普通高等教育“十四五”规划教材

冶金工业出版社

机械设备维修工程学

Engineering of Mechanical Equipment Maintenance

（中英文对照）

王立萍　廖桂兵　朴景威　张国强　编著

扫码查看
视频、课件

北　京
冶　金　工　业　出　版　社
2024

内 容 提 要

本书通过中英文对照的方式，系统地介绍了机械设备维修与维护的基本理论，机械零部件的拆卸与修复，机械设备的装配与安装，以及机械设备的润滑与密封等知识。本书内容全面，深入浅出，有助于读者同时掌握中、英文机械设备维修与维护的专业术语和方法技术等内容。书中每章后均附有思考题。

本书可作为大专院校机械类专业的教学用书，也可供从事机械设备维修和维护工作的工程技术人员参考。

图书在版编目(CIP)数据

机械设备维修工程学＝Engineering of Mechanical Equipment Maintenance：汉英对照/王立萍等编著．—北京：冶金工业出版社，2024. 3

普通高等教育“十四五”规划教材

ISBN 978-7-5024-9785-9

Ⅰ.①机…　Ⅱ.①王…　Ⅲ.①机械维修—高等学校—教材—汉、英　Ⅳ.①TH17

中国国家版本馆 CIP 数据核字(2024)第 053493 号

机械设备维修工程学
Engineering of Mechanical Equipment Maintenance

出版发行　冶金工业出版社　**电　　话**　(010)64027926
地　　址　北京市东城区嵩祝院北巷 39 号　**邮　　编**　100009
网　　址　www. mip1953. com　**电子信箱**　service@ mip1953. com

责任编辑　高　娜　美术编辑　吕欣童　版式设计　郑小利
责任校对　郑　娟　责任印制　禹　蕊
北京建宏印刷有限公司印刷
2024 年 3 月第 1 版，2024 年 3 月第 1 次印刷
787mm×1092mm　1/16；16 印张；383 千字；243 页
定价 43. 00 元

投稿电话　**(010)64027932**　**投稿信箱**　**tougao@cnmip. com. cn**
营销中心电话　**(010)64044283**
冶金工业出版社天猫旗舰店　**yjgycbs. tmall. com**
(本书如有印装质量问题，本社营销中心负责退换)

前　言

机械设备是国民经济各行各业的重要硬件工具之一，机械设备的技术指标和工作性能对企业产品的生产率、质量、成本等起着决定性的作用。机械设备在运行或停止过程中不可避免地存在磨损、变形、断裂、腐蚀等现象，因此，设备必须得到及时、良好的维修保养，使其始终处于最佳工作状态。设备维修是企业维护生产的必要手段，是节约能源和资源的重要措施和途径。

大专院校机械专业的学生在掌握了机械设备设计制造方面的专业知识后，还应该学会如何解决机械设备使用过程中可能遇到的各种问题。这些问题涉及机械设备的维修管理、故障诊断、零件失效、拆卸、清洗、检查、修复、设备安装乃至润滑密封等。

本书共分6章，第1章为机械设备维修工程学的基本理论，第2章为机械零件失效与设备故障，第3章为机械设备的拆卸，第4章为机械零件的修复，第5章为机械设备的装配与安装，第6章为机械设备的润滑与密封。

第1章、第3章和第6章由辽宁科技大学王立萍编写，第2章由鞍山钢铁有限公司张国强编写，第4章由辽宁科技大学廖桂兵编写，第5章由鞍山钢铁有限公司朴景威编写。本书由辽宁科技大学王立萍审定。

本书由辽宁科技大学教材建设基金资助项目资助出版，在此表示衷心的感谢！我们还要感谢许多朋友、领导、老师和学生为我们提供了大量的资料和建议。有些叙述、数字、表格被本书直接引用，参考文献列于书末，在此，向其作者表示最诚挚的感谢。

希望本书能够帮助读者，特别是机械专业的大学生和机械工程师们，帮助他们了解更多关于机械设备的维护知识和技巧，找到更新、更好的方法来完成工作，拓宽眼界，提高分析问题和解决问题的能力，为构建节能、低碳、环保、安全、高效的机械设备维修与维护工作作出更大贡献。

由于作者水平所限，书中不妥之处，敬请读者批评指正。

作　者

2023年6月

Preface

Mechanical equipment is hardware tools of all walks of life in the national economy. The technical index and working performance of mechanical equipment play a decisive role in product productivity, quality, cost, etc. Wear, deformation, fracture, corrosion and other phenomena inevitably exist in the process of mechanical equipment running or stopping. The equipment must be repaired and maintained in time to make it work in the best working condition for a long time. Therefore, maintenance is a necessary means to maintain production and it is an important measure and way to save energy and resources.

Students of colleges and universities and higher vocational colleges majoring in mechanical engineering have mastered a lot of professional knowledge of mechanical equipment design and manufacturing. They should know how to solve the problems appeared during mechanical equipment working. These problems involve relevant knowledge and skills of the maintenance and management, fault theory, failure, disassembly, inspection, cleaning, repair of parts, installation of mechanical equipment and lubricating and sealing.

This book consists of 6 chapters. Chapter 1 mainly introduces basic theory of mechanical equipment maintenance engineering, Chapter 2 mainly introduces failure of parts and equipment, Chapter 3 mainly introduces disassembly of mechanical equipment, Chapter 4 mainly introduces repair technologies of mechanical parts, Chapter 5 mainly introduces assembly and installation of mechanical equipment, Chapter 6 mainly introduces lubrication and sealing of mechanical equipment.

Chapter 1, Chapter 3 and Chapter 6 are written by Wang Liping of University Science and Technology Liaoning. Chapter 2 is written by Zhang Guoqiang of Anshan Iron and Steel Co. Ltd. Chapter 4 is written by Liao Guibing of University Science and Technology Liaoning. Chapter 5 is written by Piao Jingwei of Anshan Iron and Steel Co. Ltd. And the final examination of the book is finished by Wang Liping of University Science and Technology Liaoning.

The publication of this book is funded by the Textbook Construction Fund of University of Science and Technology Liaoning. The authors express their special thanks to the University of Science and Technology Liaoning! We would also like to acknowledge many friends, leaders, teachers, and many students who devoted so much materials and suggestions to us. Some narrations, figures and tables directly cited in the book are listed in references at the end of the book, here, we would like to express our sincere thanks to those original authors.

We hope this book will help readers, especially college students majoring in mechanical engineering and mechanical engineers, learn more about maintenance of mechanical equipment and find new and better ways to do their jobs, broaden their perspective as engineers, improve problem analysis and problem-solving skills, and make greater contributions to building energy-saving, low-carbon, environment-friendly, safe, and efficient mechanical equipment maintenance job.

Due to the limitation of the author's knowledge, there are inevitably shortcomings in the book. Please readers grant criticism and correction and let us know.

Author

June 2023

目　录
Contents

1　机械设备维修工程学的基本理论 …… 1
Chapter 1　Basic Theory of Mechanical Equipment Maintenance Engineering …… 1

1.1　机械设备维修工程概述
(Overview of Mechanical Equipment Maintenance Engineering) …… 2
1.1.1　机械设备维修工程学的任务
(Tasks of Mechanical Equipment Maintenance Engineering) …… 2
1.1.2　机械设备维修工程学的内容
(Contents of Mechanical Equipment Maintenance Engineering) …… 2
1.2　机械设备常用维修管理策略
(Common Maintenance Management Strategies of Mechanical Equipment) …… 4
1.2.1　故障维修(Breakdown Maintenance, BM) …… 4
1.2.2　计划预防性维修(Preventive Maintenance, PM) …… 5
1.2.3　全员生产维修(Total Productive Maintenance, TPM) …… 6
1.2.4　预知维修(Predictive Maintenance, PdM) …… 9
1.3　点检定修制(Point Checking and Regular Repair System) …… 10
1.3.1　点检系统(Point Checking System) …… 11
1.3.2　定修系统(Regular Repair System) …… 14
1.3.3　点检员的基本素质(Basic Quality of Point Inspector) …… 15
1.4　机械设备的可靠性、维修性和有效度
(Reliability, Maintainability and Efficiency of Mechanical Equipment) …… 17
1.4.1　可靠性(Reliability) …… 17
1.4.2　维修性(Maintainability) …… 21
1.4.3　有效度(Efficiency) …… 23
1.5　统筹法(Overall Planning Method) …… 24
思考题(Questions) …… 31

2　机械零件失效与设备故障 …… 33
Chapter 2　Failure of Mechanical Parts and Equipment …… 33

2.1　机械零件的失效（Failure of Mechanical Parts） …… 33
2.1.1　磨损失效（Wear Failure） …… 33
2.1.2　零件变形失效（Deformation Failure） …… 43
2.1.3　断裂失效（Fracture Failure） …… 49
2.1.4　腐蚀失效（Corrosion Failure） …… 52
2.2　机械设备的故障（Failure of Mechanical Equipment） …… 57
2.2.1　故障模式（Failure Modes） …… 58
2.2.2　故障的分类（Classification of Failure） …… 59
2.3　机械故障诊断技术
（Failure Diagnosis Technology of Mechanical Equipment） …… 61
2.3.1　电子和计算技术（Electronics and Computing Technology） …… 63
2.3.2　声振测试和分析技术（Acoustics and Vibration Testing and Analysis Technology） …… 63
2.3.3　测温技术（Temperature Measurement Technology） …… 65
2.3.4　油液分析技术（Oil Analysis Technology） …… 66
2.3.5　无损检测技术（Non-destructive Testing Technology） …… 66
思考题（Questions） …… 66

3　机械设备的拆卸 …… 68
Chapter 3　Disassembly of Mechanical Equipment …… 68

3.1　机械零件的拆卸（Disassembly of Mechanical Equipment） …… 68
3.1.1　机械零件拆卸的一般规则和要求
（Rules and Requirements of Disassembly） …… 68
3.1.2　常用的拆卸方法
（Methods of Disassembling Mechanical Parts or Components） …… 69
3.1.3　典型连接件的拆卸
（Disassembly of Typical Connection Parts or Components） …… 70
3.2　零件的清洗（Cleaning of Parts or Components） …… 74
3.2.1　拆卸前的清洗（Cleaning before Disassembly） …… 74
3.2.2　拆卸后的清洗（Cleaning after Disassembly） …… 75
3.3　零件的检验（Checking of Parts or Components） …… 78
3.3.1　检验的原则（Principles of Checking） …… 78
3.3.2　零件检验的内容（Contents of Checking） …… 80
3.3.3　检验的方法（The methods of checking） …… 81
思考题（Questions） …… 83

4 机械零件的修复 …… 84
Chapter 4 Repair Technologies of Mechanical Parts …… 84

4.1 修复前的准备工作（Preparation before Repair） …… 84
4.1.1 组织准备（Organizational Preparation） …… 84
4.1.2 技术准备（Technical Preparation） …… 84
4.2 修复技术（Repair Technology） …… 85
4.2.1 机械加工修理技术（Machining Repair Technology） …… 85
4.2.2 塑性变形修复技术（Plastic Deformation Repair Technology） …… 90
4.2.3 焊接维修技术（Welding Repair Technology） …… 96
4.2.4 热喷涂技术（Thermal Spray Repair Technology） …… 109
4.2.5 电镀修复技术（Electroplating Repair Technology） …… 113
4.2.6 工件表面强化技术（Surface Hardening Repair Technology） …… 117
4.2.7 金属扣合技术（Metal Stitch Repair Technology） …… 125
4.2.8 粘接修复技术（Adhesive Repair Technology） …… 130
4.2.9 研磨技术（Grinding Repair Technology） …… 132
4.3 零件修复技术的选择（The Choice of Part Repair Technology） …… 133
4.3.1 修复技术的选择原则
(Principles for the Selection of Repair Technology) …… 133
4.3.2 零件修复工艺规程的制订（Work Out the Repair Procedure of Parts） …… 133
4.3.3 维修费用（Maintenance Costs） …… 135
4.3.4 零部件维修更换的原则
(Principle of Parts Maintenance and Replacement) …… 136
4.3.5 典型零件修复技术的选择（Repair Technology of Typical Parts） …… 138
思考题（Questions） …… 141

5 机械设备的装配与安装 …… 144
Chapter 5 Assembly and Installation of Mechanical Equipment …… 144

5.1 机械设备的装配（Assembly of Mechanical Equipment） …… 145
5.1.1 过盈配合的装配（Assembly of Interference Fit Parts or Components） …… 147
5.1.2 联轴器的装配（Assembly of Coupling） …… 152
5.1.3 滑动轴承的安装（Assembly of Sliding Bearing） …… 165
5.1.4 滚动轴承的装配（Assembly of Rolling Bearing） …… 168
5.1.5 齿轮的安装（Assembly of Gears） …… 174
5.2 机械设备的安装（Installation of Mechanical Equipment） …… 179
5.2.1 机械安装前的准备工作（Preparation before Installation of Machines） …… 179
5.2.2 基础的设计与施工（Design and Construction of Foundation） …… 183
5.2.3 安装过程（Installation Procedure） …… 187

思考题（Questions） …… 200

6 机械设备的润滑与密封 …… 203
Chapter 6 Lubrication and Sealing of Mechanical Equipment …… 203

6.1 机械设备的润滑（Lubrication of Mechanical Equipment） …… 203
6.1.1 润滑的分类（Classification of Lubrication） …… 203
6.1.2 润滑原理（Lubrication Theory） …… 204
6.1.3 润滑材料（Lubrication Material） …… 209
6.1.4 稀油润滑系统（Thin Oil Lubrication System） …… 222
6.1.5 干油润滑系统（Dry Oil Lubrication System） …… 223
6.1.6 典型零部件的润滑（Lubrication of Typical Parts） …… 224
6.2 密封的装配（Assembly of Sealing） …… 231
6.2.1 固定连接的密封（Sealing of Fixed Connection） …… 232
6.2.2 活动连接的密封（Sealing of Active Connection） …… 233
思考题（Questions） …… 241

参考文献（References） …… 243

1 机械设备维修工程学的基本理论

Chapter 1 Basic Theory of Mechanical Equipment Maintenance Engineering

机械设备是满足企业生产经营需要的重要财产物资，是能够为企业带来经济效益的重要经济资源，工业生产过程的连续性和均衡性主要靠机械设备的正常运转来保持。只有对机械设备加强设备管理，正确地使用和真心地维护保养，适时适地地进行状态监测，科学地维修改造，保持设备处于良好的技术状态，才能保证生产连续稳定地运行。

Mechanical equipment, as an important economic resource, is very important for meeting the need of enterprise production. The continuity and balance of industrial production process mainly rely on the normal operation of machinery and equipment to maintain. It is necessary to strengthen the management of mechanical equipment, correctly operate and sincerely maintain, timely and properly monitor their state, scientifically repair and reconstruct in order that equipment can be in a good technical state. Therefore, production can be guaranteed to run continuously and stably.

目前，工业和科技的飞速发展，新产品、新技术、新材料不断涌现。设备维修与管理的理念和模式在不断更新。因此，仅仅依靠一些传统的维修技术、老化的专业知识以及在该领域处理的几个技术问题的成功案例经验，是远远不能适应快速发展的现代环境下的设备维修需求的。

With the rapid development of industry and science and technology, new products, new technologies and new materials are constantly emerging. The idea and mode of equipment maintenance and management are updating ceaselessly. Therefore, it is far from being able to adapt to the requirements of equipment maintenance in the modern environment of rapid development only relying on some traditional repair technology, aging professional knowledge and the successful case experience of solving several technical problems in the field.

随着各行各业装备水平的快速提高，装备维修工作要跟上新时代的步伐已经刻不容缓。研究维修理论，发展维修技术，高质量、高效率、低成本、安全地完成维修任务，已成为摆在广大工程技术人员面前的重要课题。

With the rapid improvement of equipment level in various industries, it is urgent for equipment maintenance to keep pace with the new times. Studying maintenance theory, developing maintenance technology, and completing maintenance tasks with high quality, high efficiency, low cost and safety, have become an important subject in front of the vast number of engineer and technical personnel.

如何应用跨学科的现代科学、技术和理论，涵盖多种学科和技术合理、全面的方式，形成不同的环境和条件下的有效的维修理论与技术已成为现代维护领域一个严峻的挑战。

It has become a serious challenge in the field of modern maintenance that how to apply the interdisciplinary modern science, technology and theory covering multiple disciplines and technologies in a reasonable and comprehensive way to form the effective maintenance theory and technology in the field with different environment and conditions.

1.1 机械设备维修工程概述（Overview of Mechanical Equipment Maintenance Engineering）

机械设备维修工程是一门综合性的技术应用课程。以具有故障的机械设备为研究对象，所研究的领域是机械设备故障的机理和维修技术（故障排除）。

Mechanical equipment maintenance engineering is a comprehensive technology application course. The mechanical equipment with failures is the object of study, and the mechanism and maintenance technology of mechanical equipment failures (troubleshooting) is its research field.

1.1.1 机械设备维修工程学的任务（Tasks of Mechanical Equipment Maintenance Engineering）

机械设备维修工程学有三项任务：

There are three tasks of mechanical equipment maintenance engineering:

（1）研究机械设备工作性能恶化的规律和机理，寻求延长机械设备寿命和改善其可靠性的途径；

Study the regularity and mechanism of the mechanical equipment performance deterioration to seek out the way for extending the service life of mechanical equipment and improving its reliability;

（2）应用现代化科学技术成果，研究适用的维修安装技术和修复工艺；

Study suitable maintenance and installation technology and repair technology according to modern science and technology achievements;

（3）以最佳经济效益为准则，研究维修管理理论和方法，为维修决策提供依据。

Study the theory and method of maintenance management, taking the best economic benefit as standard, to provide the basis for maintenance decision.

1.1.2 机械设备维修工程学的内容（Contents of Mechanical Equipment Maintenance Engineering）

机械设备维修工程学的主要内容包括：故障理论、维修技术和维修管理。机械设备维修工程学的具体内容如图1.1所示。

The main contents of mechanical equipment maintenance engineering include failure theory, maintenance technology and maintenance management. The contents of mechanical equipment maintenance engineering are shown in Figure 1.1.

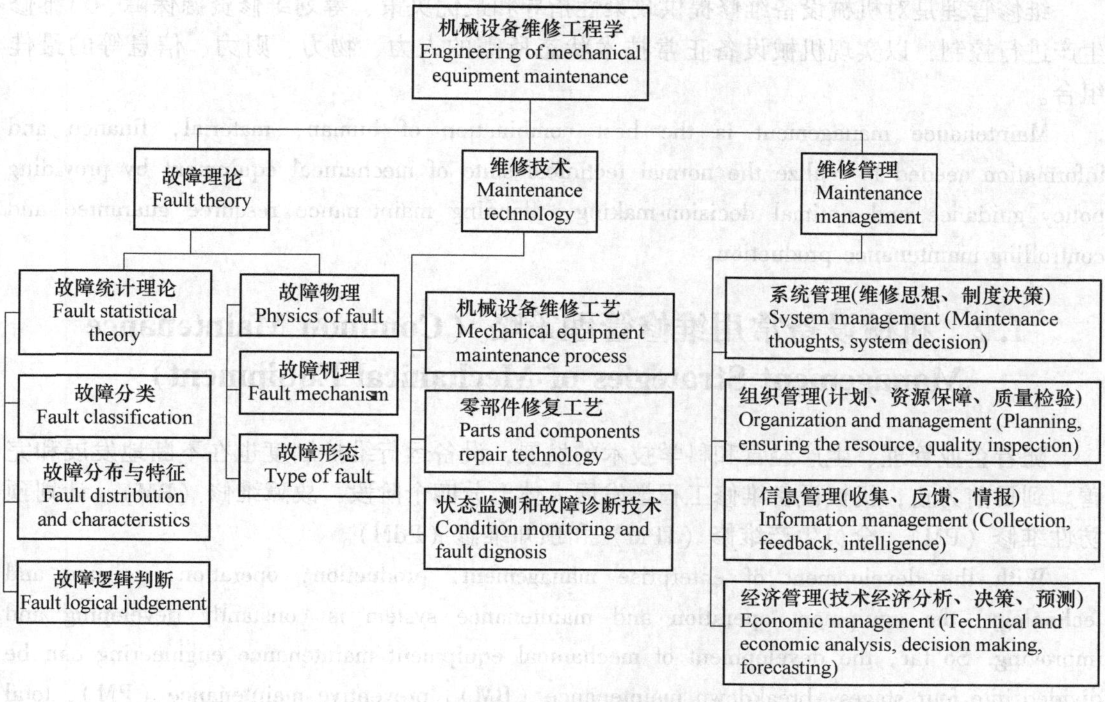

图 1.1 机械设备维修工程学的内容

Figure 1.1 Contents of mechanical equipment maintenance engineering

故障理论是机械设备维修工程学科的理论基础，它揭示了机械设备投入生产后的运动规律，是维修的决策依据。故障理论是在可靠性、维修性、摩擦、磨损、润滑学、工程诊断学等学科的基础上建立和发展起来的一门综合性理论。

Fault theory is the theoretical basis of mechenical equipment maintenance engineering discipline, which reveals the movement rules of mechanical equipment after being put into production. Fault theory is the decision basis of maintenance. Fault theory is a comprehensive theory established and developed on the basis of reliability, maintainability, friction, wear, lubricating, engineering diagnostics and other disciplines.

维修技术是指具体的生产技术，它借鉴各工程技术学科的成果，建立发展起来一套较全面的机械设备维护和零部件修复工艺体系。状态监测和故障诊断技术是根据工程诊断学原理和方法，建立发展起来的一套完善的检测和诊断系统。这种系统的建立标志着机械设备进入了现代化管理阶段，并使机械设备实现预知维修成为可能。

Maintenance technology refers to specific production technology. A comprehensive mechanical equipment maintenance and parts repair process system is established and developed learning from the achievements of various engineering and technical disciplines. State monitoring and fault diagnosis technology is a set of perfect detection and diagnosis system based on engineering diagnostics principle and method. The establishment of this kind of system marks that the mechanical equipment has entered the stage of modern management and makes it possible to realize predictive maintenance of mechanical equipment.

维修管理是对机械设备维修提供政策性指导和最优决策、筹划维修资源保障、对维修生产进行控制，以实现机械设备正常技术状态所需的人力、物力、财力、信息等的最佳组合。

Maintenance management is the best combination of human, material, finance and information needed to realize the normal technical state of mechanical equipment by providing policy guidance and optimal decision-making, planning maintenance resource guarantee and controlling maintenance production.

1.2 机械设备常用维修管理策略（Common Maintenance Management Strategies of Mechanical Equipment）

随着企业管理、生产经营和科学技术的发展，设备运行维护制度也在不断地发展和完善。到目前为止，机械设备维修工程学发展大体上分四个阶段：故障维修（BM）、计划预防性维修（PM）、全员生产维修（TPM）和预知维修（PdM）。

With the development of enterprise management, production, operation, science and technology, the equipment operation and maintenance system is constantly developing and improving. So far, the development of mechanical equipment maintenance engineering can be divided into four stages: breakdown maintenance (BM), preventive maintenance (PM), total productive maintenance (TPM) and predictive maintenance (PdM).

1.2.1 故障维修（Breakdown Maintenance, BM）

自从有了机械，维修就和机械并存，而且随生产的发展而发展。一直到19世纪，努力提高设备利用率、减少维修费用、增加利润的想法都还未出现，对于机械的故障，只是在发生了以后才去处理，即故障维修方式。

Since there have been machinery, maintenance and machinery have existed side by side, and developed with the development of production. Until the 19th century, the idea of trying to improve the utilization of equipment, reduce maintenance costs, and increase profits had not yet emerged, and mechanical failures were dealt with only after they had occurred. That is the Breakdown Maintenance method.

故障维修，是设备发生故障后再进行的修理，俗称“出了事再修”，其特征是机械损坏随时修理，维修工作处于完全被动状态，只能适应很低生产力的水平。由于不知道故障何时发生，缺乏修前的准备工作，事后维修的设备停修时间较长，停产的经济损失较大。事后维修是一种原始落后的管理方式。

Breakdown maintenance is the method of repairing equipment after failure, commonly known as “repair after failure happened”, which is characterized by mechanical damage repair at any time. Maintenance work was in a completely passive state, and can only adapt to the level of very low productivity. Because we do not know when the failure occurs, lack of preparatory work before repair, the stop and repair time of equipment is longer, so, the economic loss of production is greater. Breakdown maintenance is a primitive and backward management mode.

1.2.2 计划预防性维修（Preventive Maintenance，PM）

计划预防性维修是指定期对设备进行的维护，以减少设备发生故障的可能性。计划预防性维修是在设备仍在工作时进行的，这样就不会意外地发生故障。就维护策略的复杂性而言，计划预防性维修是介于故障维护和预知维修之间的一种维修策略。

Preventive maintenance (PM) is maintenance that is regularly performed on a piece of equipment to lessen the likelihood of failure. It is performed while the equipment is still working so that it does not break down unexpectedly. In terms of the complexity of this maintenance strategy, it locates between breakdown maintenance and predictive maintenance.

进入20世纪，生产力水平有了很大提高，出现了以福特汽车装配线为代表的流水生产作业，生产方式由单件转向批量，生产效率飞跃提高。与此同时，机械设备故障对生产的影响也显著增加。如果某一环节停止运转，则全线停工。于是出现了设备利用率的问题，产生了预防维修的观点。

In the 20th century, the productivity level has been greatly improved, and the flow production operation represented by the assembly line of Ford automobile emerged. The production mode has changed from single piece to batch, and the production efficiency has been greatly improved. At the same time, the impact of mechanical equipment failure on production also increased significantly. If one link stops running, the whole line is shut down. The problem of equipment utilization arises, and the idea of preventive maintenance arises.

计划预防维修制的具体实施可概括为定期检查、按时保养、计划修理。计划预修制主要包括日常维护保养和不同性质的定期修理（大、中、小修），它能保证机械设备经常地、长期地处于正常运行状态，能明显减少以至消除非计划停机，对连续性的大生产产生了良好的效果。

The specific implementation of PM system can be summarized as regular inspection, timely maintenance and planned repair. The PM system mainly includes daily maintenance and regular repairs of different properties (overhaul, medium repair, minor repair). It can ensure that mechanical equipment is in normal operation regularly and for a long time, and can significantly reduce or eliminate unplanned downtime, which has a good effect on continuous large-scale production.

预防维修阶段容易出现的问题是修理周期、修理工作量不能准确地反映该设备运行的客观规律、或者不能准确反映不同位置、不同负荷环境下的同一名称设备的自身状况，其结果是易造成维护不及时，更多的是造成“过剩维修”，使得备品备件储备量增大，检修工作量增大，各种费用增多（人力和停修时间），设备维修成本增高。

There are some problems in PM phase. Repair cycle, repair work can not accurately reflect the objective law of the equipment operation, or may not accurately reflect the different positions under different load conditions, the equipment condition of the same name. The result is that PM is easy to cause not timely maintenance, more "surplus" maintenance, spare parts reserves increase, maintenance workload increases, various costs increase (manpower and maintenance time), equipment maintenance costs increase.

1.2.3 全员生产维修（Total Productive Maintenance，TPM）

全员生产性维修（TPM）是一种根据设备中的每个人都应该参与维修而不仅仅是维修团队的理念来运作的策略。这种方法使用了所有员工的技能，并试图将维修与设施的日常性能结合起来。

Total productive maintenance（TPM）is a strategy that operates according to the idea that everyone in a facility should participate in maintenance，rather than just the maintenance team. This approach uses the skills of all employees and seeks to incorporate maintenance into the everyday performance of a facility.

在全员生产维修理念下，从高层管理人员到设备操作人员，每个人都应该参与维修。

Under the TPM philosophy，everyone from top-level management to equipment operators should participate in maintenance.

TPM 计划的开始将着重于建立"5S"基础和制定自主维修计划。这就解放了维修人员，使他们可以开始更大的项目并执行更有计划的维修。

The beginning of a TPM program will focus on establishing "5S" foundation and developing an autonomous maintenance plan. This frees up the maintenance staff to begin larger projects and perform more planned maintenance.

"5S"是整理（Seiri）、整顿（Seiton）、清洁（Seiketsu）、清扫（Seiso）和素养（Shitsuke）这 5 个词的缩写。"5S"起源于日本，是指在生产现场对人员、机器、材料、方法等生产要素进行有效管理。

"5S" stands for "Seiri"，"Seiton"，"Seiketsu"，"Seiso" and "Shitsuke". "5S" originated in Japan，refers to the effective management of production factors such as personnel，machines，materials and methods on the production site.

要实现 TPM，必须做到以下五件事。

Five things should be done in order to achieve TPM.

（1）整理。确定哪些物品经常使用，哪些物品不经常使用。常用的物品应就近存放，不常用的物品应存放较远。

Seiri/Sort. Determine which items are used frequently and which are not. The ones used frequently should be kept close by，and others should be stored further away.

（2）整顿。每个物品都应该有一个存放的地方，而且只有一个存放的地方。

Seiton/Systemize. Each item should have one place，and only one place to be stored.

（3）清洁。工作场所需要清洁。没有它，问题将更加难以识别，维护将更加难以执行。

Seiketsu /Shine. The workplace needs to be clean. Without it，problems will be more difficult to identify，and maintenance will be more difficult to perform.

（4）清扫/标准化。清扫就是使生产现场进入没有垃圾，没有污脏的状态，虽然已经整理、整顿过，要的东西马上就能取得，但是被取出的东西要达到能被正常使用的状态才行。而达到这种状态就是清扫的第一目的，尤其目前强调高品质、高附加价值产品的制造，更不容许有垃圾或灰尘的污染，造成品质不良。工作场所应该标准化和贴标签。

Seiso/Standardize. Cleaning is to make the production site into a state of no garbage, no dirt. Although it has been sorted out and rectified, the things to be obtained immediately, but the things to be removed to reach the state of normal use. And to achieve this state is the first purpose of cleaning, especially the current emphasis on high quality, high value-added products manufacturing, not to allow garbage or dust pollution, resulting in poor quality. Workplaces should be standardized and labeled.

（5）素养/保持。通过晨会等手段，提高员工文明礼貌水准，增强团队意识。并且只有长期坚持，才能使员工养成按规定行事的良好工作习惯。

Shitsuke/Sustain. Morning meetings and other means can be used to improve the level of civility of employees, enhance the sense of teamwork. And only these means are be done for a long term, employees can develop good work habits in accordance with the rules.

我国企业在“5S”现场管理的基础上，结合国家如火如荼的安全生产活动，在原来“5S”基础上增加了安全（safety）要素，形成“6S”。随着人们对这一活动认识的不断深入，有人又添加了节约（save）、学习（study）等内容，分别称为“7S”“8S”。

On the basis of “5S” field management, combined with national safety production activities in full swing, Chinese enterprises have added safety elements on the basis of the original “5S”, forming “6S”. With the deepening of people's understanding of this activity, some people add “Save”, “Study” and other contents, called “7S” and “8S” respectively.

安全是指保障员工的人身安全，保证生产连续安全正常进行，同时减少因安全事故而带来的经济损失。节约就是对时间、空间、能源等方面合理利用，以发挥它们的最大效能，从而创造一个高效率的、物尽其用的工作场所。学习是终身的，指社会每个成员为适应社会发展和实现个体发展的需要，贯穿于人的一生的、持续的学习过程。

Safety is to ensure the personal safety of employees, ensure the continuous safety and normal production, and reduce the economic loss caused by safety accidents. Save is the rational use of time, space and energy to maximize their efficiency, thus creating an efficient and well-used workplace. Learning is lifelong. It refers to continuous learning process throughout the life for every member of the society to adapt to social development and realize the needs of individual development.

TPM 建立在“5S”的基础上，由八大支柱支撑着它。任何一个支柱都是非常重要的，起着不同的承重责任。TPM 的八大支柱如图 1.2 所示。

TPM is built on a “5S” foundation supported by eight pillars. Any pillar is very important and carries different load-bearing responsibilities. Eight pillars of TPM is shown in Figure 1.2.

TPM 具有三个“全”特征，即全效率、全系统和全员参与。全效率是指设备生命周期成本评估和设备综合效率。这是这项工作的目标，要求设备消耗最小的总成本和最大的效益和贡献，也就是“多快好省”。全系统是指包括所有的生产方法和维护系统。

TPM is characterized by three “whole”, namely, whole efficiency, whole system and whole participation. Whole efficiency refers to equipment life cycle cost evaluation and equipment comprehensive efficiency. This is the goal of the work, requiring the equipment to consume the minimum total cost and the maximum benefit and contribution, that is, “greater, faster, better

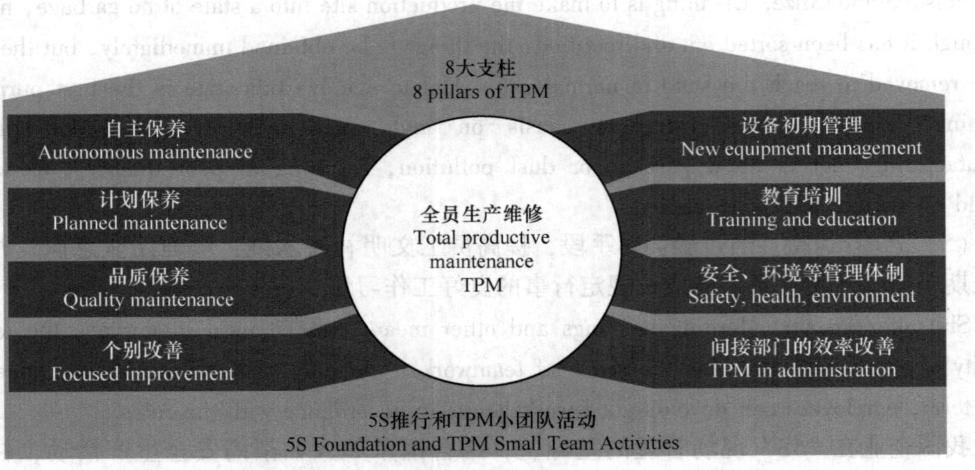

图 1.2 TPM 的八大支柱

Figure 1.2 Eight pillars of TPM

and more economical".

全系统也可称为全过程，包括与设备寿命相关的各个环节和阶段的系统。全员参与是指计划、使用、维护等与系统有关的设备的所有部门和人员的参与，包括生产工人和企业各级领导，尤其注重操作员的自主集体活动。

Whole system means that all methods of production and maintenance systems are included. The whole system can also be called the whole process, including the system related to each link and stage of the equipment life. Whole participation refers to all departments and personnel responsible for planning, using, maintenance and other equipment related to the system, to participate, including production workers and enterprise leaders at all levels, especially pay attention to the operator's autonomous group activities.

TPM 的目标可以概括为四个“零”，即停机为零、废品为零、事故为零、速度损失为零。

The objectives of TPM can be summarized as four "zeros", namely, zero stop, zero waste, zero accident and zero speed loss.

停机为零，指计划外的设备停机时间为零。

Zero stop means zero downtime for unplanned equipment.

计划外的停机对生产造成冲击相当大，使整个生产匹配发生困难，造成资源闲置等浪费。计划时间要有一个合理值，不能为了满足非计划停机为零而使计划停机时间值达到很高。

Unplanned downtime has a considerable impact on production, making it difficult to allocate the entire production, resulting in idle resources and other waste. Planned time should be a reasonable value. A high planned downtime value should not be given in order to get zero unplanned downtime.

废品为零，指由设备原因造成的废品为零。“完美的质量需要完善的机器”，机器是保证产品质量的关键，而人是保证机器好坏的关键。

Zero waste means zero waste caused by equipment. “Perfect quality needs perfect machine”. Machine is the key to ensure the quality of products, and people are the key to ensure the quality of machines.

事故为零，指设备运行过程中事故为零。设备事故的危害非常大，不仅影响生产，还可能会造成人身伤害，严重的可能会“机毁人亡”。

Zero accident means there is zero accident during the operation of the equipment. The harm of equipment accident is very big, not only affecting production, but also may cause personal injury, sven “the loss of both machine and the person”.

速度损失为零，指设备速度降低造成的产量损失为零。由于设备保养不好，设备精度降低而不能按高速度使用设备，等于降低了设备性能。

Zero speed loss means zero production loss due to reduced equipment speed. Due to the poor maintenance of the equipment, the equipment precision is reduced and the equipment cannot be used at high speed, which is equivalent to reducing the performance of the equipment.

推行 TPM 要从三大要素上下功夫：

The implementation of TPM should be carried out from the three elements:

（1）提高工作技能。不管是操作工，还是设备工程师，都要努力提高工作技能，没有好的工作技能，全员参与将是一句空话。

Improve work skills. No matter operators or equipment engineers, should strive to improve work skills. Without good work skills, TPM will be an empty talking.

（2）改进精神面貌。精神面貌好，才能形成好的团队，共同促进，共同提高。

Improve the mental attitude. Staff with good mental attitude can form a good team. They can promote and improve together.

（3）改善操作环境。通过 5S 等活动，使操作环境良好，一方面可以提高工作兴趣及效率，另一方面可以避免一些不必要设备事故。现场整洁，物料、工具等分门别类摆放，也可使设置调整时间缩短。

Improve the operating environment. Make the operating environment good through 5S and other activities. On one hand, good environment can improve work interest and efficiency of staff. On the other hand, good environment can avoid some unnecessary equipment accidents. The clean and tidy operating environment and materials and tools being put in different categories can shorten the adjustment time.

1.2.4 预知维修（Predictive Maintenance, PdM）

预知维修（PdM）的目的是预测设备何时可能发生故障，并通过安排维修来防止故障的发生。维修频率低，可靠性高。通过与软件相连接的监控设备或目视检查来识别可能的故障。

Predictive maintenance（PdM）aims to predict when asset failure might occur and prevents the failure by scheduling maintenance. It keeps maintenance frequency low and reliability high. Possible failure is identified with monitoring equipment linked to software or through visual inspections.

预知维修的目的首先是预测设备何时可能发生故障，其次是通过进行维修来防止故障的发生。对未来故障的监视允许在故障发生之前就计划维护。

The aim of predictive maintenance (PdM) is firstly to predict when equipment failure might occur, and secondly, to prevent the occurrence of the failure by performing maintenance. Monitoring for future failure allows maintenance to be planned before the failure occurs.

理想情况下，预知维护允许维修频率尽可能低，以防止计划外的反应性维修，而不会产生与进行过多预防性维修相关的成本。

Ideally, predictive maintenance allows the maintenance frequency to be as low as possible to prevent unplanned reactive maintenance, without incurring costs associated with doing too much preventive maintenance.

预知维修使用状态监视设备实时评估设备的性能。这一过程中的一个关键要素是物联网（IoT）。物联网允许不同的设备和系统连接，一起工作、共享、分析和执行数据。

Predictive maintenance uses condition-monitoring equipment to evaluate the performance of equipinent in real-time. A key element in this process is the Internet of Things (IoT). IoT allows for different equipment and systems to connect, work together, and share, analyze and action data.

物联网依靠预知维护传感器来捕获信息，使其有意义，并识别任何需要关注的领域。使用预知维修和预知维修传感器的一些例子包括振动分析、油分析、热成像和设备观察。

IoT relies on predictive maintenance sensors to capture information, make sense of it and identify any areas that need attention. Some examples of using predictive maintenance and predictive maintenance sensors include vibration analysis, oil analysis, thermal imaging, and equipment observation.

前述机械设备维修策略中，全员生产维修是目前应用最广泛、效果最好的维修策略。全员生产维修的工作内容主要包括点检定修制、设备分级管理制、设备文明生产达标管理制等，其中点检定修制是一套加以制度化的比较完善的科学管理方法，它的实质就是以预防维修为基础、以点检为核心的全员维修制。

Among the aforementioned mechanical equipment maintenance strategies, full production maintenance is currently the most widely used and effective maintenance strategy. The work content of full staff production and maintenance mainly includes the system of routine inspection and repair, equipment grading management, and equipment civilized production standard management. Among them, point checking and regular repair system is a relatively complete scientific management method that is institutionalized. Its essence is the full staff maintenance system based on preventive maintenance and centered on routine inspection.

1.3 点检定修制（Point Checking and Regular Repair System）

点检定修制以点检为核心，是通过点检员在规定的时间、按规定的路线、利用专用设备和个人经验对设备进行周期性检查，并对检查结果进行分析比较来判断设备的状态，进而采取相应措施维持提高设备能力的一种管理制度。

Point checking and regular repair system takes the point checking as the core, which is a management system in which point inspectors conduct periodic check of the equipment at the specified time, according to the specified route, using special equipment and personal experience, and analyzes and compares the check results to judge the status of the equipment, and then takes corresponding measures to maintain and improve the equipment capacity.

点检定修制是以预防维修为主，全员参与的一种计划综合维修。包括点检制和定修制两个方面。其主要优点是：降低维修费，减少故障，提高维修效率，提高设备综合效率，有利于协调组织生产。

Point checking and regular repair system is a kind of comprehensive maintenance plan with the participation of all staff, which mainly includes the point checking system and regular repair system. Its main advantages include reducing maintenance costs and failure reduction, improving maintenance efficiency, improving equipment comprehensive efficiency, being beneficial for coordinating and organizing production.

把设备管理分为三方：点检方、检修方和生产方。点检方属于管理方，履行生产设备管理的全面职能，从这个意义上而言，其处于核心地位。检修方是在管理方的业务计划指导下，完成各种设备检修或备件制造业务。生产方是指生产操作工人接受点检方关于设备日常点检业务指导和管理。

Equipment management is divided into three parties: point checking department, maintenance department and production department. Point checking department belongs to the management department, to perform the overall function of production equipment management. In this sense, point checking department is in the core position. Under the guidance of the business plan of the management, the maintenance department shall complete all kinds of equipment maintenance or spare parts manufacturing business. The production department refers to the production operation workers who receive guidance and management from the point checking department about the daily point checking of equipment.

1.3.1 点检系统（Point Checking System）

设备点检系统是以点检系统为核心的 TPM 管理系统。对设备进行规范、定期的定点检查，以准确掌握设备技术状况、设备故障早期信息和设备退化趋势。及时采取措施消除幼年故障，以提高设备的工作效率，延长设备的使用寿命，保证企业正常生产。

Equipment point checking system is TPM management system centered on the point checking system. Standard and regular, fixed point checking on equipment are taken in order to accurately master the equipment technical conditions, the early information and degradation trend of equipment failure. Timely methods can be taken to eliminate the failure in infancy in order to improve the equipment work efficiency and prolong the life of equipment, to ensure normal production enterprise.

点检是指对设备按规定周期进行预防性的检查，检查部位是某些点，故称为点检。

Point checking refers to the preventive check of the equipment according to the specified period. The parts need to be inspected are some points, so it is called point checking.

点检分为日常点检、专业点检和精密点检。

Point checking is divided into daily point checking, professional point checking and precision point checking.

（1）日常点检。日常点检也称为生产点检，由生产（操作）工人、值班维修工人（保全工、钳工、电工、马达工等）进行的点检工作。按要求按规定路线开展巡检，生产人员每班2次，维修人员每班N次（按工种、按设备特点规定检查次数）。

Daily point checking. Daily point checking is also called production point checking, which is carried out by production (operation) workers, on-duty maintenance workers (maintenance workers, fitters, electricians, motor workers, etc.). Carry out patrol check according to the requirements and the prescribed route. The production personnel work twice per shifts and the maintenance personnel work N times per shift (check the check times according to the types of work and the characteristics of equipment).

设备人员一般按经验和五感，对岗位范围内的设备的使用情况逐项进行检查，将结果记录在点检卡上或告知维修人员值班记录，重点发现异常情况。点检卡必须在当天提交。日常点检是预防事故发生的基础和第一道防线。

Production personnel generally check equipment they are responsible for item by item by experience and five senses. The results will be recorded in the point checking card or tell the maintenance personnel on duty record and focus on finding abnormal conditions. The point checking card must be submitted on the same day. Daily point checking is the basis and the first line of defense to avoid accidents.

（2）专业点检。专业点检的对象是本企业的关键设备（A类设备）或关键部位。凭经验或设计要求给本单位的设备制订出周期检查表（点检计划表），由专业技师、技术工人按表依次进行点检工作。针对日常点检发现的问题或有争议的问题开展工作。专业技师、技术工人需携带点检仪器、工量具进行检查。

Professional point checking. The object of professional point checking is the key equipment (class A equipment) or key parts of the enterprise. Work out periodic check schedule (point checking schedule) for the equipment of the unit according to experience or design requirements, and carry out point checking by professional technicians and technical workers according to the schedule. Work will be carried out on the problems or controversial issues found during daily point checking. Professional technicians and technical workers shall carry point checking instruments and measuring tools for checking.

根据显示的数据凭技术知识和工作经验判断出异常情况和隐患部位，并要组织维护人员排除这一故障。专业点检是骨干支柱，是技术性点检，大量的诊断、维护工作靠此完成。这是避免事故的第二道防线。

Abnormal conditions and hidden danger positions shall be judged based on technical knowledge and work experience according to the data displayed, and maintenance personnel shall be organized to eliminate this failure. Professional point checking is the backbone and technical point checking, and a large number of diagnosis and maintenance work is completed by this. This is the second line of defense to avoid accidents.

（3）精密点检。精密点检由专职分管技术人员进行，或外聘专业技术人员进行。这是一种不定期的点检，完全根据设备运行状况和技术需要安排。设备状况较佳，运转稳定的车间设备一般每年安排4~6次，或按季节温度变化安排点检次数。

Precision point checking. Precision point checking shall be carried out by full-time technical personnel in charge or by external professional technical personnel. This is a kind of irregular point checking, completely according to the operation condition of equipment and technical requirements. The workshop equipment with good condition and stable operation is usually arranged 4~6 times per year, or the point checking times are arranged according to the seasons and temperature change.

精密点检是技术性很强的点检，是针对设备的“疑难杂症”开展的监控和诊断。必须应用测量工具、诊断仪器比较鉴别，才能推导出结论性意见。它是日常点检和专业点检的完善与补充，是避免事故的第三道防线。

Precision point checking is a very technical point checking, which is the monitoring and diagnosis for the “difficult and complicated diseases” of the equipment. It is necessary to use measuring tools and diagnostic instruments to compare and identify, so as to derive conclusive opinions. It is the perfection and supplement of daily point checking and professional point checking. It is the third line of defense to avoid accidents.

点检制度的实质是全体员工的参与、管理和责任。它以预防为基础，以点检为中心，以定期维护为目标。及时掌握设备的运行状态，指导设备的维护保养，可以说是一种严肃、科学的管理方法。

The essence of the point checking system is the participation, management and responsibility of all staff. It is based on prevention, centered on point checking and aimed at regular maintenance. It can be said that it is a serious and scientific management method to timely master the running state of equipment and guide the maintenance of equipment.

全面实施设备点检制度，主要进行以点检为核心的以下步骤维修工作：

Carrying out the equipment point checking system comprehensively refers to mainly carrying out the maintenance with point checking as the core from the following procedures:

（1）定人。设立兼职和全职的点检人员。

Fix personnel. Part-time and full-time point inspectors are all set up.

（2）定点。确定设备的故障点、检查地点、项目和内容。

Fix the points. The failure points, the check site, items and contents of the equipment should be identified.

（3）定量。对有退化趋势的设备进行定量测定。

Fix quantitative. The equipment with deterioration tendency shall be quantitatively determined.

（4）定周期。对于不同的设备，不同的设备故障点，应该有不同的检查周期。

Fix period. There should be different checking cycle for different equipment and different equipment failure point.

（5）定标准。确定各点检点是否正常的依据。这就是判断标准。

Fix standards. The basis to determine whether each point checking site is normal or not should be sure. That is the judgment standard.

（6）定点检计划表。检查员被指示按照规定的路线行进。

Fix check schedule. The inspectors are instructed to follow the prescribed route.

（7）定记录格式。包括作业记录、异常记录、故障记录及倾向记录，都要有固定的格式。

Fix record format. There should be a fixed format for work records, abnormal records, failure records and tendency records.

（8）定点检业务流程。明确点检作业和点检结果的处理程序。

Fix check business process. The procedure of point checking operation and result should be clear.

1.3.2 定修系统（Regular Repair System）

定修是指在推行设备点检管理的基础上，根据预防维修的原则，按照设备的状态，确定设备的检修周期和检修项目，在确保检修间隔内的设备能稳定、可靠运行的基础上，做到使连续生产系统的设备停修时间最短，物流、能源和劳动力消耗最少，是使设备的可靠性和经济性得到最佳配合的一种检修方式。

The regular repair system refers to a way of repair with implementing the point checking management of equipment based on the principle of preventive maintenance, according to the state of the equipment, determining the maintenance cycle and maintenance items of equipment, equipment overhaul interval are guaranteed, ensuring stable and reliable operation of equipment during maintenance interval, and getting the shortest stop time for continuous production system of the equipment, the minimum consumption of logistics, energy and labor, the best match of the equipment reliability and economic efficiency.

定修包含以下内容：

The regular repair system includes the following contents:

（1）要有科学的定修模型。

A scientific regular repair model is required.

（2）要有完善的以网络图方式编成的检修作业标准和修理质量基准。

There should be an overall plan process flow chart with perfect maintenance operation standards and repair quality standards.

（3）要有明确的生产方、点检方、检修方、社会协作方的业务分工协议书。

There should be a clear agreement on the division of business among the production party, the point checking department, the maintenance department and the social cooperation party.

（4）要有现代化的施工机具和检测仪表等手段。

Modern construction machines and instruments and other means are necessary.

（5）推行以作业长制为中心的现代化基层管理方式。

Implement a modern primary level management mode centered on the operation head system.

定修制就是包括了上述要点、特征和内容的一种设备检修管理制度。即在定修模型的指导下，按照工程委托、接受工程、工程实施、工程记录四个步骤所形成的一整套科学而严密的管理制度。

The regular repair system is a kind of equipment maintenance management system which includes the above points, features and contents. That is, under the guidance of regular repair model, a set of scientific and strict management system is formed according to the four steps of project entrustment, acceptance, implementation and record.

定修制的具体内容包括：检修工程管理、计划值管理、修理计划设定周期与进度编制、年度计划编制和评定年度计划、月度日程计划编制、检修工程实施、检修工程委托处理、检修工程接受业务处理、检修工程实施处理、检修工程记录处理，以及科学的工稳编码管理和管理用的成套表格。

The detailed contents of the regular repair system include maintenance engineering management, plan value management, repair plan setting cycle and schedule compilation, preparing and evaluating annual plans, monthly schedule preparation, maintenance engineering implementation, commission maintenance works, maintenance engineering acceptance business processing, maintenance engineering implementation treatment, maintenance engineering record processing, as well as scientific stability coding management and management of the set of forms.

定修模型是各生产工序的定修配合，按照能源平衡、物流平衡、生产平衡与检修人员平衡，设定各项定修工程的周期和时间的标准化模式，这个标准化模式称为定修模型。

The regular maintenance model is the coordination of the regular maintenance of each production process. According to the energy balance, logistics balance, production balance and maintenance personnel balance, the standardized model of the cycle and time of each regular maintenance project is set, which is called the regular maintenance model.

1.3.3 点检员的基本素质（Basic Quality of Point Inspector）

点检工作由设备主管厂长领导，由设备职能部门管理，按生产工序和地域划分区域。每个区域设立至少 1 名机械点检员，一般每个厂设立至少 1 名设备员。

Point checking work is guided by the director in charge of equipment and managed by the functional department of equipment. Each area shall have at least one mechanical point inspector. Each plant shall have at least one mechanical engineer.

点检员的基本素质如下：

Basic qualities of point inspector are followings:

（1）有一定的设备管理理论，技术素质和实践经验，能进行一般的机械制图；

A point inspector must have the equipment management theory, the technical quality and the practical experience, can carry on the general mechanical drawing;

（2）有较强的故障判别能力，会使用简易诊断仪器；

A point inspector should have strong ability of failure discrimination, can use simple diagnostic instrument;

（3）有一定的业务协调能力，与维修站、设备组、工段进行横向联络；

A point inspector should have a certain business coordination ability, and maintenance station，equipment group，section horizontal contact；

（4）全面熟悉所管辖设备的结构、性能、设备的操作规程，检修规程。

A point inspector should be familiar with the structure，performance，operation procedure and maintenance procedure of the equipment.

点检员是设备技术状态的管理员，是设备运行、维护和检修技术管理的中坚力量，对设备的技术状态负责。

Point inspectors are the manager of the equipment technical status，and are responsible for the equipment technical status. Point inspectors are the backbone of the equipment operation，maintenance，checking and repairing technical management.

点检员职责包括：

Duties of point inspectors should include：

（1）负责对所辖区域设备技术状态管理，按点巡检要求，设备技术维护要求定期按时进行设备的现场巡视并作好记录；

Be responsible for the technical management of the equipment in certain area. According to the point checking requirements，equipment technical maintenance requires for regular check and make records；

（2）负责制定设备的检修计划，备品备件计划，设备技术维护计划的制定和现场故障处理；

Be responsible for making equipment maintenance plan，spare parts plan，and equipment technical maintenance plan and on-site failure treatment；

（3）负责设备检修计划的组织实施，质量监督验收等；

Be responsible for the organization and implementation of equipment maintenance plan，quality supervision and acceptance，etc；

（4）负责与维修工段（公司）联络，开出检修委托工作量签证；

Be responsible for the contact with maintenance section（company），issue maintenance commissioned workload visa；

（5）负责设备故障的检测分析诊断，提出处理意见和检修，改进方案；

Be responsible for the detection，analysis and diagnosis of equipment failure and put forward treatment opinions and maintenance，improvement plan；

（6）负责对设备的现场管理、指导、检查，督促生产岗位人员搞好设备操作和日常维护；

Be responsible for on－site management，guidance and check of equipment，and urge production staff to do well in equipment operation and daily maintenance；

（7）负责设备运行档案的填写与管理；

Be responsible for filling in and manage equipment operation files；

（8）负责本区域岗位人员的设备知识培训。

Be responsible for the training of equipment knowledge for the personnel in this region.

设备管理人员职责包括：

Responsibilities of equipment management personnel should include：

(1) 负责对点检员的业务指导和管理;

Be responsible for the point inspector's business guidance and management;

(2) 负责重点检修项目的管理;

Be responsible for the management of key maintenance projects;

(3) 负责备品备件的组织，管理;

Be responsible for the organization and management of spare parts;

(4) 负责点检员开出的委托和工作量复审;

Be responsible for the commission and workload review issued by the point inspector;

(5) 负责重点设备的技术改进和改造;

Be responsible for the technical improvement and transformation of key equipment;

(6) 负责技术资料的管理;

Be responsible for the management of technical data;

(7) 负责对维修公司的评价。

Be responsible for the evaluation of the maintenance company.

点检管理的要点是实行全员管理，专职点检员按区域分工管理。点检员本身是一贯制管理者。点检是按照一整套标准化、科学化的轨道进行的。

The key point of point checking management is to implement the whole staff management, full-time point checking staff according to the regional division of labor management. The point inspectors themselves are institutional managers. The point checking is carried out according to a set of standardized and scientific tracks.

1.4 机械设备的可靠性、维修性和有效度（Reliability, Maintainability and Efficiency of Mechanical Equipment）

设备的质量反映了满足规定和潜在要求的设备的总和。设备的质量特性包括可靠性、可维护性和有效性。因此，产品的可靠性、可维护性和有效性是产品内在质量的本质特征。

The quality of the equipment reflects the sum of the equipment to meet specified and potential requirements. The quality characteristics of the equipment includ reliability, maintainability and efficiency. Therefore, the reliability, maintainability and efficiency of products are the essential characteristics that characterize the internal quality of products.

1.4.1 可靠性（Reliability）

可靠性是指机械设备在规定条件下和规定时间内完成规定功能的能力。

Reliability refers to the ability of machine to complete the required function under required conditions and within required time.

可靠性是衡量机械设备质量的重要指标。机械设备不可避免地会出现故障，需要维护和修理。维修性是衡量设备维修质量的重要指标。在维修工程中，可靠性和维护性与效率密切相关。设备的有效性反映了设备整个生命周期的质量。

Reliability is an important index to measure the quality of mechanical equipment. Mechanical equipment inevitably breaks down and needs to be maintained and repaired. Maintainability is an important index to measure the quality of equipment maintenance. In maintenance engineering, reliability and maintainability are closely related to efficiency. The effectiveness of the equipment reflects the quality of the whole life cycle of the equipment.

可靠度是指机械设备在规定条件下和规定时间内完成规定功能的概率。

Reliability degree is a probability of machine to complete specified functions under specified conditions and within the specified time.

可靠性和可靠度定义基本上相同，只是一个表示定性，一个表示定量，其差别一个是“能力”，一个是“概率”。

The definition of reliability and reliability degree are basically the same, except that one represents qualitative and another represents quantitative. The difference is that reliability is capability and reliability degree is probability.

可靠性的基本目的是尽量减少故障。可靠性关系到产品的经济性。不同的对象有不同的可靠性要求。高可靠性意味着高价格。

The basic purpose of reliability is to minimize failures. Reliability is related to the economy of the product. Different objects have different reliability requirements. High reliability means high price.

规定条件是指工作环境、运行和维护条件。

Specified conditions refer to the working environment, operation and maintenance conditions.

规定的时间是指使用寿命。开始时可靠性高，之后逐渐降低。可靠性与时间有关。

Specified time refers to the service life. The reliability is high at the beginning and gradually decreases later. The reliability is related to time.

规定的功能是指确保给定的参数在技术文件规定的范围内。

Specified function refers to that ensuring given parameters are within the scope specified in the technical document.

可靠性的定量计算结果就是设备的可靠度。

The quantitative calculation result of reliability is the reliability of equipment.

用于计算可靠度的函数包括可靠度函数 $R(t)$、故障概率函数 $F(t)$、失效密度函数 $f(t)$、失效率 $\lambda(t)$ 和平均寿命 t。

Functions used to calculate reliability degree include reliability degree $R(t)$, failure probability $F(t)$, failure density $f(t)$, failure rate $\lambda(t)$, and average life time t.

（1）可靠度函数 $R(t)$（Reliability degree $R(t)$）。

可靠度函数表示在规定时间间隔内，机械设备不发生故障的概率，以 $R(t)$ 表示。

Reliability function $R(t)$ refers to the probability of mechanical failure does not occur within the given time interval.

例如：一个机械设备由 100 个零件组成，在 t 时间内一个没有坏，完全可靠，此时可靠度是 100%。如果在 t 时间内坏 100 个，完全不可靠，此时可靠度是 0。所以，$0 \leq R(t)$

≤1，如图 1.3 所示。

Give an example. A mechanical device is composed of 100 parts. If none of them are bad at a given time t, the reliability of this mechanical device is 100%. If 100 parts are all bad at a given time t, the reliability of this mechanical device is 0. Therefore, reliability is a value between 0 and 1, $0 \leqslant R(t) \leqslant 1$, as shown in Figure 1.3.

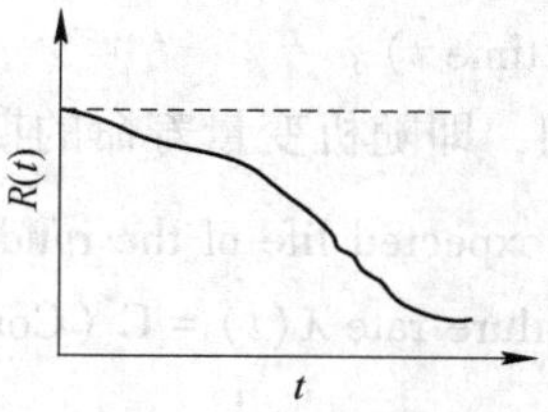

图 1.3 可靠度函数

Figure 1.3 Reliability degree

(2) 故障概率函数 $F(t)$ (Failure probability $F(t)$)。

故障概率函数又称不可靠度函数，表示在规定时间间隔内机械设备发生故障的概率，以 $F(t)$ 表示。故障概率函数也称为不可靠度函数。故障概率函数取值范围为 0~1，即 $0 \leqslant F(t) \leqslant 1$，故障概率函数与可靠性函数之和为 1，即 $R(t)+F(t)=1$。故障概率函数如图 1.4 所示。

Failure probability $F(t)$ refers to the probability of mechanical failure within the prescribed time intervals. The failure probability is also known as the unreliability. Failure probability is evaluated between 0 and 1, i. e. $0 \leqslant F(t) \leqslant 1$. And the sum of failure probability and reliability is 1, i. e. $R(t)+F(t)=1$. The failure probability is shown in Figure 1.4.

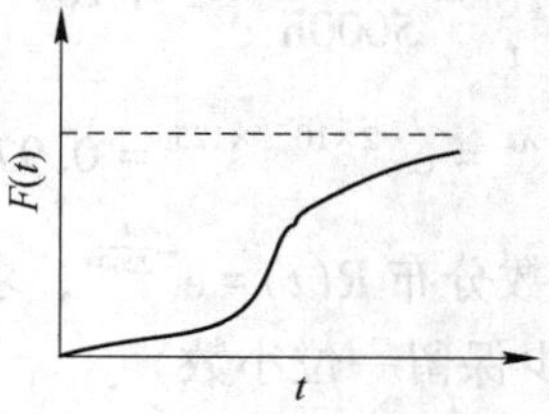

图 1.4 故障概率函数

Figure 1.4 Failure probability function

(3) 失效密度函数 $f(t)$ (Failure density $f(t)$)。

失效密度函数 $f(t)$ 为给定时刻的失效频率。它也被描述为机械故障随时间的分布。

Failure density $f(t)$ refers to the failure frequency at a given moment. It is also described as the distribution of mechanical failure over time.

对于失效密度函数有（For the failure density）：

$$\int_0^\infty f(t)\,dt = 1 \tag{1.1}$$

(4) 失效率 $\lambda(t)$ (Failure rate $\lambda(t)$)。

失效率 $\lambda(t)$ 是指机器在工作时间 t 时刻未发生故障时，经过 t 时刻后单位时间内发生故障的概率。也称为 t 时刻的失效率。

Failure rate $\lambda(t)$ refers to the probability of failure per unit time after time t when machine is not yet fail at time t during work time. It is called failure rate at time t.

$$\lambda(t) = \frac{f(t)}{R(t)} \tag{1.2}$$

（5）平均寿命 $\bar{t}$（Average life time $\bar{t}$）。

平均寿命也就是平均工作时间，即随机变量寿命的期望值。

Average life time $\bar{t}$ refers to the expected life of the random variable.

如果失效率是常数，则（If failure rate $\lambda(t)$ = C（Constant），so）：

$$\bar{t} = \frac{1}{\lambda} \tag{1.3}$$

例 1.1 设某元件失效密度函数 $f(t) = \lambda e^{-\lambda t}$，失效率 λ 为常数，若它的平均寿命 $\bar{t}$ 为 5000h，试求其失效率和使用 125h 时的可靠度 $R(125)$（百分数的小数点后至少保留一位小数）。

Example 1.1 Suppose a component failure density $f(t) = \lambda e^{-\lambda t}$, failure rate λ is a Constant. If its average $\bar{t}$ is 5000h, try to solve the failure rate λ and reliability degree $R(125)$ after working 125h（keep at least one decimal place in final result）.

解（Solution）：

$$R(t) = \frac{f(t)}{\lambda(t)} = \frac{\lambda e^{-\lambda t}}{\lambda} = e^{-\lambda t}$$

$$\lambda = \frac{1}{\bar{t}} = \frac{1}{5000\text{h}} = 2 \times 10^{-4} \frac{1}{\text{h}}$$

$$R(125) = e^{-\lambda t} = e^{-2\times 10^{-4}\times 125} = 0.975 = 97.5\%$$

例 1.2 某元件可靠度服从指数分布 $R(t) = e^{-\frac{t}{2000}}$，求平均寿命 $\bar{t}$ 和使用 200h 时的可靠度 $R(200)$（百分数的小数点后至少保留一位小数）。

Example 1.2 The reliability of a component obeys exponential distribution $R(t) = e^{-\frac{t}{2000}}$, try to solve the average life $\bar{t}$ and reliability degree $R(200)$ of the component after working 200h（keep at least one decimal in final result）.

解（Solution）：

平均寿命（Average life time）$\bar{t}$：

$$\bar{t} = \frac{1}{\lambda} = \frac{1}{\frac{1}{2000\text{h}}} = 2000\text{h}$$

使用 200h 时的可靠度（Reliability degree at t=200h）：

$$R(200) = e^{-\frac{t}{2000}} = e^{-\frac{200}{2000}} = e^{-0.1} = 0.905 = 90.5\%$$

1.4.2 维修性 (Maintainability)

评价机械设备使用性能时，一方面要考虑这些设备从开始工作到发生故障这段时间的可靠度和工作寿命，另一方面还要考虑到一旦发生故障，是否可以在较短时间内经过修理恢复到原来的工作状态，后者为机械设备所具有的维修性。

When evaluating the service performance of mechanical equipment, on the one hand, the reliability and service life of these equipment from the beginning of work to the failure should be taken into account; on the other hand, whether it can be restored to the original working state after repair in a short time in case of failure, the latter is the maintainability of mechanical equipment.

维修性是指在规定的条件下和规定的时间内，按规定的程序和方法进行维修时，保持和恢复到完成规定功能的能力，称为该设备的维修性。

Maintainability refers to the ability to maintain and restore mechanical equipment to complete specified function under specified conditions and within specified time, according to specified procedures and specified repair methods.

维修度是指在规定的条件和规定时间内，按规定程序和方法进行维修时，保持和恢复到能完成规定功能的概率，称为该设备的维修度。维修度是可维修系统维修难易的客观指标。

Maintenance degree refers to the probability to maintain and restore to complete specified function under specified conditions and within the specified time, according to the specified procedures and specified repair methods. Maintenance degree is an objective index of maintenance difficulty of repairable system.

全部维修受很多因素影响。从机械设备和零件的本身而言，能否进行维修的难易程度和性质（易接近性、标准化程度、可测试性、结构合理性）都是在设计和安装过程中所确定的。维修性设计最高目标达到无维修，实现无维修设计。

Total maintenance is affected by many factors. From the point of view of mechanical equipment and parts themselves, The ease and nature of maintenance (accessibility, standardization, testability, structural rationality) are determined during the design and installation process. The highest goal of maintainability design is to achieve no maintenance and realize no maintenance design.

维修时间是一个随机变量，描述维修时间概率分布尺度用维修度表示。维修性表示定性概念；维修度表示定量尺度。

Maintenance time is a random variable, and the probability distribution scale of describing maintenance time is expressed by maintenance degree. Maintainability means qualitative concept. Maintenance degree represents quantitative scale.

用于计算维修度的函数包括维修度函数 $M(t)$，不可维修度函数 $G(t)$，维修密度函数 $m(t)$，修复率 $\mu(t)$ 和平均修复时间（MTTR）$\bar{t}'$.

Functions used to calculate maintainability degree include maintenance degree $M(t)$, unmaintenability $G(t)$, maintenance density $m(t)$, maintenance rate $\mu(t)$, and average time to repair（MTTR）$\bar{t}'$.

（1）维修度函数 $M(t)$。维修度是表示可维修系统在规定条件下进行维修并在规定时间内完成维修的尺度。它是在 $\tau \leqslant t$ 时间内完成维修的概率，$M(t)$ 越大，越易维修。$t=0$，$M(t)=0$；$t\to\infty$，$M(t)=1$。

Maintenance degree $M(t)$. The degree of maintenance $M(t)$ refers to the scale when the repairable system carries out the maintenance under the required conditions and within the required time. It is a probability of finishing maintenance within the required time. The larger the $M(t)$ of the equipment, the easier it is to maintain. $t=0$, $M(t)=0$; $t\to\infty$, $M(t)=1$.

（2）不可维修度函数 $G(t)$。不可维修度函数 $G(t)$ 是指在规定时间内未完成维护的概率。$t=0$，$G(t)=1$，$t=\infty$，$M(t)=0$，$G(t)=1-M(t)$。

Unmaintenability $G(t)$. Unmaintenability $G(t)$ refers to the probability of finishing maintenance within the required time. $t=0$, $G(t)=1$, $t=\infty$, $M(t)=0$, $G(t)=1-M(t)$.

（3）维修密度函数 $m(t)$。维修密度 $m(t)$ 指单位时间内的维修程度。

Maintenance density $m(t)$. Maintenance density $m(t)$ refers to the maintenance degree per unit time.

$$m(t)=\frac{\mathrm{d}M(t)}{\mathrm{d}t} \tag{1.4}$$

$$M(t)=\int_0^t m(t)\,\mathrm{d}t \tag{1.5}$$

（4）修复率 $\mu(t)$。修复率 $\mu(t)$ 是指在规定的条件下、规定的时间内，产品在任何规定的维修级别上所修复的故障总数与该级别上的总维修时间的比值。

Maintenance rate $\mu(t)$. Maintenance rate $\mu(t)$ refers to the ratio of the total number of failures repaired at any required repair level of the product to the total repair time at this level under the required conditions and within the required time.

$$\mu(t)=\frac{m(t)}{G(t)}=\frac{m(t)}{1-M(t)} \tag{1.6}$$

$$M(t)=\frac{\mu(t)-m(t)}{\mu(t)} \tag{1.7}$$

若 $\mu(t)=\mu$ 为常数，则（If failure rate $\mu(t)=\mathrm{C}$（Constant），so）：

$$G(t)=\mathrm{e}^{-\mu t} \tag{1.8}$$

$$M(t)=1-G(t)=1-\mathrm{e}^{-\mu t} \tag{1.9}$$

（5）平均修复时间 $\bar{t}'$。平均修复时间等于各次修复时间总和除以修理次数。

Mean time to repair（MTTR）$\bar{t}'$. Mean time to repair（MTTR）$\bar{t}'$ is equal to the total repair time divided by the repair times.

$$\bar{t}'=\frac{\text{各次修复时间总和(total repair time)}}{\text{修理次数(repair times)}} \tag{1.10}$$

若 $\mu(t)=\mu$ 为常数，则（If failure rate $\mu(t)=\mathrm{C}$（Constant），so）：

$$\mu = \frac{1}{\bar{t}'} \tag{1.11}$$

例 1.3 维修密度函数 $m(t)$ 按指数分布 $m(t) = \mu e^{-\mu t}$，在某时间间隔内设备共进行 4 次修理，其时间为 5h，3h，6h，10h，求设备在 10h 修复的维修度 $M(10)$（百分数的小数点后至少保留一位小数）。

Example 1.3 Maintenance density $m(t)$ is distributed exponentially $m(t) = \mu e^{-\mu t}$. The equipment was repaired 4 times in a certain time interval. The time is 5, 3, 6, 10 hours. Please calculate the maintenance degree $M(10)$ of equipment repaired within 10 hours (Keep at least one decimal in final result).

解 (Solution):

求 (Calculate) $\bar{t}'$:

$$\bar{t}' = \frac{\text{各次修复时间总和(total repair time)}}{\text{修理次数(repair times)}} = \frac{5 + 3 + 6 + 10}{4} = 6$$

求 (Calculate) μ:

$$\mu = \frac{1}{\bar{t}'} = \frac{1}{6}$$

求 (Calculate) $M(10)$:

$$M(t) = \frac{\mu(t) - m(t)}{\mu(t)} = 1 - e^{-\mu t}$$

$$M(10) = 1 - e^{-\mu t} = 1 - e^{-\frac{10}{6}} = 0.811 = 81.1\%$$

1.4.3 有效度 (Efficiency)

机械设备的有效度是把设备从开始工作到发生故障（可靠性问题），从发生故障后进行维修（维修性问题）这两个阶段的可靠度和维修度结合起来，用一个统一尺度，来评价全部过程中有效工作的程度，就是机械设备的有效度（有效利用率）用 A 表示（如图 1.5 所示），用式（1.12）计算。

The efficiency of mechanical equipment is a scale that combines the reliability and maintenance of the equipment from the beginning of operation to the failure (reliability problem) and the maintenance after the failure (maintainability problem) to evaluate the effectiveness of the equipment throughout the process. The efficiency (effective utilization) of mechanical equipment is denoted by A (as is shown in Figure 1.5), and can be calculated by formula (1.12).

$$A = \frac{\text{MTBF}}{\text{MTBF} + \text{MTTR}} \times 100\% \tag{1.12}$$

式中 MTBF——平均故障间隔期（可靠度尺度）即平均工作时间（The mean failure interval (reliability scale) is the mean operating time）：

$$\text{MTBF} = \frac{\sum t_i}{n} = \frac{t_1 + t_2 + \cdots + t_n}{n} \tag{1.13}$$

t_i ——工作时间（The working time）；

MTTR——平均维修时间（维修度尺度），即平均故障时间 $\bar{t}'$（The mean maintenance time (maintenance degree scale), is the mean time to failure $\bar{t}'$;

$$\mathrm{MTTR} = \frac{\sum t_{0i}}{n} = \frac{t_{01} + t_{02} + \cdots + t_{0n}}{n} \tag{1.14}$$

t_{0i}——停机维修时间（The down time）。

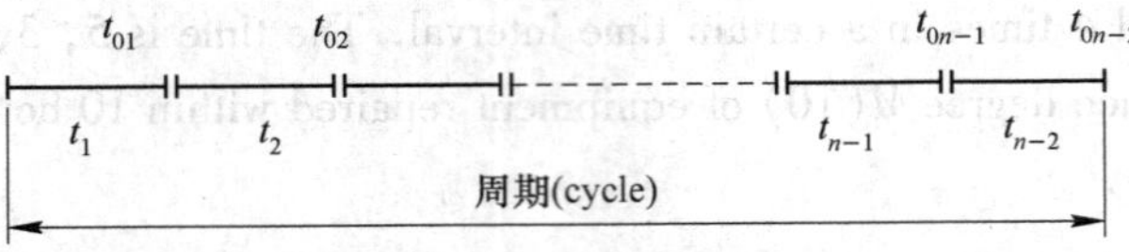

图 1.5　机械设备工作的一个周期

Figure 1.5　Cycle of mechanical equipment operation

例 1.4　某台设备运转 10000h，在此期间共发生 10 次故障，每次处理故障的时间平均 9h，计划检修时间共计 300h，试求这台设备的有效度（百分数的小数点后至少保留一位小数）。

Example 1.4　During the operation of a certain equipment for 10000h, a total of 10 failures occurred during this period. The average time for each failure was 9h, and total planned maintenance time was 300h (keep at least one decimal in final result).

解（Solution）：

$$\mathrm{MTBF} = \frac{10000}{10} = 1000\mathrm{h}$$

$$\mathrm{MTTR} = \frac{300 + 9 \times 10}{10} = 39\mathrm{h}$$

$$A = \frac{\mathrm{MTBF}}{\mathrm{MTBF} + \mathrm{MTTR}} \times 100\% = 96\%$$

1.5　统筹法（Overall Planning Method）

机械设备修理是一项复杂的工作，必须统筹安排。运用统筹法编制修理计划可以统筹全局，最优安排工作秩序，找出关键工序，从而达到缩短工期，节约人力、材力，减少投资的目的。工程负责人、施工技术人员和工人都应该掌握这种方法，用它来指导检修工作。

Mechanical equipment repair is a complex work and must be overall arranged. Overall planning method can be used to plan the overall repair project, get the optimal arrangement of work order, find out the key process, so as to shorten the time limit, save manpower, material, and reduce investment. Project leader, construction technicians and workers should master this method and use it to guide the maintenance, installation and other work.

一项工程总是包含多道工序，依照各工序间的衔接关系，用箭头表示其先后次序，画出一个表示各项任务相互关系的箭头图，注上时间，算出并标明主要矛盾线，这个箭头图称为统筹图或工序流线图。

A project always contains multiple processes. Arrows are used to indicate the order of each process according to the relationships among all processes. The time each process needs to be finished is given. Calculate and indicate the principal contradiction line. This diagram with arrows is called the overall plan process flow chart.

下面举例说明统筹图的组成及绘制方法。如大修一台机床包括十道工序：拆卸、清洗、检查、零件修复、零件加工、床身与工作台研合、变速箱组装、部件组装、电器检修和安装、装配和试车。大修一台机床的工序流线图如图 1.6 所示。

The following example illustrates the composition and drawing method of the overall plan process flow chart. Overhauling a lathe includes ten procedures: disassembly, cleaning, check, parts repair, parts processing, lathe bed and workbench to grind, gearbox assembly, component assembly, electrical overhaul and installation, assembly and test run. The overall plan process flow chart of overhauling a lathe maintenance project is drawn according the relationship of each process and shown in Figure 1.6

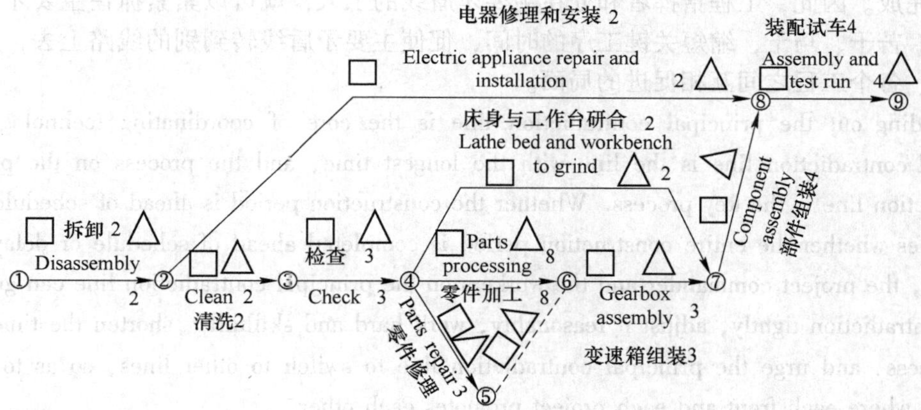

图 1.6 车床大修统筹图

Figure 1.6 The Overall plan process flow chart of overhauling a lathe

在图 1.6 中，箭尾表示工序开始，箭头表示工序完成，从箭尾到箭头表示一道工序过程。箭线把各个结点连接起来，并表明各工序先后顺序和相互关系。

In Figure 1.6, the tail of an arrow line indicates the beginning of process. The head of an arrow line indicates the completion of process. An arrow line indicates a process. Arrow line connects each node and indicates the order and relationship of each process.

拆卸工序需要 2 天，清洗需要 2 天，检查需要 3 天，零件修理需要 3 天，零件加工需要 8 天，变速箱组装需要 3 天，床身和工作台研合需要 2 天，部件组装需要 4 天，电器维修和安装需要 2 天，装配试车需要 4 天。虚线箭头表示虚设工序，需要 0 天。

The process of disassembly takes 2 days, cleaning takes 2 days, check takes 3 days, parts repair takes 3 days, parts processing takes 8 days, gear assembly takes 3 days, lathe bed and workbench to grind need 2 days, component assembly takes 4 days, electrical appliances repair and installation takes 2 days, assembly and test run takes 4 days. Dotted arrow line represents free process takes 0 days.

解题包括三个步骤：第一步，找出所有可能的路线，计算各路线工时，确定主要矛盾线和项目总工期；第二步，计算每道工序的最早可能开工时间和最迟必须开工时间；第三步，计算每道工序的时差。

Problem solving includes three steps. First step is identifying all possible routes, calculating total time of each route, and determining the main contradiction lines and the total project duration. Second step is calculating the earliest possible start time and the latest mandatory start time of each process. Third step is calculating the time difference of each process.

(1) 找出所有可能的路线，确定主要矛盾线和项目总工期（Finding out all possible line and determine the principal contradiction line and total duration of the project）。

找出主要矛盾线是统筹技术的核心。主要矛盾线是所有路线中工时最长的一条路线，处于主要矛盾线上的工序是关键工序，它的工期提前与否，决定着整个工程工期提前完成或推迟完成。因此，工程指挥者和处在主要矛盾线的工人，就可以紧紧抓住主要矛盾，合理调整，苦干、巧干，缩短关键工序的时间，促使主要矛盾线转到别的线路上去，形成各条战线、各个工程之间互相促进的局面。

Finding out the principal contradiction line is the core of coordinating technology. The principal contradiction line is the line with the longest time, and the process on the principal contradiction line is the key process. Whether the construction period is ahead of schedule or not determines whether the entire construction period is completed ahead of schedule or delayed. In this way, the project commander and the workers on the principal contradiction line can grasp the main contradiction tightly, adjust it reasonably, work hard and skillfully, shorten the time of the key process, and urge the principal contradiction line to switch to other lines, so as to form a situation where each front and each project promotes each other.

找主要矛盾线的方法是在画好工序流线图后，算出每条线路的总工期，其中工期最长的路线就是主要矛盾线。

The way to find the principal contradiction line is to work out the total construction period of each line after drawing the process flow diagram. The line with the longest time is the principal contradiction line.

运用图 1.6 数据找主要矛盾线。先计算每条路线所需要的总时间：

Figure 1.6 is used to find the principal contradiction line. Calculate the total time required for each route:

第一条路线（Route 1）：①—②—⑧—⑨　2+2+4=8（days）。

第二条路线（Route 2）：①—②—③—④—⑦—⑧—⑨　2+2+3+2+4+4=17（days）。

第三条路线（Route 3）：①—②—③—④—⑥—⑦—⑧—⑨　2+2+3+8+3+4+4=26（days）。

第四条路线（Route 4）：①—②—③—④—⑤—⑥—⑦—⑧—⑨　2+2+3+3+0+3+4+4=21（days）。

第三条路线所需时间最长，所以是主要矛盾线。项目总工期为26天。

Route 3 takes the longest days, so it is the principal contradiction line. 26 days is the total duration of the project.

值得注意的是，主要矛盾线可能不止一条，次要矛盾线也可能用其他颜色标出。对主要矛盾线进行改造，重新绘制大修计划工艺流程图。大修计划工艺流程图必须从非主要矛盾线中绘制人员以支撑主要矛盾线后重新绘制。

It is worth noting that there may be more than one principal contradiction line, and the secondary contradiction line may also be marked in other colors. The principal contradiction line can be transformed, then the overhaul plan process flow chart needs to be redrawn. The overhaul plan process flow chart must be remade after drawing personnel from the non - principal contradiction line to support the principal contradiction line.

(2) 计算每道工序的最早可能开工时间和最迟必须开工时间（Calculation of the earliest possible start time and the latest must start time of each process）。

找出主要矛盾线，可以看到非主要矛盾线上的项目是有潜力可挖的。潜力到底有多大？要靠计算时差来解决。时差实际上是指每道工序的最晚必须开始时间和最早可能开始时间之间的时差。一般情况下，非主要矛盾线上工序存在时差，主要矛盾线上工序的时差应为0。由于时差，人员可从非主要矛盾线中提取支撑主要矛盾线。为了计算时间，有必要知道每道工序最早可能开工时间和最晚必须开工时间。以图1.4为例讨论时差计算方法。

After finding out the principal contradiction line, we can see that the project on the non-principal contradiction line has potential to be arranged. How big is the potential? It depends on time difference. The time difference actually refers to the time difference between the latest start time and the earliest possible start time of each process. Generally, the time difference exists for the processes on the non-principal contradiction line, and the time difference on the principal contradiction line should be 0. Because of the time difference, personnel can be drawn from the non-principal contradiction line to support the principal contradiction line. To calculate the time, it is necessary to know the earliest possible start time and the latest necessary start time of each process. Figure 1.4 is taken as an example below of the earliest possible start time and the latest must start time of each process.

计算最早的可能开工时间应该从第一个工序沿着箭头方向，一步一步，直到最后一个工序的流程图。可能的最早开工时间为前一工序最早可能开工时间与前一工序时间之和。也就是说，第一个工序最早可能开工的时间为零，第二个工序可能最早的开工时间是第一个工序可能最早开工时间和第二个工序必须完成的时间之和。剩余工序的最早可能开工时间也根据这个规则计算。最早可能开工时间填在图中□里。

The calculation of the earliest possible start time should start from the first process along the arrow direction, step by step, until the final process of the flow chart. The earliest possible start time is the sum of earliest possible start time of preceding process and the preceding process time. That is to say, the earliest possible start time of the first process can start is zero. The earliest possible start time of second process is the sum of the earliest possible start time of the first process and the time the second process must be finished. The earliest possible start time of the rest processes are also calculated according to this rule. The earliest possible start time is filled in □ on the figure.

如果前面有多个工序，那么这个工序最早可能开工时间应该是前面所有工序的最大的可能开工时间。例如，在最后一个工序——装配试车前面有两个工序：电器维修和安装的最早可能开工时间是 2 天，部件装配的最早可能开工时间是 18 天，所以，装配试车的最早可能开工时间应该是 18 天加上部件装配这道工序的时间 4 天，等于 22 天。

If there is more than one preceding process, the earliest possible start time of this process should be the maximum possible start time of all preceding processes. For example, there are two processes before the last process—assembly and trail run process, the earliest possible start time of one of its preceding process—the electric appliance repair and installation process is 2 days, and the earliest possible start time of the other preceding process—the components assembly process is 18 days, so, the earliest possible start time for the assembly test should be 18 days plus 4 days for the assembly process, which is 22 days.

最迟必须开工时间的计算应该从最后一个工序开始，逆向计算到第一个工序。此工序的最迟必须开工时间是其后续工序的最新必须开工时间与此工序时间的差值。也就是说，项目的总持续时间是 26 天，最后一个工序必须完成的时间是 4 天，那么，最后一个工序的最迟必须开工的时间应该是 26-4=22 天。电器维修安装工序的最迟必须开工时间为最后一个工序的最迟必须开工时间 22 与该工序必须完成时间 2 的差值，故为 20 天。其余工序的最新必须开工时间也根据此规则计算。最迟必须开工时间填在图中△里。

The calculation of the latest must start time should start from the last process to the first process. The latest must start time of this process is the difference of the latest must start time of its following process and this process time. That is to say, the total duration of the project is 26 days, the last process must be finished in 4 days, then, the latest must start time of the last process should be 22 days. The latest must start time of electric appliance repair and installation process is the difference of the latest must start time of the last process, 22 days, and the time this process must be finished, 2 days, so it is 20 days. The latest must start time of the rest processes are also calculated according to this rule. The latest must start time is expressed in △.

如果有多个紧后工序，则此工序的最迟必须开工时间应为所有紧后工序的最小的最迟必须开工时间与此工序的启动时间之差。例如，第三个工序检查后面有三个工序，最迟必须开工时间分别是 16 天、7 天和 12 天，第三个工序的完成时间是 3 天，所以最迟必须开工时间应该是 7 和 3 的差，即 4 天。每个工序最早可能开工时间和最迟必须开工时间如图 1.7 所示。

If there is more than one following process, the latest must start time of this process should be the difference of the minimum latest must time of all following processes and this process. For example, there are three processes after the third process—check process, the latest must start time of them are 16, 7 and 12 respectively, the finishing time of the third process is 3 days, so the latest must time of the third process—check process should be the difference of 7 and 3, which is 4. The earliest possible start time and the latest must start time of each process are shown in Figure 1.7.

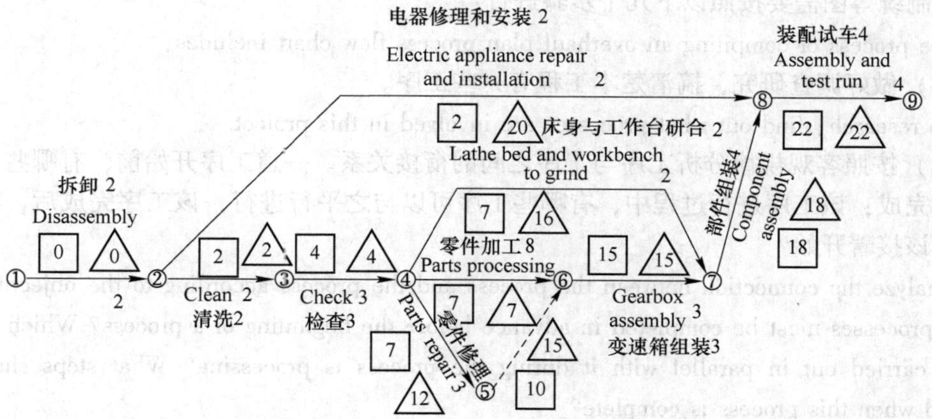

图 1.7　每道工序的最早可能开工时间和最迟必须开工时间

Figure 1.7　The earliest possible start time and the latest must start time of each process

（3）计算时差（Calculation of time difference of each process）。

计算完最早可能开工时间和最迟必须开工时间之后，就可以计算时差了。

After calculating the earliest possible starting time and the latest must starting time of each process, the time difference of each process can be calculated.

每道工序的时差等于该道工序最迟必须开工时间与最早可能开工时间之差。例如，第 3 道工序检查的时差就是这道工序最迟必须开工时间 4 与最早可能开工时间 4 之差，时差就是 0。

Time difference of each process is the difference of the latest must start time of this process and the earliest possible start time of it. For example, time difference of the third process—check process should be the difference of 4 and 4, which is 0.

每道工序的时间计算结果如表 1.1 所示。

The calculation results of time difference of each process can be expressed in Table 1.1.

表 1.1　车床大修每道工序的时差表

Table 1.1　Time difference of each procedure of lathe overhaul

①—②	②—③	③—④	②—⑧	④—⑦	④—⑥	④—⑤	⑥—⑦	⑦—⑧	⑧—⑨
0	0	0	18	9	0	5	0	0	0

有时差的工序，也就是有支援其他任务的潜力。主要矛盾线的工序时差必须等于零，否则就会延误整个任务完成期限。凡是时差为零的工序连接起来，就是主要矛盾线，这是要特别重视的线路，要加强控制，加强调度。

A process with time difference has the potential to support other tasks. The process time difference of the main contradiction line must be zero. Otherwise, the whole task will be delayed. All the working procedures with zero time difference are connected, which is the main contradiction line. Special attention should be paid to this line, and control and scheduling should be strengthened.

编制统筹图需要按照以下几个步骤进行：

The process of compiling an overhaul plan process flow chart includes：

（1）做好调查研究，搞清楚本工程有哪些工序。

Do research；find out what processes are involved in this project.

（2）按照客观规律分析工序与工序之间的衔接关系。一道工序开始前，有哪些工序必须先期完成；该工序进行过程中，有哪些工序可以与之平行进行；该工序完成后，有哪些工序应该接着开始？

Analyze the connection between the process and the process according to the objective law. Which processes must be completed in advance before the beginning of a process? Which process can be carried out in parallel with it during the process is processing? What steps should be followed when this process is complete?

（3）确定完成各工序所需的时间。

Determine the time required to complete each process.

第一，单一时间估计法（肯定型）。这种方法就是在估算各项工序时间时，已有定额资料可供参考，或有先例可循，这时只需要确定一个时间值。

First，single-time estimation（sure type）. In this way，a quota of data is available or precedents are followed when estimating the time of each process，and we only need to determine the time.

第二，三种时间估计法（非肯定型）。如果该项工作以前没有做过，或做的次数很少，估计一个时间定额难以估准，此时即可先预计三个时间值，然后再求可能完成时间的平均值，这三个时间是：最乐观时间、最保守时间和最可能时间。

Second，three kinds of time estimation methods（unsure type）. If the work has not been done before or has been done very few times，it is difficult to estimate a time quota. At this time，three time can be estimated first，and then the average of the possible completion time can be calculated. The three time are the most optimistic time，the most conservative time，and the most probable time.

最乐观时间，是指在顺利情况下，完成某工序可能出现的最短时间，用 a 表示。

The most optimistic time refers to the shortest time possible to complete a certain process under smooth conditions，which is denoted by a.

最保守时间，是指在不利情况下，完成某工序可能出现的最长时间，用 b 表示。

The most conservative time is the maximum time possible to complete a certain process under difficult conditions，denoted by b.

最可能时间，是指在正常情况下，完成某工序最可能出现的时间，用 m 表示。

The most probable time is the most likely time to complete a certain process under normal circumstances，denoted by m.

然后，按以下公式求出平均值，即：

Then，calculate the average value according to the following formula，namely：

$$t_{\Sigma} = \frac{a + 4m + b}{6} \tag{1.15}$$

这样就可以把非肯定型化为肯定型。

So you can convert the unsure type to the sure type.

（4）把施工任务分配到各施工单位，做好人力、设备和原材料的安排。

Assign construction tasks to the construction units, make a good arrangement of manpower, equipment, and raw materials.

（5）订好施工方案。

Work out the construction plan.

（6）绘统筹图。

Draw an overhaul plan process flow chart.

（7）时间计算。计算每项工序最早开工时间、最迟必须开工时间和时差，确定主要矛盾线并用红线（次主要矛盾线用其他颜色）标明。

Times calculation. Calculate the earliest starting time, the latest must start time and time difference of each process, determine the principal contradiction line, and mark it with red line (the secondary principal contradiction line is marked with other colors).

（8）根据主要矛盾线的长度和时差，对箭头图加以调整。

Adjust the arrow chart according to the length and time difference of the principal contradiction line.

思考题（Questions）

1.1　名词解释（Nouns explanation）：

可靠性（reliability），可靠度（reliability degree），维修性（maintainability），维修度（maintenance degree），点检（point checking）

1.2　设某元件失效密度函数 $f(t) = \lambda e^{-\lambda t}$，失效率 λ 为常数，若它的平均寿命为4000h，试求其失效率和使用150h后的可靠度（百分数的小数点后至少保留一位小数）。

Suppose the failure density of a component $f(t) = \lambda e^{-\lambda t}$, failure rate λ is constant. If its average life is 4000h, please calculate its failure rate and reliability degree after working for 150h (the percentage has at least one decimal).

1.3　某元件可靠度服从指数分布 $R(t) = e^{-\frac{t}{3000}}$，求平均寿命和使用后300h的可靠度（百分数的小数点后保留一位小数）。

The reliability of a component obeys exponential distribution $R(t) = e^{-\frac{t}{3000}}$, please calculate the average life and the reliability degree after working 300h (the percentage has at least one decimal).

1.4　目前，机械设备维修工程提出了哪些维修模式？

How many kinds of mode of mechanical equipment maintenance engineering have been proposed up to now?

1.5　机械设备维修工艺学分哪三部分内容？

What are the three contents of mechanical equipment maintenance engineering?

1.6　全员生产维修的特点、核心和目标分别是什么？

What are the characteristics, core and objectives of TPM?

1.7 点检的种类有哪些？

How many types of point checking are there?

1.8 推行 TPM 有哪三个关键因素？

What are the three key factors to implement TPM?

1.9 根据图 1.8，回答下列问题。

According to Figure 1.8, answer the following questions according to the process time（unit：days）in the plan below.

（1）找出图 1.8 中所有可能的路线并计算每条路线的时长；

Find out all possible process routes in the Figure 1.8 and calculate the work hours required for each route;

（2）确定主要矛盾线和总工期；

Determine the principal contradiction line and the total duration of the project;

（3）计算每道工序最早可能开工时间填在图 1.8 的□中，计算每道工序最晚必须开工时间填在图 1.8 的△中。

Calculate the earliest possible start time of each process and fill in □ of Figure 1.8. Calculate the latest must start time of each process and fill in △ of Figure 1.8.

（4）计算每道工序的时差填在表 1.2 中。

Calculate the time difference of each procedure and express it in Table 1.2.

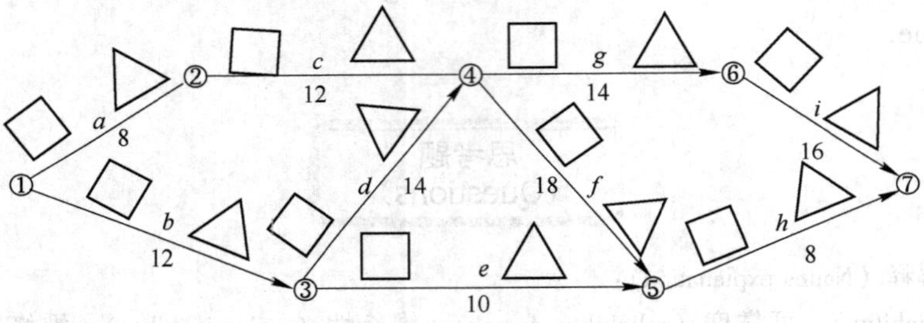

图 1.8　题 1.9 的图

Figure 1.8　Picture of question 1.9

表 1.2　每道工序的时差

Table 1.2　Time difference of each procedure

a	*b*	*c*	*d*	*e*	*f*	*g*	*h*	*i*

2 机械零件失效与设备故障

Chapter 2 Failure of Mechanical Parts and Equipment

2.1 机械零件的失效（Failure of Mechanical Parts）

机械设备的宏观性能的变化经常是由于零件或材料的失效引起的。机械零件常见的失效类型有：磨损失效、变形失效、断裂失效和腐蚀失效。

Changes in the macroscopic performance of mechanical equipment are often caused by the failure of parts or materials. Common failure types of mechanical parts include wear failure, deformation failure, fracture failure and corrosion failure.

2.1.1 磨损失效（Wear Failure）

据统计，机械设备故障模式中60%~80%是由磨损失效造成的，所以磨损失效最为常见。磨损是指表面材料因摩擦而产生的小损失现象。

Statistically, 60%~80% of mechanical equipment failure mode is caused by wear failure, so wear failure is one of the most common failure. Wear refers to the phenomenon of small loss of surface materials caused by friction.

2.1.1.1 摩擦磨损理论（Theory of Friction and Wear）

摩擦是指两个相互接触的物体在外力作用下相对运动或有相对运动趋势时，接触面之间产生的切向运动阻力现象。摩擦力是指摩擦时产生的切向阻力的大小。

Friction refers to the phenomenon of tangential motion resistance between contact surfaces when two bodies in contact with each other move relatively or have a relative movement trend under the action of external force. Friction refers to the tangential resistance caused by friction.

摩擦学是研究摩擦、磨损与润滑的学科。摩擦通常用分子力学的术语来解释，即摩擦由机械阻力和分子引力组成，这是粗糙表面和光滑表面摩擦的主要组成部分。

Tribology is a discipline of friction, wear, and lubrication. Friction is usually explained in molecular-mechanical terms, that is, friction consists of mechanical resistance and molecular attraction, which is the main component of friction for rough contact surfaces and for smooth surfaces.

当两个不平整的表面接触时发生机械阻力，凹面和凸面相互啮合。两个不平整的表面相对滑动并沿凸起部分反复抬升，凸起部分变形或破碎。此时所显示的阻力称为机械阻力。机械阻力和分子引力的大小是此消彼长的，因此在确定载荷条件下，常常存在一个最佳的粗糙度。表面粗糙度高于或低于最佳粗糙度，摩擦力都是增大的。

Mechanical resistance occurs when two uneven surfaces contact, and the concave and convex on the surface mutually mesh. Two uneven surfaces relatively slide and are repeatedly lifted along the convex part, so the convex part deforms or is broken. The resistance shown at this time is known as mechanical resistance. Mechanical resistance and molecular attraction rise and fall, so there is always an optimal roughness under certain loading conditions. Whether the surface roughness is higher or lower than the optimum roughness, the friction force increases.

2.1.1.2 摩擦的分类（Classification of Friction）

摩擦有很多不同的分类方法，其中常见的是按运动方式或润滑状况来分类的。

Friction can be classified by many different ways, most commonly by mode of motion or condition of lubricating.

按摩擦副的运动方式，摩擦分三类：滑动摩擦、滚动摩擦和混合摩擦。滑动摩擦是指摩擦副之间作相对滑动时所表现出的摩擦。滚动摩擦是指摩擦副之间作相对滚动时所表现出的摩擦。混合摩擦是指摩擦副之间既有相对滑动，又有相对滚动时所表现出的摩擦。如齿轮的啮合面之间的摩擦就是混合摩擦。

According to the motion mode of friction pair, there are three types of friction: sliding friction, rolling friction and mixed friction. Sliding friction is a kind of friction between friction pairs when they slide relative to each other. Rolling friction is a kind of the friction between the friction pairs when they are rolling relative to each other. Mixed friction is a kind of friction there is both relative sliding and relative rolling friction between the friction pairs. For example, the friction between the meshing surfaces of gears is mixed friction.

按摩擦副间的润滑状态，摩擦分为五类：干摩擦、半干摩擦、边界摩擦、半液体摩擦和液体摩擦。干摩擦是指摩擦副表面间完全没有润滑介质或其他杂质存在时所表现出的摩擦。液体摩擦是指摩擦副两表面间完全被润滑介质隔开，彼此间不发生直接接触的摩擦。边界摩擦是指摩擦副两表面间只有一层很薄（0.1μm 以下）的连续油膜存在时所产生的摩擦。半干摩擦是指摩擦副两表面间部分被很薄的油膜分开，部分直接接触，介于干摩擦和边界摩擦之间的一种摩擦方式。其特性取决于边界摩擦和干摩擦所占的比例。半液体摩擦是指介于液体摩擦与边界摩擦之间的一种摩擦形式，称为半液体摩擦。其特性取决于液体摩擦和边界摩擦所占的比例。

According to the lubricating state between friction pairs, there are five categories of friction: dry friction, semi-dry friction, boundary friction, semi-liquid friction and liquid friction. Dry friction occurs when there is no lubricating medium or no other impurities on the surface of the friction pairs. Liquid friction refers to the friction of the friction pairs which are completely separated by lubricating medium between the two surfaces and are without direct contact. Boundary friction refers to the friction generated when there is only a thin continuous oil film (less than 0.1μm) between the two surfaces of the friction pair . Semi-dry friction refers to a friction mode between two surfaces of a friction pair, which is separated by a very thin oil film and partially directly contacted, and is between dry friction and boundary friction. Its characteristics depend on the proportion of boundary friction and dry friction. Semi-liquid friction refers to a form of friction between liquid friction and boundary friction. Its properties depend on the proportion of liquid and boundary friction.

磨损会降低机械设备的运动精度，甚至使机器完全失去工作能力。磨损会缩短机器的寿命，增加维护时间和成本，增加材料消耗。磨损并不总是坏事。例如机器跑合阶段的磨损、利用磨损原理进行的机械加工（如研磨、抛光、磨削）等，都是利用磨损为生产服务的。常用的机械磨损理论有黏着理论和分子-机械理论。

Wear will reduce the movement accuracy of mechanical equipment, or even make the machine completely lose its working ability. Wear and tear will shorten the life of the machine, increase maintenance time and cost. and increase material consumption. Wear is not always a bad thing. For example, wear in the running-in stage of the machine, mechanical processing (such as grinding, polishing, grinding) using wear principle, etc. are all usage of wear to serve the production. The commonly used mechanical wear theory includes adhesion theory and molecular-mechanical theory.

2.1.1.3 磨损的分类（Classification of Wear）

根据导致磨损的主要原因，可将磨损分为以下四类：磨料磨损、黏着磨损、疲劳磨损和腐蚀磨损。

According to the main causes of wear, wear can be divided into the following four categories: abrasive wear, adhesive wear, fatigue wear and corrosion wear.

A 磨料磨损（Abrasive Wear）

磨料与零件表面相对运动，作用在磨料上的力可以分解为垂直于表面的分力和平行于表面的分力。垂直于表面的分力使磨料嵌入表面。对于塑性好的材料表面，产生大量密集的压痕，反复作用后，产生疲劳破坏；对于脆性材料，表面不发生变形就产生脆性破坏。平行于表面的分力使磨料产生切向运动，导致表面被刻划、切削而留下沟槽：对于塑性材料，磨料切削会在摩擦表面上切下一条切屑；对于脆性材料，一次就从表面上切下许多碎屑。

The abrasive moves relative to the surface of the part. The force acting on the abrasive can be decomposed into the component force perpendicular to the surface and the component force parallel to the surface. The component force perpendicular to the surface embeds the abrasive into the surface. For a plastic material surface, it produces a large number of dense indentations, and after repeated application, fatigue failure occurs. For the brittle material surface, failure occurs without surface deformation. The component force parallel to the surface of the components makes the abrasive move tangentially. As a result, the surface is carved, cut and grooves are left. For plastic materials, abrasive cutting will cut a chip on the friction surface. For brittle materials, many fragments are cut from the surface at one time.

磨料磨损的主要影响因素有磨料粒度、磨料的几何形状、磨料的硬度和压力等。磨损量随粒度增大而增加。但增大到临界尺寸以后，磨损速率保持不变。尖锐的磨料磨损速率高，但当磨料被磨钝后，磨损速率下降，若磨料被磨碎，磨损速率又会增加。

The main influencing factors of abrasive wear are abrasive particle dimension, abrasive geometry, abrasive hardness, and pressure. Wear increases with the increase of particle dimension. However, after increasing to the critical dimension, the wear rate remains unchanged. Sharp abrasive wear rate is high, but when the abrasive is blunt, wear rate decreases. If the abrasive is ground up, wear rate increases.

磨料的硬度远远超过零件的硬度时，影响不大。硬度相当时有影响，磨料硬度低于零件的硬度，差值增大，磨损速率下降；略高于零件的硬度，磨损严重。因此，用表面强化的办法使零件的硬度达到或超过磨料的硬度，可以提高零件的耐磨性。但硬度高的材料，韧性下降，脆性增加，会带来不良后果。

If abrasive hardness is far larger than the hardness of the parts, the influence is small. When the hardnesses of the two are similar, the influence occurs. When abrasive hardness is lower than the hardness of parts, the difference of hardness increases, and the wear rate decreases. When abrasive hardness is slightly higher than the hardness of the parts, wear is serious. Therefore, the hardness of the parts can reach or exceed the hardness of the abrasive by means of surface strengthening in order that the wear resistance of the parts can be improved. But, as material with high hardness, the toughness of the material will decrease and brittleness will increase. Bad consequence will occur.

磨损速率与压力成正比。常用的减轻磨料磨损措施包括减少磨料的进入、增强零件的抗磨性。

The rate of wear is proportional to the pressure. Common methods to reduce abrasive wear include reducing the entry of abrasive and enhancing the wear resistance of parts.

为了减少磨料的进入，可以配备高效、高容的空气滤清器及燃油、机油滤清器；增加用于防尘的密封装置如毡圈密封件等；在润滑系统中装入吸铁石、集屑房及油污染程度指示器；经常清理更换空气、燃油、机油滤清装置。为了增强零件的抗磨性，可以采用热处理和表面处理方法改善零件材料的性质，提高表面硬度，尽可能使表面硬度超过磨料硬度。

In order to prevent abrasive from entering into the contact surfaces, high-efficient, high-capacity air filter and fuel, oil filter can be equipped. Sealing devices such as felt ring seals for dustproof add can be used. Magnet, chip collecting room and oil pollution level indicator can be installed in the lubricating system. The air, fuel and oil filter should be clean and replaced frequently. In order to enhance the wear resistance of parts, heat treatment and surface treatment can be used to improve the properties of parts, improve the surface hardness, as far as possible to make the surface hardness more than the abrasive hardness.

选用耐磨性能好的材料。对于要求耐磨又有冲击载荷作用的零件，可采用中碳钢淬火，低温回火，得到马氏体钢的办法使零件既具有耐磨性，又具有较好的韧性。选用一硬一软的摩擦副，使磨料被软材料所吸收，减少磨料对重要、价高材料的磨损。例如蜗轮蜗杆一般采用铸铁、铸青铜、锡青铜、铝青铜等软材料，而蜗轮蜗杆一般采用碳钢或合金钢等硬材料。

Materials with good wear resistance should be chosen. For the parts which require not only wear resistance but also impact load, the method of quenching medium carbon steel, tempering at low temperature and obtaining martensite steel can be used to make the parts have both wear resistance and good toughness. A hard-soft friction pair should be chosen. Therefore, the abrasive can be absorbed by the soft material to reduce the wear of the abrasive to the important and expensive material. For example, worm gear generally uses soft materials such as cast iron, cast bronze, tin bronze, and aluminum bronze, while worm gear generally uses hard materials such as carbon steel or alloy steel.

注意在磨料磨损和腐蚀磨损同时存在的摩擦副中，仅用提高摩擦副的硬度，而较少考虑提高耐蚀性的情况下，往往会使磨损加剧。如通过热处理或选用中高碳淬火钢作酸性砂泵的耐磨材料，其磨损速率比低硬度耐蚀性较好材料的磨损速率更大。

It should be noted that in the friction pair where abrasive wear and corrosion wear exist simultaneously, only the hardness of the friction pair being improved, and less consideration being given to improving corrosion resistance will often aggravate the wear. For example, the wear rate of choosing high carbon quenching steel as acid sand pump wear resistant material or adopting heat treatment is greater than that of materials with low hardness and better corrosion resistance.

B 黏着磨损（Adhesive Wear）

黏着磨损是指构成摩擦副的两个摩擦面，在相对运动时，由于黏着作用，一个表面上的材料转移到另一个表面上所引起的磨损。

Adhesive wear refers to the wear caused by the transfer of materials that make up the friction pair transfer from one surface to the other during the relative movement of the two friction surfaces.

摩擦副在重载条件下工作，由于润滑不良，相对运动速度高、摩擦产生的热量来不及散掉，摩擦副表面产生极高的温度，材料表面强度降低，承受高压的表面凸起部分相互黏着，继而在相对运动中被撕裂下来，材料从强度低的表面上转移到材料强度高的表面上，造成摩擦副的灾难性破坏，如咬死或划伤。

Friction pair works under overload conditions. The friction pair surface will produce high temperature due to bad lubricating, high relative speed, the heat generated by friction has no time to dissipate. Therefore, the surface temperature of friction pair increase to a high degree. The strength of the material surface decrease. The convex parts of the surface under high pressure stick to each other and then was torn down in relative motion. The material transfers from the surface with low intensity to the surface with high strength. The friction pair suffers catastrophic damage, such as scuffing or scratch.

黏着磨损的主要影响因素有摩擦副的表面状态、摩擦副材料表面成分与组织等。摩擦表面洁净，无吸附膜，易发生黏着磨损。可使用适当的润滑剂。

The main influencing factors of adhesive wear are surface state of friction pair, composition and structure of friction pair material surface. When the friction surface is clean and there is no adsorption film, it is easy to occur adhesive wear. Appropriate lubricant should be used.

构成摩擦副的两摩擦表面的材料相互间形成固溶体的趋势，直接和黏着磨损有关，两者越易形成固溶体或金属间化合物，越易发生黏着磨损。所以同类金属或原子结构、晶体结构相近的材料，比性质有明显差异的材料构成的摩擦副更易发生黏着磨损。

The tendency of solid solution formation between the materials on the two friction surfaces forming the friction pair is directly related to adhesive wear. Therefore, similar metals or materials with similar atomic and crystal structures are more likely to have adhesive wear than friction pairs made up of materials with obvious differences in properties.

因此选用性质差异大的材料构成摩擦副，是降低黏着磨损的有效途径。在摩擦副的一个表面上覆盖铅、锡、银、铟或者软的合金都可以提高抗黏着磨损的能力（如图 2.1 所示）。如巴氏合金等常用作轴承衬瓦的表面材料，就是为了提高其抗黏着磨损的能力。

Therefore, it is an effective way to reduce adhesive wear to choose materials with large difference in properties to form friction pair. Coating a surface of a friction pair with lead, tin, silver, indium, or a soft alloy improves the ability to resist adhesive wear (as shown in Figure 2.1). For example, Babbitt alloy is commonly used as material of bearing lining surface because it can improve the ability of bearing lining surface to resist adhesive wear.

图 2.1 表面涂层

Figure 2.1 Surface Coating

C 疲劳磨损（Fatigue Wear）

疲劳磨损是指摩擦副材料表面上局部区域在循环接触应力作用下产生疲劳裂纹，由于裂纹扩展并分离出微片和颗粒的一种磨损形式。

Fatigue wear occurs in local areas on the surface of friction pair materials under the action of cyclic contact stress. As a result, crack grows and separates microchips and particles.

根据摩擦副间的接触和相对运动方式，可将疲劳磨损分为：滚动接触疲劳磨损和滑动接触疲劳磨损。滚动轴承、传动齿轮等有相对滚动摩擦副表面间出现的麻点和脱落现象都是由滚动接触疲劳磨损造成的。滚动接触疲劳磨损的特点是经过一定次数的循环接触应力的作用，麻点或脱落才会出现。实际上，纯滚动接触疲劳磨损是比较少见的，大多数情况下含有滑动成分。滑动接触疲劳磨损是指作用于表面上污秽载荷会使表面产生压平或压入，触点区产生相应的应力和应变，摩擦运动时的反复作用造成了触点处结构、应力状态的不均匀和应力集中，从而造成裂纹的萌生与扩展，最终使部分表面材料以微粒形式从表面脱落，形成磨屑。

The fatigue wear can be divided into rolling contact fatigue wear and sliding contact fatigue wear according to the contact and relative motion modes between friction pairs. The pitting and dropping phenomenon between the surface of rolling bearing and transmission gear are caused by rolling contact fatigue wear. Rolling contact fatigue wear is characterized by pitting or shedding after a certain number of cycles of contact stress. In fact, pure rolling contact fatigue wear is relatively rare and sliding components are contained in most cases. Sliding contact fatigue wear refers to normal load applied to a surface causes the surface be flat or pressed in and the corresponding stress and strain are produced in contact area. The repeated actions of contact friction motion cause uneven structure, stress condition and stress concentration, lead to crack initiation and propagation, eventually make part of the surface material fall off from the surface in the form of particles, wear debris form.

凡是影响裂纹萌生的扩展的因素都对接触疲劳磨损有影响，影响滚动接触疲劳磨损的主要因素包括材质、粗糙度、表面应力状态、配合精度的高低、润滑油的性质等。钢中非金属夹杂物的存在易引起应力集中，在这些夹杂物的边缘形成裂纹，从而使疲劳磨损易于发生。材料的组织状态、内部缺陷等对磨损也有重要影响。通常晶粒均匀、细小、碳化物体小且分布均匀的组织抗疲劳裂纹产生的能力强。

All the factors that affect crack initiation and propagation have influence on contact fatigue wear. The main factors that affect rolling contact fatigue wear include material, roughness, surface stress state, matching accuracy, and properties of lubricating oil. The existence of nonmetallic inclusions in steel tends to cause stress concentration, and cracks are formed at the edges of these inclusions, so fatigue wear is easy to occur. The microstructure and internal defects of materials also have an important influence on wear and tear. Generally, the structure with small and uniform grain dimension, small carbonized object and uniform distribution have strong ability to resist fatigue crack.

硬度在一定范围内增加，其抗疲劳磨损的能力增加，因此对轴承钢的表面和传动齿轮啮合面都要求其硬度在 HRC60 左右，摩擦副各表面的最佳硬度应根据工况和运动方式不同通过试验来确定，对于已磨损的摩擦副的修复，用试验的方法来确定各表面的最佳硬度值，对于多数机修部门来说是有困难的。多数情况下，使修复后各表面的硬度与原始值相近，即可认为这种修复方式是可行的。

When the hardness of material increases within a certain range, the ability of the material to resist fatigue wear increases. So, the hardness of bearing steel surface and the mesh surface of driving gear are required about HRC60. The optimum hardness on the surface of the friction pair should be determined according to different working condition and movement way through the experiment. It is difficult to determine the best the surface hardness value using experimental method for most of the engineering department. In most cases, if the hardness of each surface after repaired is close to the original value, it can be considered that this repair method is feasible.

实践表明，在一定范围内降低表面粗糙度是提高抗疲劳磨损能力的有效途径。例如，滚动轴承表面粗糙度 *Ra* 由 0.4μm 降低到 0.2μm 时，使用寿命提高 2~3 倍；粗糙度 *Ra* 由 0.2μm 降低到 0.1μm 时，使用寿命提高 1 倍。粗糙度 *Ra* 继续减小，对使用寿命的影响较小。

The practice shows that reducing the surface roughness in a certain range is an effective way to improve the anti-fatigue wear ability. For example, when the surface roughness of rolling bearing *Ra* decreases from 0.4μm to 0.2μm, its service life increases by 2~3 times. When the roughness *Ra* decreases from 0.2 micron to 0.1 micron, the service life increases by 1 times. If roughness *Ra* decreases continuously, there will be little impact on service life.

D 腐蚀磨损(Corrosive Wear)

腐蚀磨损是指在腐蚀性环境中，摩擦表面上发生的比单纯机械摩擦损失与单纯腐蚀损失之和高得多的一种破坏形式。腐蚀磨损的主要特点是磨损过程中既有机械摩擦起作用，又有腐蚀破坏起作用，表现出极高的磨损速率。通常是摩擦为腐蚀发生提供了新鲜的金属表面，腐蚀改变了金属表面的性质，使摩擦造成的磨损速度加快。

Corrosion wear refers to a form of damage that occurs on the friction surface in corrosive environment, which is much higher than the sum of the pure mechanical friction loss and the pure corrosion loss. The main characteristics of corrosion wear are the concurrence of mechanical friction and corrosion damage, which show high wear rate. It is usually friction provides fresh metal surfaces for corrosion to occur. Corrosion changes the properties of metal surfaces and accelerates the rate of wear caused by friction.

根据腐蚀介质的特性，通常将腐蚀磨损分为：氧化磨损和特殊介质中的腐蚀磨损。氧化磨损是指在摩擦过程中摩擦表面在空气中氧或润滑剂中氧的作用下所生成的氧化膜很快被机械摩擦去除的磨损形式。特殊介质中的腐蚀磨损是指在摩擦过程中，环境中的酸、碱等电解质作用于摩擦表面上所形成的腐蚀产物迅速被机械摩擦所除去的磨损形式，叫做特殊介质中的腐蚀磨损。

According to the characteristics of corrosive medium, corrosion wear is usually divided into oxidation wear and corrosion wear in special medium. Oxidation wear refers to a form of wear in which the oxide film formed on the friction surface under the action of oxygen in the air or in the lubricant is quickly removed by mechanical friction. Corrosion wear in special medium refers to a form of wear in the process of friction the corrosion products formed by the action of environment electrolytes such as acid and alkali on the friction surface are quickly removed by mechanical friction.

2.1.1.4 磨损的规律(Wear Law)

机器在工作中各机件磨损发展情况随工作条件而异，但所有机件磨损的发展却有着共同的规律。图2.2是零件的磨损量 W 随时间 t 增长的变化曲线，称磨损特性曲线。它表示了机件在设计制造合理，且在正常工作条件下，磨损增长的情况。

The development of machine wear varies with the working conditions, but the development of all machine wear has a common rule. Figure 2.2 is the wear characteristic curve describe the change of wear quantity W with time t. It shows that wear of the machine parts designed and manufactured reasonably increases under normal working conditions.

从图2.2可以看出，曲线分三个明显的部分，根据曲线的特性可将磨损分为三个阶段。OA 段为磨合（初期）磨损时期：磨损速率下降。由于新加工零件表面比较粗糙，因此零件的磨损十分迅速，随着时间的延长，表面粗糙度下降，实际接触面增大，凸起部分磨平所造成的塑性变形导致冷作硬化，所以磨损速率逐渐下降，且具有最低的数值。A 点

越靠近 O 点越好，这样就可以在短时间内以最低的磨损量达到磨合要求。选用合适的磨合载荷，相对运动速度、润滑条件等参数是尽快达到正常磨损的关键。AB 段为正常（工作）磨损时期（或叫自然磨损时期）：磨损成直线缓慢上升，磨损速率小且稳定。该时间的长短称为零件修理前的使用寿命。一切可延长该阶段持续时间的措施都有利于减轻零件的磨损，提高零件的使用寿命。合理地保养与维护是延长零件使用寿命的关键。BC 段为强烈（事故或急剧）磨损时期：磨损速率迅速增加。这是由于工作条件恶化、零件几何尺寸改变、配合间隙增大，润滑条件改变、并附加冲击作用存在造成的。在这种情况下，磨损加剧，零件迅速破坏，有可能出现大事故。因此及时发现和修理即将进入该阶段工作的零部件具有十分重要的意义。

As can be seen from Figure 2.2, the curve is divided into three sections, and the wear can be divided into three stages according to the characteristics of the curve. Section OA is the run-in (initial) wear period. The wear rate decreases in this section. As the surface of the newly machined parts is relatively rough, the wear of the parts is very rapid. As time goes on, the surface roughness decreases and the actual contact surface increases. The plastic deformation caused by the flattening of the bulge part leads to cold hardening. The closer point A is to point O, the better is the situation. In this way, it can meet the requirements of running in a short time with the minimum wear. It is the key to achieve normal wear as soon as possible to choose proper running load, relative moving speed, lubricating condition, and other parameters. Section AB is the normal (working) wear period (or natural wear period). Wear rises slowly in a straight line with a small and stable wear rate. The length of time is called the service life of the part before repair. All methods that can extend the duration of this stage are beneficial to reduce the wear and improve the service life of parts. Reasonable maintenance is the key to extend the service life of parts. Section BC is a period of intense (accident or sharp) wear. Wear rate increases rapidly in this section. This is due to the deterioration of working conditions, geometric dimensions of parts change, fit gap increase, lubricating conditions change, and additional impact. In this case, wear speeds up and parts are quickly destroyed. There may be a big accident. Therefore, it is of great significance to timely discover and repair the parts that are about to enter this stage.

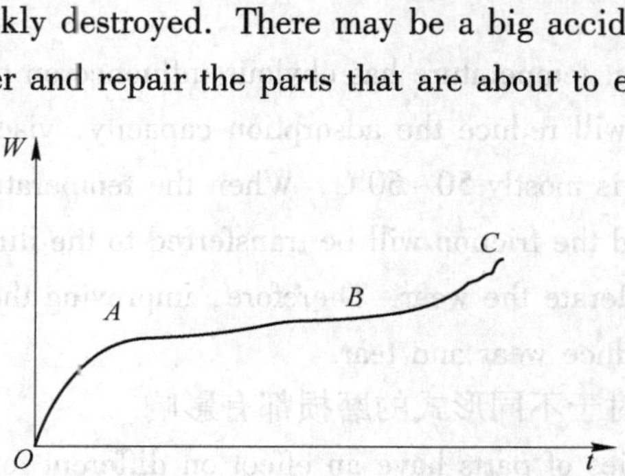

图 2.2 磨损特性曲线

Figure 2.2 Wear characteristic curve

2.1.1.5 影响磨损的主要因素（The Main Factors Affecting the Wear）

影响磨损的主要因素有：摩擦副的相对运动速度及压力、润滑情况、工作温度、零件材质、表面加工质量及配合间隙等。因此，为了减轻磨损，就必须采取措施来保证上述参数在合理的设计和使用范围内。

The main factors affecting the wear are the relative speed and pressure of friction pair, lubricating, working temperature, parts material, surface processing quality and fit gap, etc. Therefore, in order to reduce wear, methods must be taken to ensure that the above parameters are in a reasonable range of design and use.

在润滑状态良好的情况下，零件相对运动速度越高，越易形成液体摩擦，磨损越轻微。若润滑不良，速度越高，磨损越严重。因此，避免长时间在高速、过载及有冲击载荷条件下工作，可减轻零件的磨损。润滑状况的好坏直接影响零件的磨损行为，若能保证零件在液体摩擦状态下工作，则可以不考虑磨损问题，但实际上，零件工作时，绝大多数是介于边界摩擦和液体摩擦之间，因此，改善润滑使摩擦副间形成具有一定厚度和承载能力的油楔，对于促进载荷均匀分布，降低微凸处的单位压力大有好处，从而可以减轻磨损。

Under the condition of good lubricating, the higher the relative moving speed of the parts, the easier the formation of liquid friction, the less wear. The poorer lubricating, the higher the speed, the more serious wear. Therefore, avoiding the machine working under high speed, overload and impact load for a long time can reduce the wear of parts. The lubricating conditions directly affect the wear behavior of parts. If parts work can be guaranteed under liquid friction state, the problem of wear can be ignored. But in fact, the great majority parts work between boundary friction and fluid friction. Therefore, improving the lubricating by forming certain thickness between the friction pair and oil wedge with carrying capacity is of great benefit to promote the uniform distribution load, reduce the unit pressure of micro convex place, moreover, reduce wear.

工作环境温度对润滑油的油性有明显的影响，温度升高会使油的吸附能力下降，黏度降低，化学稳定性变差。安全温度多为 50~60℃，当温度升高到 150~200℃时，油膜会破坏，摩擦会向边界摩擦或干摩擦方向转化，使磨损加快。因此，改善零件的工作环境，对减轻磨损很有好处。

The working environment temperature has obvious influence on the oiliness of lubricating oil. The increase of temperature will reduce the adsorption capacity, viscosity, and chemical stability of oil. The safe temperature is mostly 50~60℃. When the temperature rises to 150~200℃, the oil film will be damaged, and the friction will be transferred to the direction of boundary friction or dry friction, which will accelerate the wear. Therefore, improving the working environment of the parts is very beneficial to reduce wear and tear.

零件材料的力学性能对于不同形式的磨损都有影响。

The mechanical properties of parts have an effect on different forms of wear.

零件表面的加工质量包括宏观几何形状和表面粗糙度甚至刀痕方向都对磨损有影响。对于每一种载荷，都存在一个最佳的粗糙度，粗糙度过低或过高，都会使磨损增加。

The machining quality of the surface of parts, including macroscopic geometry, surface roughness and even the direction of cutter marks, all affect the wear. For each type of load, there is an optimal roughness, and too little or too much roughness will increase wear.

配合间隙对磨损量的影响也很大，一般来说，间隙不应过小，也不应过大。间隙过小，不易形成液体摩擦，摩擦热也不易散去，易出现黏着磨损和咬死现象；间隙过大也不易形成液体摩擦，还会由于出现振动而引入冲击载荷。因此当配合间隙达到一定程度时，应及时修理，恢复原有的配合间隙。

Fit clearance has a great impact on wear. Generally speaking, fit clearance should not be too small and should not be too large. If fit clearance is too small, it is not easy to form liquid friction and friction heat is not easy to dissipate. Adhesive wear and bite death phenomenon is easy to appear. Too large fit clearance is not easy to form liquid friction, and will introduce impact load due to vibration. Therefore, when the fit clearance reaches a certain extent, the parts should be repaired in time to restore the original fit clearance.

2.1.2 零件变形失效（Deformation Failure）

金属受力产生的变形，可分为两个阶段：弹性变形阶段和塑性变形阶段。在弹性变形阶段，应变与应力之间呈线性关系；应力消失后，变形完全消失。在塑性变形阶段，应变与应力呈非线性关系；应力消失后，变形不能完全消失，总有一部分被保留下来；此时，材料的组织和性能都会发生相应的变化。

The deformation of metal under stress can be divided into two stages: elastic deformation stage and plastic deformation stage. In elastic deformation stage, there is a linear relationship between strain and stress. After the stress disappears, the deformation completely disappears. In plastic deformation stage, the relationship between strain and stress is nonlinear. After the stress disappears, the deformation cannot disappear completely, and some deformation are always retained. At this point, the structure and properties of the material will change accordingly.

2.1.2.1 金属的弹性变形（Elastic Deformation of Metal）

关于弹性变形问题，应了解两种效应：包申格效应和弹性后效。

Two kinds of effects should be known about elastic deformation: Bauschinger's effect and elastic aftereffect.

A 包申格效应（Bauschinger's Effect）

当对一个试件预先加载变形，然后再同向加载变形时，弹性极限升高；反向加载变形时，弹性极限降低。这种现象被称为包申格效应。例如：退火黄铜轧材试样，最初拉伸弹性极限为240MPa，最初压缩弹性极限为176MPa；经最初压缩后的试样，得到0.17%的残余压缩变形。

When a specimen is pre-loaded and deformed, and the load in the same direction is applied on it then, the elastic limit of the specimen increases. Or, the load in the opposite direction is applied on it, the elastic limit decreases. This phenomenon is known as the Bauschinger's effect. For example, when rolled brass specimen is annealed, the initial tensile elastic limit of brass specimen is 240MPa, the initial compression elastic limit of brass specimen is 176MPa. After initial compression of the brass specimens, the residual compression deformation of 0.17% is obtained.

若再给予第二次压缩，则压缩弹性极限升为287MPa；若对试样第二次加载不是压缩，而是拉伸时，则拉伸弹性极限大大降低，只有 85MPa。包申格效应在研究金属的疲劳问题很重要。给予较大的残余塑性变形，或是在引起金属恢复或再结晶的温度下退火（钢在 400~500℃以上，铜合金在 250~270℃以上），可以消除包申格效应。

If the second compression is given to the brass specimen, the compression elastic limit of brass specimen rises to 287MPa. If the specimen is stretched instead of compressed as the second loading, the tensile elastic limit is greatly reduced to only 85MPa. Bauschinger's effect is very important in the study of metal fatigue. The elimination of Bauschinger's effect can be achieved by giving large residual plastic deformation, or by annealing at the temperature that causes metal recovery or recrystallization (above 400~500℃ for steel and 250~270℃ for copper alloy).

B 弹性后效（Elastic Aftereffect）

把弹性极限以内一定大小的力，骤然加到金属试样上或骤然去除时，试样立即产生或立即消失的应变总落后于应力的现象，就叫作弹性后效。

When a force of a certain value within the elastic limit is suddenly added to a metal specimen or be suddenly removed, the strain immediately produced or disappeared from the metal specimens always lags behind the stress. This phenomenon is called elastic aftereffect.

图 2.3 中，坐标 σ 表示拉应力，坐标 ε 表示应变，坐标 t 表示时间，OA 表示拉伸应力在弹性极限内从 0 增加到某一值 a 并从 A 到 B 保持不变，当应力从 O 增加到 A 时，应变是 0，当应力从 A 到 B 保持不变停止增加时，应变从 D 增加到 E。应变 OE 表示拉应力作用下的应变 ε 与时间 t 的变化关系，OK 为应变达到总应变时持续时间长短，EG 表示拉应力去除后，在恢复过程中应变 ε 与时间 t 的变化关系，KG 为弹性恢复过程持续时间的长短。当拉应力从 B 减小到 0 时，应变保持不变，但当拉应力停止减小时，应变开始从 M 减小到 G。

In Figure 2.3, the coordinate σ is tensile stress, the coordinate ε is strain, the coordinate t is time. OA is the tensile stress increase from 0 to a certain value within elastic limit and keep the value from A to B. The stain is 0 when the tensile stress increase from O to A, and increase from D to E when the tensile stress stop increasing and remain unchanged from A to B. OE is the

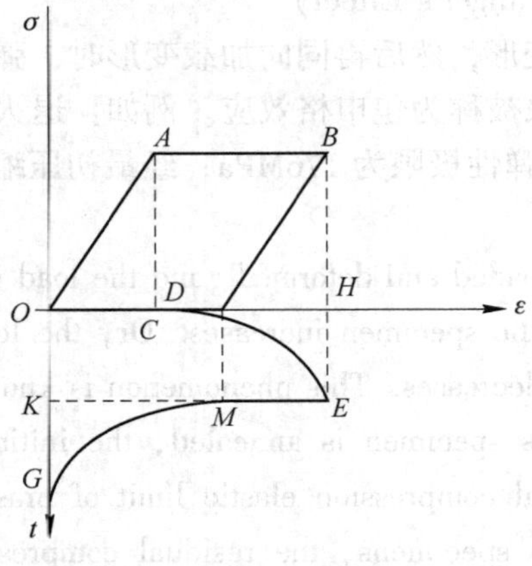

图 2.3 弹性后效示意图

Figure 2.3 Schematic diagram of elastic aftereffect

relationship of strain ε and time t under the action of tensile stress, OK is the length of time when strain reaches total strain, EG is the relationship of strain ε and time t in the process of recovery after tensile stress is removed, KG is the length of the elastic recovery process duration. When the tensile stress decrease from B to 0, the strain keeps unchanged from E to M, but when the tensile stop decreasing, the strain begins to decrease from M to G.

弹性后效的影响因素包括材料的性质，应力的大小及应力状态，温度等。金属组织结构越不均匀、作用力越大、温度越高，则弹性后效越大。剪应力引起的弹性后效最显著。

Influence factors of elastic aftereffect are material properties, magnitude of stress and stress state, temperature, etc. The more uneven the structure of the metal, the greater the force and the higher the temperature, the greater the elastic aftereffect. The elastic aftereffect caused by shear stress is the most significant.

单晶体金属不存在弹性后效或其作用很小，实际的金属都是多晶体。长时间回火可以消除弹性后效，钢的回火温度是300~450℃。弹性后效在工程修理中有着实际意义，如经过校直的零件（如曲轴），经过一段时间之后，又会发生弯曲，就是弹性后效的表现，所以校直后的零件应进行回火处理。

Single crystal metals have no elastic aftereffect or have little effect. The metals are multicrystal actually. Method to eliminate elastic aftereffect can be finished by long time tempering. The tempering temperature of steel is 300 ~ 450℃. Elastic aftereffect has practical significance in engineering repair. For example, the parts having be straightened (such as crankshaft) bend again after a period of time. This phenomenon is the performance of elastic aftereffect. So, straightened parts should be tempered.

2.1.2.2 金属的塑性变形（Plastic Deformation of Metal）

单晶体在剪切力作用下才产生塑性变形，其变形过程有两种方式：滑移和孪生。多晶体由于相邻晶粒的位向不同，又存在晶界，其塑性变形比较复杂。总之，晶粒越细，金属的强度和硬度越高，塑性和韧性也越好。

The plastic deformation of a single crystal occurs under shear force, and there are two ways of deformation: slip and twinning. Due to the different orientation of adjacent grains and grain boundary, the plastic deformation of polycrystals is complex. In general, the finer the grain, the stronger and harder the metal, the better its plasticity and toughness.

塑性变形对金属组织和性能的影响表现为两种现象：加工硬化和残余内应力。

The effect of plastic deformation on metal structure and properties is manifested in two phenomena: work hardening and residual internal stress.

A 加工硬化（Work Hardening）

在塑性变形过程中，金属随变形程度的增加，强度和硬度提高，而塑性和韧性下降，这种因冷变形加工而引起金属强化的现象，称为加工硬化。

In the process of plastic deformation, the strength and hardness of metal increase with the increase of deformation degree, while the plasticity and toughness decrease. This phenomenon of metal strengthening caused by cold deformation processing is called work hardening.

除力学性能外，在冷变形过程中，还发生某些物理、化学性能的变化，如电阻增加、耐蚀性降低等。加工硬化在实际生产中的意义包括：可利用加工硬化提高金属的强度。加工硬化可以保证金属结构和零件的工作安全：金属构件在承受载荷时，其内部应力的分布由于种种原因不可能很均匀，往往在局部区域产生应力集中，它与工作应力叠加起来所形成的应力峰值，可能大大超过材料的屈服极限，并引起塑性变形。

Besides mechanical properties, in the process of cold deformation, some physical and chemical properties will change. For example, resistance increase, corrosion resistance decrease, etc. The significances of work hardening in practical production include: Work hardening can be used to improve the strength of metal. The work hardening can ensure the safety of the metal structure and parts of work. The internal stress distribution of metal parts may not very evenly due to various reasons. Stress concentration tends to produce in local area. The peak stress overlaid with working stress could significantly exceed the material yield limit, and cause plastic deformation.

如果没有加工硬化效应，该部位可能因不断地产生变形而失效或断裂。因为有了加工硬化，阻止了局部区域变形的发展，同时又由于塑性变形使应力松弛（改变了应力的分布状态），从而减少了应力集中。加工硬化的不利方面是消耗的动力增加，变形量受限制。

If there is no work hardening effect, the area may be broken due to continuous deformation. Due to work hardening, the development of local area deformation is prevented, and due to plastic deformation, the stress is relaxed (the distribution of stress is changed), thus the stress concentration is reduced. The disadvantages of work hardening are increasing power consumption and limiting deformation.

加工硬化的消除办法是进行中间退火，以便继续加工，这就增加了成本，延长了生产周期。

Work hardening can be eliminated by intermediate annealing to allow further processing, but this increases costs and prolongs production cycles.

B　残余内应力（Residual Internal Stress）

残余内应力是指金属经过冷变形加工后，除了发生组织和性能的变化外，由各部分不均匀的变形引起的应力。残余内应力可分为以下三类：宏观内应力、晶间内应力和晶格畸变内应力。

Residual internal stress refers to the stress caused by uneven deformation of various parts after cold deformation of metal, except for changes in structure and properties. Residual internal stress can be divided into the following three categories: macroscopic internal stress, intercrystalline internal stress, and lattice distortion internal stress.

因金属各部分变形不均匀引起，发生在较宽范围内的内应力，称为宏观内应力。如冷轧板表面变形量比内部变形量大，但由于钢板内外层是相互牵制的，因此，表面和内部产生了方向相反、大小相等的内应力。

Macroscopic internal stress is the internal stress that occurs in a wide range due to the uneven deformation of various area of part. For example, the surface deformation of the cold-rolled plate is larger than the internal deformation. But, the internal stress of surface and the internal area have the opposite direction and the same size because the inner and outer layers of the steel plate are mutually dependent.

晶间内应力是由金属冷塑性变形、因晶粒或亚晶粒变形不均匀引起。

Intercrystalline internal stress is caused by uneven deformation of grain or subgrain during cold plastic deformation of metal.

晶格畸变内应力是由于在晶体缺陷附近引起晶格畸变而产生，它是金属塑性变形时所产生的最主要的内应力。

Lattice distortion internal stress is the most important internal stress caused by plastic deformation of metal due to lattice distortion near crystal defect. It is the main internal stress produced by plastic deformation of metal.

由于残余内应力的存在，金属弹性内能增加，金属处于不稳定状态。后续加热和切割时，由于内应力释放而产生变形或开裂。残余内应力的消除办法是进行时效处理。时效处理包括自然时效处理和人工时效处理。

The existence of residual internal stress increases the elastic internal energy of metal, and the metal is in an unstable state. The part will deform or crack when later be heat and cut due to the release of internal stress. The method of eliminating residual internal stress is aging treatment. Aging treatment includes natural aging treatment and artificial aging treatment.

自然时效处理是指铸造件或焊接件，放置一年或更长时间，使内应力消除。因时间长，生产上较少采用。人工时效处理是指退火或振动。铸件的退火处理：把铸件加热到接近塑性状态的温度，保温一段时间，使铸件在残余应力作用下发生塑性变形，达到消除内应力的目的，然后再缓慢冷却。

Natural aging treatment is used for casting or welding parts which are placed for one year or longer to eliminate the internal stress. Because the time of natural aging treatment is long, it is less used on production. Artificial aging treatment refers to annealing or vibration. Annealing treatment of casting refers to heating the casting to a temperature close to the plastic state, keeping it warm for a period of time, making the casting plastic deformation under the action of residual stress, achieving the purpose of eliminating internal stress, and then slowly cooling the casting.

零件在使用中变形的原因主要表现为以下几个方面。

The main reasons for the deformation of parts in use are as follows.

（1）毛坯制造方面：高温冷却使零件产生变形。

For blank manufacturing, parts deform after high temperature cooling.

（2）机械加工方面：切去表面后破坏了内应力的平衡，由于内应力重新分布，零件将发生变形；在切削过程中，由于切削力和切削热作用，零件的表面层会发生极大的塑性变形，因而产生内应力。

For mechanical processing, after cutting the surface of the parts, the balance of internal stress is destroyed. Due to the redistribution of internal stress, the parts will be deformed. In the process of cutting, because of the cutting force and the cutting heat, the surface layer of the parts will have a great plastic deformation, thus generating internal stress.

（3）修理质量方面：没有考虑已发生的变形而增大变形；机械修理中许多修复工艺，如焊接、堆焊、压力加工等，都能产生新的应力和变形。如：堆焊中热应力和相变应力的作用，压力加工时弹性后效和内应力松弛等。

For quality of repair, increase the deformation without considering the deformation of the parts. Many repair processes in mechanical repair, such as welding, surfacing, pressure processing, etc., can produce new stress and deformation. For example, the role of thermal stress and phase change stress in surfacing welding, elastic aftereffect during pressure processing and internal stress relaxation.

（4）使用方面：环境恶劣；极限载荷；温度过高；超负荷；或者操作不当引起温升过高，也会增加蠕变变形。

For application, creep deformation will also be increased due to harsh environment, limit load, excessive temperature, overload or improper operation that causes excessive temperature rise.

（5）设计方面：零部件布置不合理，也会造成变形。

For design, unreasonable parts layout will cause deformation.

引起零件变形的主要原因有：外载荷、温度、残余内应力、材料内部缺陷等。减轻变形危害可以从设计、制造、修理、使用几方面入手。

The main causes of parts deformation are: external load, temperature, residual internal stress, material internal defects and so on. The reduction of deformation hazards can be started from the following aspects.

（1）设计。充分考虑如何实现机构的动作和保证零件的强度；充分考虑零件的刚度、变形以及所设计的零件在制造、装配和使用中会发生什么问题。

Design. Fully consider how to achieve the action of the mechanism and ensure the strength of parts; Full consideration is given to the stiffness and deformation of the parts and what problems will occur in the manufacturing, assembly and use of the designed parts.

合理地布置各零部件，改善零件的受力状况，可以减少变形；尽量使零件壁厚均匀，可以减少热加工时的变形。

Reasonably arrange all parts and components to improve the stress condition of parts and reduce deformation; Make the wall thickness of parts uniform as far as possible to reduce the deformation during thermal processing.

（2）制造。制订毛坯制造工艺时，要重视变形问题，采取各种措施以减少毛坯的残余应力；毛坯制成后以及在以后的机械加工过程中，必须安排足够的消除内应力的工序。

Manufacturing. When formulating the blank manufacturing process, attention should be paid to the deformation problem, and various measures should be taken to reduce the residual stress of the blank; After the blank is made and in the subsequent mechanical processing, adequate procedures to eliminate internal stress must be arranged.

在机械加工中，把粗加工和精加工分开，留出存放时间；机械加工中，尽量保留工艺基准，留给修理时使用，可以减少修理加工中因基准不一而造成的误差。

In machining, separate roughing and finishing to allow storage time; In mechanical processing, try to keep the process standard for use in repair, which can reduce the error caused by different standards in repair and processing.

（3）修理。大修时不能只检查配合面的磨损情况，对于相互位置精度也必须认真检查和修复；应制订出合理的检修标准，并且应该设计出简单、可靠、容易操作的专用量具和专用卡具；大修时合理选择定位基准；注意热加工或压力加工时，采取措施减少应力和变形。

Repair. During overhaul, the wear condition of the mating surface cannot be checked only, and the mutual position accuracy must also be carefully checked and repaired; Reasonable maintenance standards shall be worked out, and special measuring tools and special clamping tools shall be designed that are simple, reliable and easy to operate; Reasonable selection of positioning datum during overhaul; Take measures to reduce stress and deformation during hot or pressure processing.

（4）使用。正确安装设备，使各部件尽可能地承受不应有的持续应力的作用。如：安装机床时，要检测几个方向的水平，否则，会由于存在不应有的安装应力，导致部件变形。

Use. Properly install the equipment so that each component can bear the undeserved continuous stress as much as possible. For example: the installation of machine tools requires detecting the level of several directions, otherwise, due to the existence of undue installation stress, parts will deform.

移动部件停机时的位置应适当。如：车床拖板停机时应放在尾部、冲床的冲头停机时落在垫块上都有利于减轻变形。

When the moving parts stop, the position should be appropriate. For example, the lathe should be put on the tail when the towing board stops, punch when the machine stops falling on the pad is conducive to reducing deformation.

精密机床不能用于粗加工，以防由于过载引起精密零件变形，导致精度下降。恰当存放备品备件。如：丝杆、轴等零件应垂直存放，这样可以防止变形的产生。

Precision machine tools cannot be used for rough machining, in case of deformation of precision parts caused by overload, resulting in decreased precision. Properly store spare parts. Such as: screw, shaft and other parts should be stored vertically, so as to prevent deformation.

2.1.3 断裂失效（Fracture Failure）

断裂常见的分类形式有以下四种：韧性断裂与脆性断裂、解理断裂与切变断裂、穿晶断裂和晶界断裂、一次加载断裂与反复加载断裂。

There are four common types of fracture classification: ductile fracture and brittle fracture, cleavage fracture and shear fracture, transgranular fracture and grain boundary fracture and primary loading fracture and repeated loading fracture.

2.1.3.1 韧性断裂与脆性断裂（Ductile Fracture and Brittle Fracture）

零件在断裂之前有明显的塑性变形并伴有颈缩现象的一种断裂形式，称为韧性断裂。引起金属韧性断裂的实质是实际应力超过了材料的屈服强度。分析金属韧性断裂的失效原因应从设计、材质、工艺、使用载荷、环境等角度考虑问题。

Ductile fracture is a form of fracture in which the part has obvious plastic deformation and necking phenomenon before fracture. The substance of metal ductile fracture is caused by the fact that the actual stress exceeds the yield strength of the material. When analyzing failure reasons, we should consider design, material, process, service load and environment.

零件在断裂之前无明显的塑性变形，发展速度极快的一种断裂形式，称为脆性断裂。由于试件在外力作用下，位错移动到晶界处受阻而在晶界前堆积起来，并在晶界前形成显微裂纹。在这种裂纹的尖端会形成很大的应力集中，由于材料的塑性差，集中的应力不能通过塑性变形而释放，结果迫使裂纹不断扩大，最终导致材料的脆性断裂。

Brittle fracture is a form of fracture in which the parts have no obvious plastic deformation before fracture and the development speed is very fast. The dislocation moves to grain boundary and accumulates in front of grain boundary due to external force. Microcracks were formed before grain boundaries. A large stress concentration will be formed at the crack tip. Due to the poor plasticity of the material, the concentrated stress cannot be released through plastic deformation, which forces the crack to expand continuously and finally leads to brittle fracture of the material.

脆性断裂的基本特征有下列三点:

Basic features of brittle fracture accidents are as followings:

(1) 脆断时结构承受的工作应力较低，通常不超过材料的屈服强度，甚至不超过常规设计所确定的许用应力。有时也称这类破坏为低应力脆性破坏。

The working stress of the structure under brittle fracture is relatively low, which usually does not exceed the yield strength of the material or even the allowable stress determined by the conventional design. It is sometimes called low-stress brittle failure.

(2) 断口平直、光亮，并与拉应力方向相垂直，没有或只有微小的屈服及减薄现象。

The fracture is straight, shiny and perpendicular to the direction of tensile stress, with no or only slight yield and thinning phenomenon.

(3) 断裂前无征兆，断裂是瞬时发生的。

No warning before the fracture, the fracture is instantaneous.

出现上述现象的实质是：这种低应力脆断大多是由于在构件内部存在的具有宏观尺寸的裂纹源的扩展造成的。这些裂纹源可能是由材料中早已存在的各种缺陷（裂纹、夹渣）、结构的不连续处（如形状、板厚突变）或不良的表面加工质量所产生，这些裂纹源表现出明显的缺口效应。在这种情况下，整个结构所受的平均应力虽然不大，但在裂纹尖端可产生很大的应力集中，局部应力大大超过材料的强度极限，使裂纹迅速扩展，最终导致整个构件的迅速破坏。

The essence of the above phenomenon is that the low-stress brittle fracture is mostly caused by the expansion of the crack source with macroscopic size in the component. These crack sources may be caused by various existing defects in the material (cracks, slag inclusion), structural discontinuities (such as shape, plate thickness mutation) or poor surface processing quality, and these crack sources show obvious notch effect. In this case, although the average stress on the whole structure is not large, great stress concentration can be generated at the crack tip, and the local stress greatly exceeds the strength limit of the material, making the crack expand rapidly, and eventually leading to the rapid destruction of the entire component.

2.1.3.2 解理断裂与切变断裂 (Cleavage Fracture and Shear Fracture)

通常在拉应力作用下，严格地沿着某些结晶学平面(解理面)发生的断裂，称为解理断裂。

Cleavage fracture occurs usually under the action of tensile stress, strictly along some crystallographic plane (cleavage plane) fracture, known as cleavage fracture.

通常在切应力作用下，伴随有大量塑性变形，沿着滑移面且顺着滑移方向发生滑移而使晶粒分离的一种断裂形式，称为切变断裂。多呈韧性断裂特征。

Shear fracture is a type of fracture in which grain separation occurs along the slip surface and along the direction of slip, usually accompanied by a large amount of plastic deformation under the action of shear stress, is called shear fracture. Most of them are characterized by ductile fracture.

2.1.3.3 穿晶断裂和晶界断裂（Transgranular Fracture and Grain Boundary Fracture）

穿晶断裂是指裂纹割断晶粒的断裂形式，韧性断裂或脆性断裂都是穿晶断裂。

Transgranular fracture refers to a fracture form in which a crack cuts off a grain. Ductile or brittle fractures are transgranular.

晶界断裂是指沿着晶界断裂。晶界断裂往往是脆性断裂，韧性的晶界断裂只有在高温蠕变中才能发生。

Grain boundary fracture is a form of fracture occurs along grain boundary. Grain boundary fracture is usually brittle fracture, and ductile grain boundary fracture can only occur in high temperature creep.

2.1.3.4 一次加载断裂与反复加载断裂（Primary Loading Fracture and Repeated loading Fracture）

每一个零件都必然带有从原子位错到肉眼可见宏观裂纹等不同大小、不同性质的缺陷，根据不同的定义，这些缺陷都可以叫作裂纹。

Every part must have defects of different sizes and properties, from atomic dislocation to macroscopic cracks visible to the naked eye. According to different definitions, these defects can be called cracks.

一次加载断裂是指零件或试样在一次静载荷（缓慢递增的或恒定的载荷）作用下或在一次冲击能量作用下发生断裂的现象，包括静拉伸、静压缩、静弯曲、静扭转、静剪切、高温蠕变和一次冲击断裂等。

Primary loading fracture refers to the phenomenon that a part or specimen breaks under the action of a static load (slowly increasing or constant load) or under the action of a shock energy, including static stretching, static compression, static bending, static torsion, static shear, high-temperature creep and primary impact fracture.

反复加载断裂是指零件或试样在经历反复多次的应力或能量负荷循环作用后才发生断裂的现象，一般称为疲劳断裂。这一类断裂的类型很多，包括反复拉压疲劳、弯曲疲劳、扭转疲劳、疲劳点蚀、热应力疲劳、共振疲劳以及多次冲击断裂等。

Repeated loading fracture refers to the phenomenon that the parts or samples break after experiencing repeated stress or energy load cycle, which is generally called fatigue fracture. There are many types of such failures, including repeated tension and compression fatigue, bending fatigue, torsion fatigue, fatigue pitting, thermal stress fatigue, resonance fatigue and multiple impact fracture.

2.1.4 腐蚀失效（Corrosion Failure）

金属零件的腐蚀失效是指金属材料与周围介质发生化学或电化学作用而导致的破坏。金属零件的腐蚀失效分两类：化学腐蚀和电化学腐蚀。

Corrosion failure of metal parts means the damage caused by chemical or electrochemical effects between metal material and the surrounding medium. Corrosion failure of metal parts is divided into chemical corrosion and electrochemical corrosion.

2.1.4.1 金属的化学腐蚀（Chemical Failure of Metal）

金属在高温气体（含氧）介质下的氧化是典型的化学腐蚀。

The oxidation of metals in high temperature gaseous (oxygen containing) media is typical of chemical corrosion.

(1) 钢铁的高温腐蚀，通常按以下方式进行：高温氧化、脱碳、氢蚀和铸铁的肿胀。

High-temperature corrosion of steel is usually carried out in the following ways: high-temperature oxidation, decarburization, hydrogen corrosion and swelling of cast iron.

1) 高温氧化。钢铁在空气中加热，在较低温度（200～300℃）下表面便出现可见的氧化膜。当温度高于800℃时，表面上开始形成多孔、疏松的“氧化皮”。这些氧化皮与基体结合松散，稍受震动便会一层层地脱落下来。

High-temperature oxidation. When steel is heated in air, visible oxide film appears on the surface at a lower temperature (200～300℃). When the temperature is higher than 800℃, the surface begins to form porous, loosen “oxide scale”. These oxide scales are loosely bound to the substrate, and will be peeled off layer by layer upon slight vibration.

气相的组成，对铁的高温腐蚀有着强烈的影响，特别是水蒸气和硫的化合物的影响最大。在烟道中，若含有过量的空气，对钢铁的腐蚀有很大影响。氧的含量越高，腐蚀速度越大，而一氧化碳却具有相反的作用。

The composition of the gas phase has a strong influence on the corrosion of iron at high temperatures, especially the compounds of water vapor and sulfur. In the flue, if there contains the excess air, it will have a very big influence on the steel corrosion. The higher the oxygen content, the greater the corrosion rate, while carbon monoxide has the opposite effect.

2) 脱碳。钢在气体腐蚀过程中，通常总是伴随有脱碳现象。脱碳是指在腐蚀过程中，除了生成氧化皮外，与氧化皮层相连的内层发生渗碳体减少的现象，这是由于渗碳体 Fe3C 与介质中的氧、氢、二氧化碳、水等作用的结果。

Decarburization. Decarburization usually occurs in the process of gas corrosion of steel. In the process of corrosion, besides the formation of oxide skin, the inner layer connected with the oxide layer is reduced by cementite, which is the result of the interaction between cementite Fe3C and oxygen, hydrogen, carbon dioxide and water in the medium.

脱碳作用生成的气体，使表面膜的完整性受到破坏，从而降低了膜的保护作用加快了腐蚀的进行。同时，碳钢表面渗碳体减少（即表面层已变成铁素体组织），使表面层的硬度和强度都大幅度下降，降低了零件的耐磨性和疲劳极限，从而降低了设备或零件的使用寿命。

The decarburization produces gases that destroy the integrity of the surface film, thereby reducing the protective effect of the film and speeding up the corrosion. At the same time, carbon steel surface cementite reduces (that is, the surface layer has become ferrite tissue), so that the surface layer of hardness and strength are greatly reduced, the wear resistance and fatigue limit of parts reduce, thereby the service life of equipment or parts reduce too.

实践证明，增加气体介质中的一氧化碳和甲烷含量，将使脱碳作用减小，钢中添加铝和钨也可降低钢的脱碳倾向。

It has been proved that increasing carbon monoxide and methane content in gas medium can reduce decarburization, and adding aluminum and tungsten to steel can also reduce decarburization tendency.

3）氢蚀。氢气在常温常压下对碳钢不会产生明显的作用，氢气在温度高于200～300℃，压力高于30.4MPa时，氢对钢材作用显著，使钢剧烈脆化，这就叫氢蚀。在合成氨、合成甲醇、石油加氢及其他一些化学和炼油工业中，氢蚀常发生。

Hydrogen corrosion. Hydrogen at normal temperature and pressure on carbon steel will not have a significant role. When the temperature is higher than 200～300℃, pressure is higher than 30.4MPa, hydrogen will make steel occur severe embrittlement, this is called hydrogen corrosion. Hydrogen corrosion often occurs in ammonia synthesis, methanol synthesis, petroleum hydrogenation and other chemical and oil refining industries.

钢材发生氢蚀有两个阶段：逆氢脆阶段和氢蚀阶段。

There are two stages of hydrogen corrosion in steel: reverse hydrogen embrittlement and hydrogen corrosion.

逆氢脆阶段是指氢和钢材接触时，氢即被钢表面所吸收，并以原子状态沿晶界向钢材内部扩散。溶解于钢中的氢虽不和钢材组成起任何化学变化，也没有改变钢的组织，但是却使钢变脆，韧性降低，如果将钢材在200～300℃低温下加热或常温下长时间静置，其韧性又可以部分或全部恢复，这一阶段，称为逆氢脆阶段。

In reverse hydrogen embrittlement stage, when hydrogen and steel contact, hydrogen is absorbed by the steel surface and diffuses to the steel interior in the atomic state along the grain boundary. Although hydrogen dissolved in steel does not form any chemical changes with the steel, nor does it change the structure of the steel, it makes the steel brittle and reduces its toughness. If the steel is heated at 200～300℃ or left standing for a long time at normal temperature, its toughness can be partially or completely restored. This stage is called reverse hydrogen brittleness stage.

氢蚀阶段是指高温高压下，侵入并扩散到钢中的氢与不稳定的碳化物发生反应生成了甲烷，产生脱碳，并且反应的气体生成物在钢材内部积聚，产生很大的内应力，使晶界产生裂纹，内部出现龟裂，而在表面则出现许多鼓泡。这就使钢的强度和韧性大大降低，发生永久脆化。

Hydrogen corrosion stage occurs under high temperature and pressure, hydrogen invaded and diffused into the steel reacts with unstable carbides to form methane and produce decarbonization, and the gas products of the reaction accumulate in the steel, generating great internal stress, causing cracks in grain boundary and internal cracking, and many bubbles appear on the surface. This greatly reduces the strength and toughness of the steel, leading to permanent embrittlement.

氢与硫化氢共存时，硫化氢与铁发生反应在钢表面生成硫化铁层，它在高温高压中是多孔性物质，且易剥落。剥落后活性铁表面又与高温下的硫化氢作用而腐蚀，如此反复进行，所以，当氢和硫化氢共存时，硫化氢对钢的腐蚀起着促进作用。

When hydrogen and sulfuretted hydrogen coexist, hydrogen sulfide reacts with iron to form an iron sulfide layer on the steel surface and the iron sulfide layer is easy to peel off. After peeling off, the surface of active iron is corroded by hydrogen sulfide at high temperature. This process is repeated. Therefore, when hydrogen and hydrogen sulfide coexist, hydrogen sulfide promotes the corrosion of steel.

防止氢蚀的措施是降低钢的含碳量或在钢中加入铬、钛、钼、钨、钒等合金元素，以形成稳定的碳化物，能提高钢抗氢蚀的能力。

Methods to avoid hydrogen corrosion is reducing the carbon content of steel or adding chromium, titanium, molybdenum, tungsten, vanadium and other alloy elements in the steel to form a stable carbide.

4）铸铁的肿胀。铸铁的肿胀是一种晶间气体腐蚀，腐蚀性气体沿着晶粒边界，石墨夹杂物和细微裂缝渗入到铸铁内部，发生氧化作用。由于所生成的氧化物体积较大，从而加大了铸铁的尺寸。铸件发生肿胀后，强度将大大降低。

Swelling of cast iron. The swelling of cast iron is a kind of intercrystalline gas corrosion. When corrosive gas goes along the grain boundary, graphite inclusions and fine cracks enter into the cast iron, oxidation occurs. The size of cast iron is increased by the large size of the oxide produced. When the casting is swollen, the strength is greatly reduced.

防止铸铁肿胀的措施是在生铁中加入适量（5%~10%）的硅，形成 SiO_2，提高氧化膜的保护性能，阻止氧的渗入，可使肿胀现象不发生。但如果硅的添加量过低（<5%），由于硅促进铸铁的石墨化，反而会使肿胀更加严重。

Methods to avoid the swelling of cast iron is adding appropriate amount of silicon in pig iron (5%~10%) to form SiO_2, improving the protective properties of the oxide film, preventing the infiltration of oxygen. These can make the swelling phenomenon does not occur. However, if the amount of silicon added is too low (<5%), the swelling will be more serious because silicon promotes graphitization of cast iron.

（2）防止钢铁气体腐蚀的方法。防止钢铁气体腐蚀的方法有很多：合金化、改善介质、应用保护性覆盖层，还可以涂耐高温的涂料。近年来的新方法有物理气相沉积（PVD）和化学气相沉积（CVD）方法。

Methods to prevent steel gas corrosion. There are many methods to prevent steel gas corrosion: alloying, improve medium, application of protective coating, and apply high temperature resistant paint. In recent years, the new methods physical vapor deposition (PVD) and chemical vapor deposition (CVD) are developed.

1）合金化。合金化是指改善钢铁抗氧化性能最有效的合金元素是 Cr、Al、Si，它们与氧的亲和力比铁强，在氧化性介质中首先与氧结合形成极稳定的 Cr_2O_3、Al_2O_3 和 SiO_2，这些氧化物结构致密，能够牢固地与金属基体结合，形成有效的保护层，阻止金属离子和氧离子的扩散，大幅度提高钢的抗氧化性。

Alloying. The most effective alloying element can improve steel oxidation resistance is Cr, Al, Si and their affinity with the oxygen stronger than iron, in oxidizing medium first combines with oxygen to form extremely stable Cr_2O_3, Al_2O_3, and SiO_2. The compact structure of the oxide, can be firmly combined with metal substrate and form effective protective layer to prevent the spread of the metal ions and oxygen ions and increase the oxidation resistance of steel.

2）改善介质。改善介质是指降低和清除有害气体，应用保护气体。

Improve medium. Harmful gas should be reduced and removed, and protective gas is applied.

3）应用保护性覆盖层。应用金属或非金属涂层，把金属和气体介质隔离。最常采用热扩散法，钢铁材料可以渗铬、渗铝、渗硅及铬-铝共渗等。

Apply protective coating. Application of protective coating, such as metal or non-metallic coating, shall be applied to isolate the metal from the gaseous medium. The most commonly used way is thermal diffusion method Steel materials can be chrome, aluminum, silicon and chromium-aluminum co-infiltration.

2.1.4.2 金属零件的电化学腐蚀（Electrochemical Corrosion of Metal Parts.）

金属的电化学腐蚀在工程界（特别是化工设备）大量存在，涉及面最广，危害也很严重。这是因为金属电化学腐蚀的条件容易形成和存在。

Electrochemical corrosion of metals exists in a large number of engineering fields (especially chemical equipment), involving the widest range and causing serious harm. Because of metal electrochemical corrosion conditions, electrochemical corrosion of metals is easy to form and exist.

局部腐蚀最普遍，典型的局部腐蚀有：接触腐蚀、小孔腐蚀、缝隙腐蚀、晶间腐蚀、应力腐蚀开裂和腐蚀疲劳。

Local corrosion is most common. The typical local corrosion includes contact corrosion, pinhole corrosion, crevice corrosion, intergranular corrosion, stress corrosion cracking and corrosion fatigue.

A 接触腐蚀（Contact Corrosion）

接触腐蚀是指在电解质溶液中，两种具有不同性质的金属或合金相互接触时，电位较负的金属腐蚀速度增加，电位较正的金属腐蚀速度减小，甚至停止。这类腐蚀现象就是接触腐蚀，或称为电位腐蚀。

Contact corrosion usually occurs in the electrolytic solution. When two metals or alloys with different properties contact each other, the corrosion rate of the metal increases with a negative potential while the corrosion rate of the metal with a positive potential decreases or even stops. This type of corrosion is called contact corrosion or potential corrosion.

B 小孔腐蚀（Pitting Corrosion）

小孔腐蚀是指金属件的大部分表面不发生腐蚀或腐蚀很轻微，但是局部地方出现腐蚀小孔并向深处发展的腐蚀现象，称为小孔腐蚀，简称点蚀或孔蚀。

Pitting corrosion refers to phenomenon that most of the surfaces of metal parts are not corroded or slightly corroded, but corrosion holes occur in local places and develop to deep depth.

C 缝隙腐蚀（Crevice Corrosion）

缝隙腐蚀是指金属与金属连接处或非金属连接处，由于存在一定的缝隙（0.025～0.1mm），当溶液进入并处于常留状态时所引起的一种局部腐蚀。

Crevice corrosion refers to the local corrosion caused by the presence of a certain crevice (0.025～0.1mm) at the metal-to-metal or nonmetallic connections when solution enters and stays in a constant state.

D 晶间腐蚀（Intercrystalline Corrosion）

晶间腐蚀是指沿着金属晶界或它的近旁发生的腐蚀现象。

Intercrystalline corrosion refers to the corrosion phenomenon occurring along or near the grain boundary of the metal.

E 应力腐蚀开裂（Stress Corrosion Cracking）

应力腐蚀开裂是指在特定腐蚀环境和机械拉应力共同作用下的一种极为隐蔽的局部腐蚀形式，而且往往事先无明显预兆，常常造成灾难性的事故。

Stress corrosion cracking is in a specific corrosion environment and mechanical tensile stress under the connection action of a very hidden form of local corrosion, and often without obvious warning, often causes catastrophic accidents.

应力腐蚀发生的条件包括：有一定大小的拉应力作用于零件，这个拉应力除了构件工作应力外，还可能是残余应力，有时残余拉应力占主导地位；腐蚀环境（包括介质、浓度、温度等）是特定的；金属材料本身对应力腐蚀开裂有敏感性。它取决于金属材料的化学成分和组织结构。

Conditions for stress corrosion are as followings: There is a certain size of tensile stress for parts. The tensile stress may be residual stress in addition to the members of the working stress. Sometimes residual tensile stress is dominant. The corrosive environment (including medium, concentration, temperature, etc.) is specified. The metal material itself is sensitive to stress corrosion cracking. It depends on the chemical composition and organization of the metal material.

应力腐蚀开裂都呈脆性断裂。防止金属腐蚀的主要方法有：合理选择材料、合理设计结构、覆盖保护层、添加缓蚀剂、电化学保护、改变环境条件。

Stress corrosion cracking belongs to brittle fracture. Corrosion fatigue refers to the phenomenon of reduced fatigue strength or fatigue life of metal materials under the combined action of corrosive medium and alternating stress.

F 腐蚀疲劳（Corrosion Fatigue）

腐蚀疲劳是指金属材料在腐蚀介质和交变应力共同作用下产生疲劳强度或疲劳寿命降低的现象。

Corrosion fatigue refers to the phenomenon that the fatigue strength or fatigue life of metal materials decreases under the combined action of corrosive medium and alternating stress.

防止腐蚀疲劳首先防止腐蚀介质的作用，若必须在腐蚀介质下工作，则采用耐腐蚀材料，以及根据不同的介质条件分别采用阴极保护或阳极保护；或者采取表面涂防腐覆盖层等表面处理方法。

First of all, the function of corrosive medium should be prevented. If it is inevasible to work in corrosive medium, corrosion resistant materials should be adopted, and cathodic protection or anodic protection should be adopted according to different media conditions. Or surface treatment methods such as coating anticorrosive covering layer should be adopted.

此外，在某些机械设备零部件上，还会出现堵塞、黏附和老化变质等常见现象。对于这类机械零件，除了改进结构设计，选用适当材质和制造工艺外，在使用和维护上，分别采取净化、避免有害介质的作用等措施来避免或延缓失效过程，也是十分重要而有效的。

In addition, in some mechanical equipment parts, there will be blockage, adhesion and aging deterioration and other common phenomena. For this kind of mechanical parts, the selection of appropriate materials and manufacturing process should be done in addition to the improvement of structural design. Moreover, in the use and maintenance, purified and no harmful media should be respectively taken to avoid or delay the failure process.

2.2 机械设备的故障（Failure of Mechanical Equipment）

故障理论就是综合分析机械设备全寿命期各阶段有关维修的不同矛盾，研究它们之间的有机联系，把许多分支科学地统一起来，从而构成一个完整的知识体系。它具有自己特有的研究对象——有故障的机械设备。

Failure theory is a complete knowledge system formed by comprehensive analyzing the different contradictions of maintenance in mechanical equipment life stages, studying their organic links. Failure theory has its own special research object, failure mechanical equipment.

故障理论主要应用了可靠性理论、维修性理论、摩擦、磨损和润滑、工程诊断学、金属物理、断裂力学等学科理论，以及用先进的测试技术同维修实践相结合，从而揭示机械系统进入生产过程后的运动规律，这些构成维修的决策依据。

Failure theory mainly consist of the theory of reliability, maintainability, friction, wear and lubricating, engineering diagnostics, metal physics, such as fracture mechanics theory. Advanced testing technology are combined with maintenance practice so as to reveal mechanical system into the production process. These theories constitute maintenance decision basis.

一切机械设备丧失规定功能的现象，统称故障。从微小的机器工作偏差，到严重的机械设备事故，都是机械故障的表现形式。

The phenomenon of all mechanical equipment loss function (out of order) is called failure or fault. Those from the tiny deviation while machine working to severe mechanical equipment accident are all forms of mechanical failure or fault.

机械在使用过程中，承受各种能量作用（力、力矩、温度、腐蚀、冲击等），这些能量一方面使机械按规定要求进行工作，另一方面造成机械性能及状态变化，随着时间的推移，机械性能下降，输出参数变化，最后发生故障。

While mechanical equipment working, various kinds of energy are exerted on the machine, including force, torque, temperature, corrosion, impact, etc. The energy, on the one hand, make machine work as required regulations, on the other hand, cause the change of mechanical properties and the state. The performance of machine degrades as time goes by. The output parameters change. Finally, machine failure occurs.

例如，轧钢机一方面在电机传动下完成钢材的轧制，另一方面其中的各个零件、部件会发生一些变化，如机架会在轧制力作用下发生周期性的变形，轧辊在轧制一定重量的钢材后会发生磨损、热膨胀，有些零件在冷却水的作用下会发生腐蚀，当这些磨损、变形或腐蚀达到一定程度，轧钢机就会出现故障。

Rolling mill, for example, on the one hand, completes steel rolling under the motor driven, on the other hand, all parts and components of rolling mill will change, such as frame may deform under periodic rolling force. The deformation of the roll will happen after a certain weight of rolling steel wear and thermal expansion. Some parts will occur corrosion under the action of cooling water. Rolling mill will be fail when the wear, deformation or corrosion reaches a certain degree.

故障过程是一个发展过程，例如，疲劳裂纹需交变应力达到一定的循环次数，材料损伤是材料检验性能偏离初始性能。

Failure process is a developing process. For example, fatigue crack requires alternating stress to reach a certain times of cycle, and material damage is the deviation of material check performance from initial performance.

机械设备的故障必定表现为一定的物质状况及特征，它们反映出物理的、化学的异常现象，并导致功能的丧失，这些物质状况的特征称为故障模式。机械设备的故障需要通过人的感官或测量仪器得到，相当于医学上的“病症”。

Mechanical equipment failure must show the certain material condition and characteristics. They reflect the physical and chemical anomalies, leading to loss of function. These characteristics of the material conditions are called fault mode. Mechanical equipment failure needs to be got through the human senses or measuring instruments, equivalent to the medical “disease”.

2.2.1 故障模式（Failure Modes）

常见的故障模式按以下几个方面进行归纳：

Common failure modes are summarized in the following aspects:

(1) 属于机械零部件材料性能方面的故障，包括疲劳、断裂、裂纹、蠕变、过度变形、材质劣化等。

Failure modes belonging to the mechanical parts material performance failure include excessive fatigue, creep, fracture, crack, deformation, material degradation, etc.

(2) 属于化学、物理状况异常方面的故障，包括腐蚀、油质劣化、绝缘绝热劣化、导电导热劣化、熔融、蒸发等。

Failure modes belonging to the chemical, physical anomalies of failure condition include corrosion, degradation of oil, degradation of insulating/resistance heat and insulation, degradation of electric conduction and thermally conductive, melt, evaporation, etc.

(3) 属于机械设备运动状态方面的故障，包括振动、渗漏、堵塞、异常噪声等。

Failure modes belonging to the mechanical equipment failure motion state include abnormal vibration, leakage, congestion/block/chock, noise, etc.

（4）多种原因的综合表现，如磨损等。

There is some comprehensive performance of a variety of reasons such as wear and tear, etc.

此外，还有配合件的间隙增大或过盈丧失、固定和紧固装置松动与失效等。

In addition, there is increased clearance or interference loss of mating parts, loosing and failure of fastening and fixed device and so on.

2.2.2 故障的分类（Classification of Failure）

故障可以按发生速度、后果和危害程序等进行理论上的分类，企业的故障分类更结合实际。

Failures can be classified according to the speed of occurrence, consequences and degree of harm. The failure classification principle of enterprises is more practical.

A 故障按发生速度分类（Classification According to the Failure rate）

根据故障发生速度，故障分为渐发性故障、突发性故障、复杂性故障。

According to the speed of the failure, failure is divided into gradual failure, sudden failure, complex failure.

a 渐发性故障（Gradual Failure）

如图2.4（a）所示，渐发性故障的主要特征是在给定时间段 $t_1 \to t_2$ 内，发生故障的概率 $F(t)$ 与设备已经工作过的时间 t_1 有关，使用过的时间越长，发生故障的概率就越高。它与材料的磨损、腐蚀、疲劳、蠕变等过程有密切的关系。大多数机器零部件的故障都属于这种类型，通常用监控手段可以预测这类故障。

As shown in Figure 2.4 (a), the main feature of the gradual failure is that the probability $F(t)$ of failure is related to the time t_1, which has been used for a given period of time from t_1 to t_2. The longer the time used, the higher the probability of failure. It is closely related to the wear, corrosion, fatigue, creep, and other processes of materials. This is the type of failure that most machine parts and components have, and it is usually predicted by monitoring.

如图2.4（a）所示，若用 U 表示零件损伤量，t 为零件使用时间，v 表示损伤过程速度，则损伤速度 $v = v(t)$ 是时间的函数。

As shown in Figure 2.4 (a), if U is used to represent the damage quantity of parts, t is the service time of parts, and v is the damage process speed, then damage velocity $v = v(t)$ is a function of time.

$$v(t) = \frac{dU}{dt} \tag{2.1}$$

而且，损伤是从设备一开始运行就开始的，所以，这类故障的无损时间间隔 t 等于零。但是只有当 U 达到 U_{max} 时故障才发生。

Moreover, the damage starts from the beginning of the operation of the equipment, so the lossless time interval t of this kind of failure is equal to zero. However, the failure occurs only when U reaches U_{max}.

b 突发性故障（Sudden Failure）

突发性故障是由各种不同因素及偶然性的外界影响共同作用的结果。这种作用已超过零件所能承受的限度。如热变形、超负荷等。如图 2.4（b）所示，无损时间间隔 t 间隔是随机量；损伤速度进行得非常快，损伤速度趋近于无穷大，发生故障的概率 $F(t)$ 与设备使用时间 t_1 无关。由于故障是突然发生的，事先无任何征兆，所以不能通过监控测试手段进行预测。

Sudden failure is the result of various factors and accidental external influences. This action has gone beyond the limits of the parts such as thermal deformation, overload, etc. As shown in Figure 2.4 (b), the lossless time interval t is a random quantity; The damage velocity is very fast, the damage velocity tends to be infinite, and the probability of failure $F(t)$ is independent of the service time t_1 of the equipment. Since the failure occurs suddenly and without warning, it cannot be predicted by means of monitoring and testing.

例如，因润滑油中断而使零件产生热变形，因使用不当出现超负荷而引起零件折断，因各项参数都达到极限值（载荷最大、温度最高、材料硬度最低等），而引起零件的变形和断裂。

For example, because of the lubricating oil interruption and causes the part to produce the hot deformation, because uses unjustly appears the overload but causes the part to break, because each parameter all achieves the limit value (the load is the biggest, the temperature is the highest, the material hardness is the lowest and so on), but causes the part the deformation and the fracture.

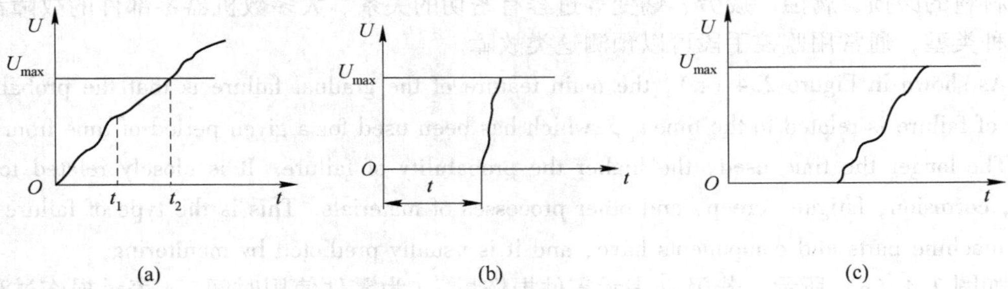

图 2.4 故障类型

Figure 2.4 Type of failure

（a）渐发性故障（gradual failure）；（b）突发性故障（sudden failure）；（c）复合型故障（complex failure）

c 复合型故障（Complex Failure）

复合型故障的发生时间是个随机量，与使用过的时长无关，损伤速度 $v = v(t)$ 则是时间 t 的函数，如图 2.4（c）所示。由于零件内部存在着应力集中，在受到外界对其作用的巨大冲击（可能是产生疲劳裂纹的根源）后，随着机器的继续使用，裂纹逐渐发展导致折断。

The occurrence time of complex failures is a random quantity, independent of the elapsed time, and the damage velocity $v = v(t)$ is a function of time t, as shown in Figure 2.4 (c). Due to the internal stress concentration of the parts, the cracks develop gradually and break off as the machine continues to use, after being subjected to the huge impact of external influences (which may be the source of fatigue cracks).

B 按故障的后果分类（Classification According to Consequences of Failure）

根据故障后果，故障可分为功能故障和参数故障。

Failure is classified by the consequences of failure as functional failure and parametric failure.

功能故障常是机器的个别零件损坏或卡滞而造成的，因此使机器不能继续完成自己的功能。例如，减速器不能旋转和传递运动，内燃机不能发动，油泵不能供油等。

Function failure is often caused by damage or clamping stagnation from the individual parts of the machine. So, the machine can't continue to complete their functions. For example, a reducer can't rotate and can't transfer movement, the internal combustion engine won't start, the pump can't supply oil, etc.

参数故障表现为机器的输出参数超过极限值。例如：加工精度不高，传动效率低，速度达不到标准值等，可能造成严重经济损失，制造出低劣产品。

Parametric failures show the output parameters of the machine exceed the limit value. For example, the machining accuracy is not high, the transmission efficiency is low, the speed cannot meet the standard, etc. Parametric failures may cause serious economic losses and produce inferior products.

C 按故障的危害程度分类（Classification According to the Harm Degree of Failure）

按危害程度，故障可分为灾难性故障、使用性故障和经济性故障。

According to the harm degree of failure classification, failure is grouped into disastrous failure, usability of failure and economic failure.

D 企业故障分类方法（Other Classification of Failure）

企业中采用故障频繁程度等级、影响程度等级和紧急程度三个方面进行综合评定。

The frequent degree of failure, degree of influence of failure and emergency degree of failure are usually used in enterprises for synthetically evaluation.

2.3 机械故障诊断技术（Failure Diagnosis Technology of Mechanical Equipment）

设备诊断技术是一种监测设备现有状况参数，分析故障原因和异常情况，预报设备未来情况的技术。诊断技术应该包括对设备的过去到现在，从现在到将来的一系列信息资料进行的预测工作。机械设备诊断技术的任务包括：提出制定诊断判定依据的理论与方法，指出技术状态识别的途径，提供技术决策（故障决策）原理，改进并总结诊断技术。

Equipment diagnosis technology is a technology to monitor the existing condition parameters of equipment, analyze the reasons for failure and abnormal conditions, and predict the future condition of equipment. Equipment diagnosis technology should include the prediction of a range of information from the past to the present, from the present to the future. The tasks of mechanical equipment diagnosis technology include putting forward the theory and method of making diagnosis decision basis, pointing out the way of technical state identification, providing the principle of technical decision (failure decision) and improving and summarizing diagnostic technology.

设备诊断技术是设备综合工程的一个组成部分，因此必须在设备寿命全过程中发挥作用。在设计、制造阶段必须考虑和制定使设备具有良好的可监控性、可诊断性。在使用阶段应充分有效地利用监测设备和诊断技术来研究和处理设备的故障。在事后阶段应注意对资料的积累、统计分析、资料库数据库的建立；为故障早期症状的正确诊断和预报及维修决策提供依据，并制定相应的标准、条例、规程等。状态监测是故障诊断的初级阶段，根据状态监测进行预知维修。

Equipment diagnosis technology is an integral part of equipment comprehensive engineering, so it must play a role in the whole life of equipment. In the design and manufacturing stage, the equipment must be considered and developed to have good monitoring and diagnostic capability. In the using stage, monitoring equipment and diagnostic technology should be fully and effectively used to study and deal with equipment failures. In the post-event stage, attention should be paid to the accumulation of data, statistical analysis, and the establishment of database, so as to provide a basis for the correct diagnosis and prediction of early failure symptoms and maintenance decisions, and formulate the corresponding standards, regulations and procedures. Status monitoring is the primary stage of failure diagnosis, according to the status monitoring for predictive maintenance.

目前故障诊断技术已经形成一门既有理论（诊断理论），又有方法（分析方法），既有实验（故障机理实验研究），又有手段（现代监测仪表与诊断系统），工程应用性强，技术背景牢固，与高技术发展密切相关的一门现代新兴学科。

The failure diagnosis technology has formed a combined theory and methods, the experimental failure mechanism and modern monitoring and diagnosis system. Engineering applicability and technical background of failure diagnosis technology are all strong. Failure diagnosis technology is new and high discipline which is closely related to a modern technology.

机械设备故障诊断技术发展方向是，在频谱分析和时间序列分析技术深入研究的基础上，进一步应用数学工具研究新的理论和技术。

The development of mechanical equipment failure diagnosis technology aims to further apply mathematics tool to research the new theory and technology based on the frequency spectrum analysis and time series analysis technology.

机械设备故障诊断技术的主要特点表现为：应用计算机进行在线监测和辅助诊断，建立具有人工智能的诊断专家系统，声发射技术、铁谱分析技术、油液分析技术、光导技术等的应用和推广。

Main features of failure diagnosis technology include establishing the diagnosis expert system with artificial intelligence acoustic emission technology, on-line monitoring and auxiliary diagnosis with the help of application of computer, applying iron spectral analysis technology, the application of oil analysis technology and optical technology and promotion.

机械故障诊断学是识别机器或机组运行状态的科学。它研究的对象是机器和机组运行状态的变化在诊断信息中的反映。它的内容包括对机器运行状态的识别、预测和监视三个方面。

Mechanical failure diagnosis is a science and technology to identify the running state of machines or unit. The research object of mechanical failure diagnosis is the reaction in the diagnostic information when the running state of machine and the unit change. The content of mechanical failure diagnosis includes the identification, predict and monitor of the machine running state.

设备故障诊断方法有许多，包括基于统计理论的故障诊断方法、基于模糊理论的故障诊断方法、基于故障树分析的故障诊断方法、基于专家系统原理的故障诊断方法、基于神经网络的故障诊断方法、基于数据融合的故障诊断方法以及基于相关技术集成的故障诊断方法。

There are a lot of equipment failure diagnosis methods, such as failure diagnosis method based on statistical theory, failure diagnosis method based on fuzzy theory, failure diagnosis method based on failure tree analysis, failure diagnosis method based on the principle of expert system and failure diagnosis method based on neural network, the failure diagnosis method based on data fusion and failure diagnosis method based on the related technology integration.

设备故障诊断技术主要涉及电子和计算技术、声振测试和分析技术、测温技术、油液分析技术、应力应变测试技术和无损检测技术等。

Mechanical failure diagnosis mainly involves electronics and computing technology, acoustics and vibration testing and analysis technology, temperature measurement technology, oil analysis technology, stress and strain measurement technology, and non-destructive testing technology.

2.3.1 电子和计算技术（Electronics and Computing Technology）

电子和计算技术是一门公共的基础技术，专用仪器和新的信号采集、分析及处理的方法都基于该两项技术。

Electronics and computing technology are common basic technologies for mechanical failure diagnosis. Some special instruments and new methods of signal acquisition, analysis and processing are based on these two technologies.

2.3.2 声振测试和分析技术（Acoustics and Vibration Testing and Analysis Technology）

用振动分析方法监测设备状态就是对现场设备进行振动信号采集、信号处理与分析，从而进行设备状态判别和诊断。

Using acoustics and vibration testing and analysis technology to monitor the state of equipment refers to collecting, processing, and analyzing the vibration signal of the field equipment, so as to discriminate and diagnose the state of equipment.

现场简易振动系统早期多采用模拟式测振仪，模拟式测振仪简易振动诊断系统包括传感器、电缆、测振仪、示波器和频率分析仪。其基本功能主要是测量机器的振动参数值，对设备做出有无故障判断。便携式多功能测振仪侧重数据采集器，操作方法简便、功能丰富，逐步取代了模拟式测振仪。

In the early stage, simple vibration system, such as analog vibrometers, are mostly used. The simple vibration diagnosis system of analog vibrometers includes sensors, cables, vibrometers, oscilloscopes and frequency analyzers. The basic function of simple vibration system is mainly measuring the vibration parameters of the machine, and making failure judgment of the equipment. Portable multifunction vibrometer focuses on data collector. The operation method of portable multifunction vibrometer is simple and the functions are rich.

便携式测振系统就是一台手持式数据采集器通过电缆直接和固定在设备上的压电式加速度传感器相连，除了测各种振动参数外，还可以在现场作时域、频域、相域等多种分析，并兼有数据存储功能。

The portable multifunction vibrometer will gradually replaces the analog vibrometer. Portable vibration measurement system is a handheld data collector connected directly with the piezoelectric acceleration sensor fixed on the equipment through cables. In addition to measuring various vibration parameters, it can also perform various analysis in time domain, frequency domain, phase domain and other fields, and have data storage functions.

振动测试系统包括经典测试系统和计算机测试系统。振动测试系统的主要部件和仪表包括传感器、信号调节器和记录显示仪表。

Vibration test system includes classical test system and computer test system. The main parts and instruments of the vibration test system include sensors, signal regulators and recording and display instruments.

传感器具有敏感功能以及将非电量变换成电量的功能。传感器的种类很多，往往可以应用多种类型的传感器来检测一种物理量，而同一种传感器也可以测量多种物理量。传感器根据被测对象分为：力传感器、位移（角位移、线位移）传感器、温度传感器、加速度传感器等。

The sensor is sensitive and have the function of converting non-electric quantity into electric quantity. There are many kinds of sensors, and one kind of parameter can be measured by multiple types of sensors, and the same kind of sensor can also measure a variety of physical quantities. Sensors are divided into force sensors, displacement (angular displacement, linear displacement) sensors, temperature sensors, acceleration sensors, etc.

信号调节器由各种电桥、谐振电路、放大、调制、解调、滤波电路、阻抗变换及标度变换等电路构成。其作用是将传感器输出的电信号变换成记录、显示仪器所需要的标准电压（或电流）信号，然后输送到记录、显示仪器中进行记录与显示，或者送入伺服驱动机构进行记录、指示与控制。

The signal regulator is composed of various bridges, resonance circuits, amplifying, modulating, demodulating, filtering circuits, impedance transformation, scale transformation and other circuits. The function of signal regulator is to convert the electrical signal output by the sensor into the standard voltage (or current) signal required by the recording and display instrument, and then transmit it to the recording and display instrument for recording and display, or send it to the servo drive mechanism for recording, indicating and controlling.

将表示被测量大小的电量进行记录与显示。通常的记录、显示仪器有光线示波器、*x-y*记录仪、电子示波器、磁带记录仪等。

The recording and display instrument records and displays the measured electric quantity. The usual recording and display instruments are ray oscilloscope, *x-y* recorder, electronic oscilloscope, tape recorder and so on.

2.3.3 测温技术（Temperature Measurement Technology）

温度是表征物体冷热程度的物理量，是国际单位制中七个基本物理量之一，它与人类生活、工农业生产和科学研究有着密切关系。随着科学技术水平的不断提高，温度测量技术也得到了不断的发展。

Temperature is a physical quantity that indicates how hot or cold an object is. As one of the seven basic physical quantities in the International System of Units, temperature is closely related to human life, industrial and agricultural production and scientific research. With the continuous improvement of science and technology, temperature measurement technology has also been developing.

温度测量方法有很多，也有多种分类，由于测量原理的多样性，很难找到一种完全理想的分类方法。目前温度测量的方法可以归纳为接触式测温和非接触式测温两大类，接触式测温方法通常包括膨胀式测温、电量式测温、接触式光电热色测温等；非接触式测温方法一般有辐射式测温、光谱法测温、激光干涉测温、声波微波法测温等。

There are many methods of temperature measurement and many kinds of classification. Due to the diversity of measurement principles, it is difficult to find a completely ideal classification method. The current temperature measurement methods can be classified into two categories: contact temperature measurement and non-contact temperature measurement. Contact temperature measurement methods usually include expansion temperature measurement, electric quantity temperature measurement, contact photoelectrothermal color temperature measurement, etc. Non-contact temperature measurement methods generally include radiant temperature measurement, spectral temperature measurement, laser interference temperature measurement, acoustic wave microwave temperature measurement and so on.

具体的接触式测温方法有玻璃液体测温计测温、双金属温度计测温、压力式温度计测温、热电偶测温、铂电阻测温、半导体测温、集成芯片测温、石英晶体测温、光纤测温、示温漆测温、液晶测温、黑体空腔测温。

The detailed contact temperature measurement methods include glass liquid thermometer temperature measurement, bimetal thermometer temperature measurement, pressure thermometer temperature measurement, thermocouple temperature measurement, platinum resistance temperature measurement, semiconductor temperature measurement, integrated chip temperature measurement, quartz crystal temperature measurement, optical fiber temperature measurement, temperature indicator paint temperature measurement, liquid crystal temperature measurement, blackbody cavity temperature measurement.

具体的非接触式测温方法有全辐射测温、亮度式测温、比色式测温、热像仪测温、多光谱测温、瑞利拉曼散射光谱测温、CARS（相干反斯托克斯喇曼散射）测温、受激荧光光谱测温、光谱吸收法测温、干涉仪测温、纹影法测温、激光散斑照相法测温、激光全息照相法测温、超声波法测温、微波衰减法测温等。几乎所有的温度测量技术都是在这些原理的基础上发展起来的。在线、非接触式、远距离测试的红外线测温技术运用比较普遍。

Specific non-contact temperature measuring methods have total radiation temperature measurement, brightness temperature, color temperature, thermal imaging temperature measurement, multi-spectral temperature measurement, Rayleigh Raman scattering spectra, CARS (coherent anti-Stokes Raman scattering), excited fluorescence spectrometric methods, spectral absorption, interferometer temperature measurement, schlieren method of temperature measurement, laser speckle photographic method of temperature measurement, laser holographic method of measuring temperature, ultrasonic method of measuring temperature, microwave attenuation method measuring temperature, etc. Almost all temperature measurement technology are developed on the basis of these principles. On-line, non-contact, infrared temperature technology for far distance measurement is widely used.

2.3.4 油液分析技术（Oil Analysis Technology）

用来研究摩擦磨损的油液分析技术在设备故障诊断中也被广泛地应用着，常用的油液分析技术包括磁塞检查法、液体颗粒计数器法、铁谱分析技术、光谱分析技术与同位素示踪技术。

Oil analysis technology used to study friction and wear is also widely used in equipment failure diagnosis. The commonly used oil analysis technology includes magnetic plug check, liquid particle counter, ferrography analysis technology, spectrum analysis and isotope tracer technology.

2.3.5 无损检测技术（Non-destructive Testing Technology）

无损检测技术（NDT）是一门独立的技术，近年来被引用到设备的故障诊断中，超声波探伤、磁粉探伤、渗透液表面裂纹探伤等技术被用来对大型固定或运动着的机器和设备进行监测和诊断，已越来越受到人们的重视。

Non-destructive testing technology (NDT) is an independent technology. In recent years, the ultrasonic flaw detection, magnetic particle flaw detection, penetrant surface crack detection and other technologies mentioned in the section of part testing have been used to monitor and diagnose large fixed or moving machines and equipment, which has attracted more and more attention.

思考题
(Questions)

2.1 机械零件常见失效形式有哪几种？

How many common failure modes of mechanical parts are summarized?

2.2 影响磨损的主要因素有哪些？

What are the main causes of wear?

2.3 磨料磨损的主要影响因素有哪些？

What are the main factors affecting abrasive wear?

2.4 黏着磨损的主要影响因素有哪些？

What are the main influencing factors of adhesive wear?

2.5 接触疲劳磨损的主要影响因素有哪些？

What are the main influencing factors of contact fatigue wear?

2.6 如何减轻磨料磨损？

How to reduce abrasive wear?

2.7 画出磨损特性曲线，并说明磨损过程分哪几个阶段？

Please draw the wear characteristic curve and indicate each stage of wear process.

2.8 金属受力产生的变形，可分为哪两个阶段，各有什么特点？

What are the two stages of metal deformation under stress? What are their characteristics?

2.9 弹性变形问题的两种效应是什么？

What are the two effects of elastic deformation?

2.10 消除加工硬化的方法是什么？

What is the way to remove work hardening?

2.11 消除内应力的方法有哪些？

What are the ways to relieve internal stress?

2.12 如何减轻变形危害？

How to reduce deformation of parts?

2.13 断裂常见的分类形式有哪四种？

What are the four common types of fracture classification?

2.14 脆性断裂的事故的基本特征有哪些？

What are the basic characteristics of brittle fracture accidents?

2.15 金属零件的腐蚀失效分哪两类？

What are the two types of corrosion failure of metal parts?

2.16 钢铁的高温腐蚀，按哪些方式进行？

What are the methods for high temperature corrosion of steel?

2.17 如何减轻钢的脱碳倾向？

How to reduce the decarbonization tendency of steel?

2.18 如何减轻钢的氢蚀？

How to reduce hydrogen corrosion of steel?

2.19 如何阻止铸铁的肿胀？

How to prevent the swelling of cast iron?

2.20 防止钢铁气体腐蚀的方法有哪些？

What are the methods to prevent gas corrosion in steel?

2.21 设备诊断技术的概念是什么？

What are the contents of mechanical failure diagnostics?

2.22 设备故障诊断中运用的主要技术有哪些？

What are the main technologies used in equipment failure diagnosis?

3 机械设备的拆卸

Chapter 3 Disassembly of Mechanical Equipment

3.1 机械零件的拆卸（Disassembly of Mechanical Equipment）

3.1.1 机械零件拆卸的一般规则和要求（Rules and Requirements of Disassembly）

拆卸的目的是便于检查和维修。由于机械设备的构造各有其特点，零部件在质量、结构、精度等各方面存在差异，因此若拆卸不当，将使零部件受损，造成不必要的浪费，甚至无法修复。为保证维修质量，在解体之前必须周密计划，对可能遇到的问题有所估计，做到有步骤地进行拆卸，一般应遵循下列规则和要求。

The purpose of disassembly is to facilitate inspection and maintenance. Because the structure of mechanical equipment has its own characteristics, and the parts are different in quality, structure, accuracy, etc. If the disassembly is improper, the parts will be damaged, resulting in unnecessary waste, or even can not be repaired. In order to ensure the maintenance quality, it is necessary to plan carefully before disassembly, estimate the problems that may be encountered, and disassemble step by step. Generally, the following rules and requirements should be followed.

（1）拆卸前必须先弄清楚构造和工作原理，做到心中有数，不能粗心大意。

Before disassembly, the structure and working principle must be clarified, so as to be clear and not careless.

（2）拆卸前做好准备工作，包括场地的选择、清理，拆前断电、擦拭、放油，对电气件和易氧化、易腐蚀的零件进行保护等。

Preparations shall be made before disassembly, including site selection and cleaning, power off, wiping and oil draining before disassembly, and protection of electrical parts and parts easy to oxidize and corrode.

（3）从实际出发，可不拆的零件尽量不拆，但需要进行必要的测试和诊断，确保无隐蔽缺陷，需要拆的一定要拆。

Proceeding from the actual situation, the parts that may not be disassembed should not to disassembled, but necessary tests and diagnosis should also be carried out to ensure that there are no hidden defects, and the parts that need to be disassembed must be disassembed.

（4）使用正确的拆卸方法，保证人身和机械设备安全。拆卸的顺序一般与装配的顺序相反，先拆外部附件，再将整机拆成总成、部件，最后全部拆成零件，并按部件汇集放置。有的拆卸需要采取必要的支承和起重措施。

Correct disassembly methods should be used to ensure personal and mechanical equipment safety. The disassembly sequence is generally the reverse of the assembly sequence. The external accessories are disassembled first, then the whole machine is disassembled into assemblies and components, and finally all of them are disassembled into parts, which are assembled and placed according to the components. Some disassembly requires necessary support and lifting measures.

（5）对轴孔装配件应坚持拆与装所用的力相同原则。

For shaft hole assembly, the principle of the same force used for disassembly and assembly shall be adhered to.

（6）拆卸应为装配创造条件，对不能互换的零件要成组存放或打标记。

Disassembly shall create conditions for assembly. Parts that cannot be exchanged shall be stored in groups or marked.

拆卸主要是指连接件的拆卸，除应遵守上述规则以外，还应掌握拆卸的方法。

Disassembly mainly refers to the disassembly of connecting pieces. In addition to the above rules, the disassembly method shall also be mastered.

3.1.2 常用的拆卸方法（Methods of Disassembling Mechanical Parts or Components）

机械零件常用的拆卸方法有击锤法、拉拔法、顶压法、温差法和破坏法。

The methods of disassembling mechanical parts or components usually include: hamming method, drawing method, top pressing method, temperature method and destroy method.

（1）击锤法是利用锤子或其他重物在敲击或撞击零件时产生的冲击能量把零件拆下。

The hammering method is using the impact energy generated by hammers or other heavy objects when striking or hitting parts to remove the parts.

（2）拉拔法适用于对精度要求较高，不允许敲击或无法用击卸法拆卸的零部件。该方法采用专门拉器进行拆卸。

The drawing method is used for parts with high accuracy that cannot be knocked or disassembled by knocking. It is disassembled with a special puller.

（3）顶压法是利用螺旋 C 型夹头、机械式压力机、液压压力机或千斤顶等工具和设备进行拆卸。它适用于形状简单的过盈配合件。

The top pressing method is to disassemble with tools and equipment such as screw C-type collet, mechanical press, hydraulic press or jack. It is applicable to interference fit parts with simple shape.

（4）温差法适用于拆卸尺寸较大、配合过盈量较大或无法用击卸、顶压等方法拆卸时，或为使过盈较大、精度较高的配合件容易拆卸。温差法是利用材料热胀冷缩的性能、加热包容件，使配合件在温差条件下失去过盈量，实现拆卸。

The temperature method can be used when the size is large, the fit interference is large, or it can not be disassembled by means of hammering or top pressing, or to make the fit parts with large interference and high precision easy to disassemble. The temperature method is using the property of thermal expansion and cold contraction of materials to heat the containing parts, so that the mating parts will lose the interference under the temperature difference condition and realize disassembly.

（5）破坏法是必须拆卸焊接、铆接等固定连接件，或轴与套互相咬死，或为保存主件而破坏副件时，可采用车、锯、錾、钻、割等方法进行破坏性拆卸。

The destroy method is dismantling the fixed connecting pieces such as welding and riveting, or the shaft and sleeve bite each other, or when the secondary parts are damaged to preserve the main parts. Destructive methods such as turning, sawing, chiseling, drilling and cutting can be used.

3.1.3 典型连接件的拆卸（Disassembly of Typical Connection Parts or Components）

螺纹连接件、过盈连接件、滚动轴承和固定连接件都属于典型的连接件，机械设备维修工作中经常涉及这些典型连接件的拆卸。

Threaded connectors, interference fitting connectors, rolling bearings and fixed connections are typical connectors, mechanical equipment maintenance work often involved in the removal of these typical connectors.

3.1.3.1 螺纹连接件的拆卸（Disassembly of Threaded Connectors）

螺纹连接件应用广泛，它具有简单、便于调节和可多次拆卸装配等优点。虽然它拆卸较容易，但有时因重视不够或工具选用不当、拆卸方法不正确而造成损坏，应特别引起注意。

The threaded connection is widely used. It has the advantages of simplicity, easy adjustment, and multiple disassembly and assembly. Although it is easy to disassemble, special attention should be paid to the damage caused by insufficient attention, improper tool selection and incorrect disassembly method.

A 一般拆卸方法（General Disassembly Method）

首先要认清螺纹旋向，然后选用合适的工具，尽量使用呆扳手或螺钉旋具、双头螺栓专用扳手等。拆卸时用力要均匀，只有受力大的特殊螺纹才允许用加长杆。

First, recognize the thread rotation direction, and then select appropriate tools. Try to use a solid wrench or screw driver, a special wrench for stud bolts, etc. The force shall be even when disassembling, and only the special thread with large force can use the extension rod.

B 特别情况的拆卸方法（Dismantling Method under Special Circumstances）

a 断头螺钉的拆卸（Disassembly of Threaded with a Broken Head）

机械设备中的螺钉头有时会被打断，断头螺钉在机体表面以下时，可在断头端的中心钻孔，攻反向螺纹，拧入反向螺钉旋出（如图 3.1（a）所示）。在机体表面以上时，可在螺钉上钻孔，打入多角淬火钢杆，再把螺钉拧出（如图 3.1（b）所示）。也可在断头上锯出沟槽，用一字形螺钉旋具拧出；或用工具在断头上加工出扁头或方头，用扳手拧出。或在断头上加焊弯杆拧出；也可在断头上加焊螺母拧出（如图 3.1（c）所示）。当螺钉较粗时，可用扁錾沿圆周剔出。

Screw head in the machines can sometimes be broken. When the head of broken screw is beneath the surface of the body, a hole can be drilled in the center of broken screw. Reverse tapping on the hole of broken screw is done. Reverse screw is driven in the interrupted screw head, the broken screw can be driven out (as shown in Figure 3.1 (a)). When the head of broken screw is higher than the surface of the body, a hole can also be drilled in the center of broken screw. A polygonal quenching steel rod can be hit into the hole. Then the broken screw can be driven out (as shown in Figure 3.1 (b)). Also, a groove can be sewed on the head of broken head, a slotted screwdriver is used to screw the broken screw out. Or, a bent rod can be welded on the broken screw head to screw out the broken screw. Moreover, a nut can also be welded on the broken head to screw out the broken screw. When the screw is big, a flat chisel can be used to winkled out along the circumference (as shown in Figure 3.1 (c)).

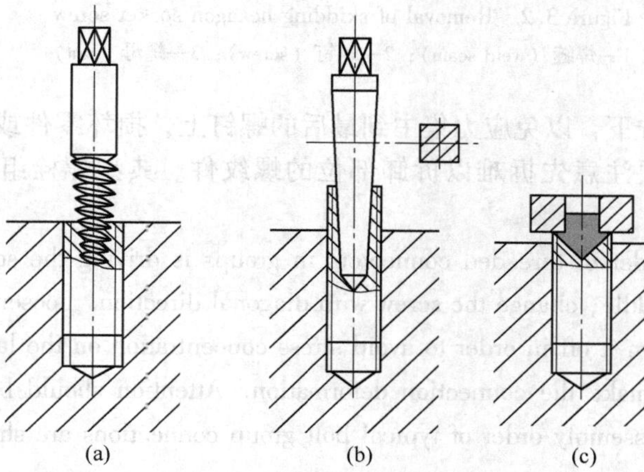

图 3.1　断头螺钉的拆卸
Figure 3.1　Removal of broken screw

b　打滑内六角螺钉的拆卸（Disassembly of Hexagon Socket with a Slide Head）

当内六角磨圆后出现打滑现象时，可用一个孔径比螺钉头外径稍小一点的六方螺母，放在内六角螺钉头上，将螺母和螺钉焊成一体，用扳手拧螺母即可把螺钉拧出（如图 3.2 所示）。

The hexagon socket head cap screw cannot be driven out when the hexagon socket head is worn and slide. A nut with a little smaller inner diameter than the outer diameter of the hexagon socket head can be welded on the hexagon socket head use an aperture hexagon nut. Then the hexagon socket head cap screw can be driven out (as shown in Figure 3.2).

c　锈死螺纹的拆卸（Disassembly of Rusty Thread）

可向拧紧方向拧动一下，再旋松，如此反复，逐步拧出；用手锤敲击螺钉头、螺母及四周，锈层震松后即可拧出；

When the rusty thread needs to be removed, the rusty thread should be driven to the tighten direction firstly and then to the loosen direction repeatedly and the rusty thread may be driven out. Knocking the head of the rusty thread all around can loosen the rust to drive the rusty thread.

d　成组螺纹连接件的拆卸（Disassembly of Threaded Connectors in Groups）

成组螺纹连接件的拆卸顺序一般为先四周后中间，对角线方向轮换。先将其拧松少许

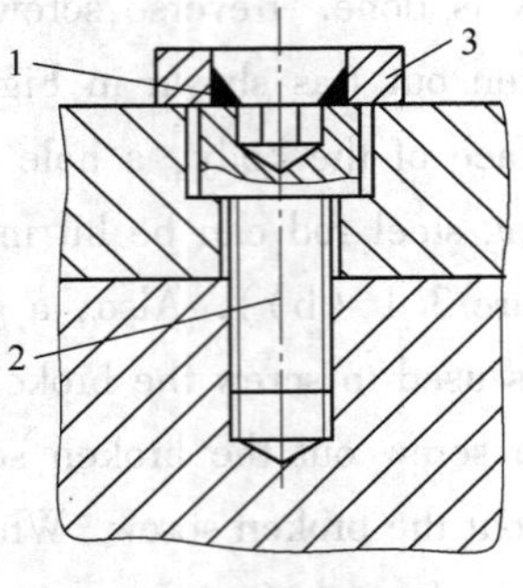

图 3.2　打滑内六角螺钉的拆卸

Figure 3.2　Removal of skidding hexagon socket screw

1—焊缝（weld seam）；2—螺钉（screw）；3—螺母（nut）

或半周，然后再顺序拧下，以免应力集中到最后的螺钉上，损坏零件或使结合件变形，造成难以拆卸的困难。要注意先拆难以拆卸部位的螺纹件。典型螺栓组连接的拆卸顺序如图 3.3 所示。

The disassembly order of threaded connectors in groups is driving the screw one by one from four edges before the middle, change the screw with diagonal direction. Loosen it a little or half way round first, and then turn it off in order to avoid stress concentration on the last screw, damage the parts or components or make the connection deformation. Attention should be paid to remove the difficult parts first. Disassembly order of typical bolt group connections are shown in Figure 3.3.

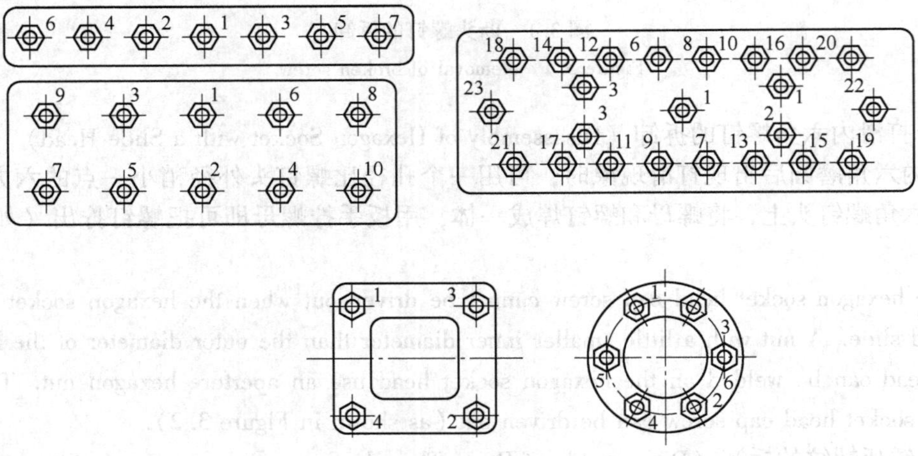

图 3.3　典型螺栓组连接的拆卸顺序

Figure 3.3　Disassembly order of typical bolt group connections

3.1.3.2　过盈连接件的拆卸（Disassembly of interference fitting connectors）

拆卸过盈件，应按零件配合尺寸和过盈量大小，选择合适的拆卸工具和方法。视松紧程度由松至紧，依次用木锤、铜棒、手锤或大锤、拉器、机械式压力机、液压压力机、水压机等进行拆卸。过盈量过大或为保护配合面，可加热包容件或冷却被包容件后再迅速压出。

When removing interference fitting parts or components, appropriate disassembly tools and methods should be selected according to the matching size of parts or components and magnitude of interference. Wood hammer, copper bar, hand hammer or sledgehammer, puller, mechanical press, hydraulic press, and hydraulic press are used depending on the degree of tightness from loose to tight to disassemble the parts or components with interference fitting. If the magnitude of interference is too large or fcr the mating surface should be protected, and enveloped parts (shaft) can be heated or the enveloping parts (hole) cooled and then the part is quickly pressed out.

无论使用何种方法拆卸，都要检查有无定位销、螺钉等附加固定或定位装置，若有必须先拆下。施力部位要正确，受力要均匀，方向要无误。

Whatever method is adopted, locating pin, alignment pin, assembling pin and such position devices should be checked and be removed firstly. The place where the force is applied should be right. The force applied on the part or component should be even. The direction the force applied should be correct.

3.1.3.3 滚动轴承的拆卸（Disassembly of Rolling Bearings）

拆卸滚动轴承时，除按过盈连接件的拆卸要点进行外，还应注意尽量不用滚动体传递力。拆卸末端的轴承时，可用小于轴承内径的铜棒或软金属、木棒抵住轴端，在轴承下面旋转垫铁，再用手锤敲击（如图 3.4 和图 3.5 所示）。

While removing the rolling bearing, rollers should not transfer force as far as possible. While removing the rolling bearing on the end of shaft, copper bar, soft metal, wood bar which is smaller than inner diameter of bearing car be used to support the end of shaft. Sizing blocks are put under the bearing. Then hand hammer is used to knock to remove the rolling bearing (as shown in Figure 3.4 and Figure 3.5).

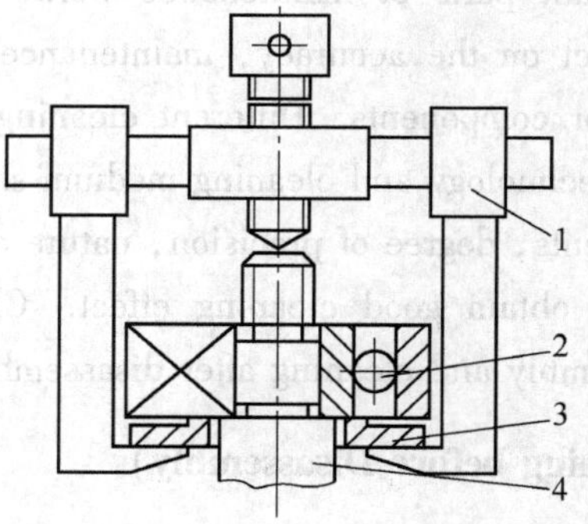

图 3.4　轴承拆卸方法之一

Figure 3.4　Method 1 of bearing disassembly

1—拆卸器（remover）；2—轴承（bearing）；3—环形件（ring）；4—轴（shaft）

3.1.3.4 固定连接的拆卸（Disassembly of Fixed Connections）

焊接和铆接属于固定连接，焊接件的拆卸可用锯割、扁錾切割、小钻头钻一排孔后再錾或锯，以及气割等，铆接件的拆卸可錾掉、锯掉、气割铆钉头或用钻头钻掉铆钉等。

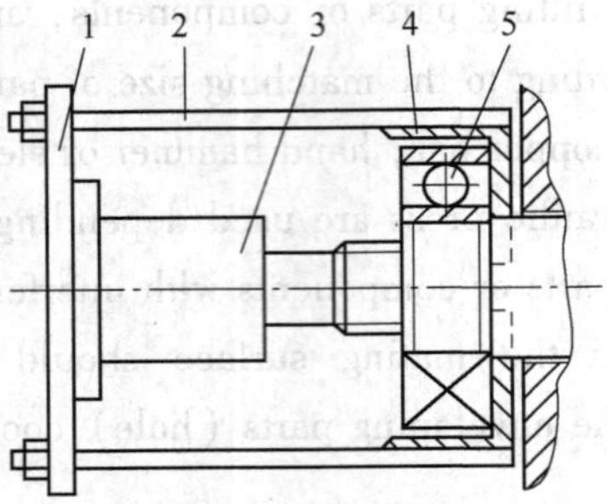

图 3.5 轴承拆卸方法之二

Figure 3.5 Method 2 of bearing disassembly

1—底板（bottom plate）；2—拉杆（pull rod）；
3—液压千斤顶（hydraulic jack）；4—角钢（angle steel）；5—轴承（bearing）

Welding and riveting are fixed connections. Sawing, flat chisel cutting and gas cutting and so on are available for disassembly of weldment after drill a row of holes with a small drill. Disassembly of riveting pieces can be chiseled out, cut out or the rivet head is cut with a drill out rivets.

3.2 零件的清洗（Cleaning of Parts or Components）

在维修过程中搞好清洗是做好维修工作的重要一环。清洗方法和清洗质量对鉴定零件的准确性、维修质量、维修成本和使用寿命等均产生重要影响。根据零件的材质、精密程度、污物性质和各工序对清洁程度的要求不同，必须采用不同的清洗方法，选择适宜的设备、工具、工艺和清洗介质，以便获得良好的清洗效果。零件的清洗包括拆卸前的清洗和拆卸后的清洗。

Cleaning is one of important parts of maintenance work. Cleaning method and cleaning quality have an important impact on the accuracy, maintenance quality, maintenance cost and service life of identified parts or components. Different cleaning methods must be adopted and appropriate equipment, tools, technology and cleaning medium should be chosen according to the material of the parts or components, degree of precision, nature of dirt and different requirements of each procedure in order to obtain good cleaning effect. Cleaning of parts or components includes cleaning before disassembly and cleaning after disassembly.

3.2.1 拆卸前的清洗（Cleaning before Disassembly）

拆卸前的清洗主要是指拆卸前的外部清洗。其外部清洗的目的是除去机械设备外部积存的大量尘土、油污、泥砂等脏物，以便于拆卸和避免将尘土、油泥等脏物带入厂房内部。外部清洗一般采用自来水冲洗，即用软管将自来水接到被清洗部位，用水流冲洗油污，采用刮刀、刷子配合进行。高压水冲刷即采用 1～10MPa 压力的高压水流进行冲刷。对于密度较大的厚层污物，可加入适量的化学清洗剂并提高喷射压力和水的温度。

Cleaning before disassembly mainly refers to cleaning the external surface of the machine before it is disassembled. The purpose of external cleaning is to remove a large amount of dirt, grease, mud and sand accumulated on the surface of the machine. Therefore, dirt, oil sludge and other dirt cannot be brought into the plant. Water is usually used to clean the external surface of the machine. Scraper and brush can be used to wipe out oil. The high-pressure water with the pressure of 1 ~ 10MPa can be used to wipe out the dirt. An appropriate amount of chemical cleaning agent can be added and the injection pressure and temperature of water can be improved to wipe out the dirt with dense thick layer.

常见的外部清洗设备有单枪射流清洗机和多喷嘴射流清洗机。单枪射流清洗机是靠高压连续射流或汽水射流的冲刷作用或射流与清洗剂的化学作用相配合来清除污物。多喷嘴射流清洗机有门框移动式和隧道固定式两种，喷嘴安装位置和数量，根据设备的用途不同而异。

Common external cleaning equipment includes single gun jet cleaning machine and multi-nozzle jet cleaning machine. Single gun jet cleaning machine can clean up dirt depending on high pressure continuous jet washing effect or air water jet washing effect, or a combination action of jet washing effect and chemical cleaning agent. There are two kinds of multi-nozzle jet cleaning machine: door frame movable type and tunnel fixed type. The location and number of nozzles installed vary according to the purpose of the equipment.

3.2.2 拆卸后的清洗 (Cleaning after Disassembly)

拆卸后的清洗包括清除油污、水垢、积炭、锈层和旧漆层等。

Cleaning of parts or components after disassembly includes removal of oil, removal of scale, removal of carbon deposits, removal of rust and removal of lacquer.

3.2.2.1 清洗油污 (Removal of Oil)

凡是和各种油料接触的零件在解体后都要进行清除油污的工作，即除油。油可分为两类：可皂化的油和不可皂化的油。可皂化的油，就是能与强碱起作用生成肥皂的油，如动物油、植物油，即高分子有机酸盐。还有一类是不可皂化的油，它不能与强碱起作用，如各种矿物油、润滑油、凡士林和石蜡等。它们都不溶于水，但可溶于有机溶剂。去除这些油类，主要是用化学方法和电化学方法。

All parts or components that come in contact with various oil must be cleaned up after disassembly, that is, remove oil. Oil can be divided into two categories: saponifiable oils and non-saponifiable oil. Saponifiable oil, such as animal oil, vegetable oil, namely polymer organic salts, can act with strong alkali to produce soap. Non-saponifiable oils, such as mineral oils, lubricants, vaseline, paraffin waxes and so on, do not act with strong alkali. They are both insoluble in water, but soluble in organic solvents. These oils are removed mainly by chemical and electrochemical methods.

常用的清洗液为有机溶剂、碱性溶液和化学清洗液等。清洗方式则有人工清洗和机械清洗两种方式。

The commonly used cleaning solution is organic solvent, alkaline solution and chemical cleaning solution. Cleaning methods include manual cleaning and mechanical cleaning.

常见的有机溶剂有煤油、轻柴油、汽油、丙酮、酒精和三氯乙烯等。有机溶剂除油是以溶解污物为基础。它对金属无损伤，可溶解各类油脂，不需加热，使用简便，清洗效果好。但有机溶剂多数为易燃物，成本高。主要适用于规模小的单位和分散的维修工作。

Common organic solvents are kerosene, light diesel, gasoline, acetone, alcohol and trichloroethylene. Organic solvent is used to remove oil based on dissolved dirt. It has no damage to metal. It can dissolve all kinds of grease, needn't to heat the part. Cleaning effect is good. But most organic solvents are flammable with a high cost. They are mainly suitable for small units and scattered maintenance work.

碱性溶液是碱或碱性盐的水溶液。利用碱性溶液和零件表面上的可皂化油起化学反应，生成易溶于水的肥皂和不易浮在零件表面上的甘油，然后用热水冲洗，很容易除油。

Alkaline solution is an aqueous solution of alkali or alkaline salt. The alkaline solution reacts with the saponifiable oil on the surface of the part to produce soap that dissolves easily in water and glycerin that does not float easily on the surface of the part. They can be washed away with hot water. Therefore, the oil is easy to be removed.

对不可皂化油和可皂化油不容易去掉的情况，应在清洗溶液中加入乳化剂，使油垢乳化后与零件表面分开。常用的乳化剂有肥皂、水玻璃（硅酸钠）、骨胶、树胶等。清洗不同材料的零件应采用不同的清洗溶液。碱性溶液对于金属有不同程度的腐蚀作用，尤其是对铝的腐蚀较强。用碱性溶液清洗时，一般需要将溶液加热到 80~90℃。除油后用热水冲洗，去掉表面残留碱液，防止零件被腐蚀。碱性溶液应用最广。

For non-saponifiable and saponifiable oils that are not easily to be removed, emulsifier should be added to the cleaning solution to separate the oil from the surface of the parts or components after emulsification. Commonly used emulsifiers are soap, sodium silicate, bone glue, gum, etc. Different cleaning solutions should be used to clean parts or components of different materials. Alkaline solution has different corrosion effects on metal, especially on aluminum. When cleaning the parts or components with an alkaline solution, the solution usually needs to be heated to 80~90℃. The residual alkaline solution on the surface should be washed away with hot water to prevent corrosion of parts or components. Alkaline solutions are most widely used.

化学清洗液是一种化学合成水基金属清洗剂，以表面活性剂为主。由于其表面活性物质降低界面张力而产生湿润、渗透、乳化、分散等多种作用，具有很强的去污能力。它还具有无毒、无腐蚀、不燃烧、不爆炸、无公害、有一定防锈能力、成本较低等优点，目前已逐步替代其他清洗液。

Chemical cleaning agent is a kind of water-based chemical synthesis metal cleaning agent. The composition of chemical cleaning liquid is mainly surfactant. Chemical cleaning agent has a strong decontamination ability because its surface active substances can reduce the interfacial tension and produce wetting, permeability, emulsification, dispersion and other functions. It also has many advantages, such as non-toxic, non-corrosive, non-burning, non-explosion, non-pollution, a certain anti-rust ability, low cost. Chemical cleaning agent has gradually replaced other cleaning agent.

油污的清洗方法主要有：擦洗、煮洗、喷洗、振动清洗和超声清洗。

The main cleaning methods to remove oil include scrubbing, boiling, spraying, vibration cleaning and ultrasonic cleaning.

3.2.2.2 清洗水垢（Removal of Scale）

机械设备的冷却系统长期使用硬水或含杂质较多的水，就在冷却器及管道内壁上沉积一层黄白色的水垢。它的主要成分是碳酸盐、硫酸盐，有的还含有二氧化硅等。

When mechanical equipment cooling system work for a long time using hard water or those water with more impurities, the cooler and pipe wall deposits a layer of yellow and white scale. The main components of scale are carbonate, sulfate, silica and so on.

水垢使水管截面缩水，热导率降低，严重影响冷却效果，从而影响冷却系统的正常工作，必须定期清除。水垢的清除方法可用化学去除法：酸盐清除水垢、碱溶液清除水垢和酸洗清除水垢。

The scale decreases the diameter of water pipe, reduces the thermal conductivity, seriously affects the cooling effect so as to affect the normal work of the cooling system. Scale must be removed regularly. The methods can be used to remove scale: acid-salt to remove scale, alkali solution to remove scale, and pickling to remove scale.

3.2.2.3 清除积炭（Removal of Carbon Deposits）

在维修过程中，常遇到清除积炭的问题，如发动机中的积炭大部分积聚在气门、活塞、汽缸盖上。积炭的成分与发动机的结构、零件的部位、燃油、润滑油的种类、工作条件以及工作时间等有很大的关系。

In the maintenance process, problems of removing carbon often arise. For example, most of the carbon in the engine accumulates on valves, pistons, and cylinder heads. The composition of carbon deposits has a great relationship with the structure of the engine, the parts or components, the types of fuel oil, lubricating oil, working conditions, and working hours.

积炭是由于燃料和润滑油在燃烧过程中不能完全燃烧，并在高温作用下形成的一种由胶质、沥青质、油焦质、润滑油和炭质等组成的复杂混合物。这些积炭影响发动机某些零件散热效果，恶化传热条件，影响其燃烧性，甚至会导致零件过热，形成裂纹。目前，经常使用机械清除法、化学法和电解法等进行积炭清除。

Carbon deposits is a kind of complex mixture composed of colloid, asphaltene, oil coke, lubricating oil and carbonaceous, which is formed under the action of high temperature due to the incomplete combustion of fuel and lubricating oil in the combustion process. The accumulation of carbon deposits affects the heat dissipation effect of some parts or components of the engine, worsens the heat transfer conditions, affects its combustion, and even causes parts or components to overheat and form cracks. At present, mechanical, chemical, and electrolytic methods are often used to remove carbon deposits.

3.2.2.4 除锈（Removal of Rust）

锈是金属表面与空气中氧、水分以及酸类物质接触而生成的氧化物，如 FeO、Fe_3O_4、Fe_2O_3 等，通常称为铁锈。去锈的主要方法有机械法、化学酸洗法和电化学酸蚀法。

The rust is the oxide formed on the metal surface when the metal contact with oxygen, water and acid substances in the air. The iron rust is FeO, Fe_3O_4, Fe_2O_3, etc. The main methods to remove rust are mechanical removal, chemical pickling and electrochemical acid etching.

3.2.2.5 清除漆层（Removal of Lacquer）

零件表面的保护层需根据其损坏程度和保护涂层的要求进行全部或部分清除。清除要先冲洗干净，然后再准备喷刷新漆。

The protective layer on the surface of the parts or components shall be completely or partially removed according to the damage degree and the requirements of the protective coating. The lacquer layer should be clean up and clear. Then new lacquer will be sprayed later.

清除方法一般用手工工具，如刮刀，砂纸，钢丝刷或手提式电动、风动工具，进行刮、磨、刷等。有条件的也可以用各种配制好的有机溶剂、碱性溶液等作退漆剂，涂刷在零件的漆层上，使之溶解软化，再借助手工工具去除漆层。为完成各道清洗工序，可使用一整套各种用途的清洗设备，包括喷淋清洗机、浸浴清洗机、喷枪机、综合清洗机、环流清洗机、专用清洗机等。究竟采用哪一种设备，要考虑其用途和生产场所。

The hand tools, such as scraper, sandpaper, wire brush or portable electric, pneumatic tools are generally used for scraping, grinding, brush the lacquer layer. Conditional also can use all sorts of organic solvent that have prepared, alkalescent solution to make paint withdrawal agent, besmear is brushed on the lacquer layer of the part, make it dissolve bate, remove lacquer layer with the aid of manual tool again. In order to complete each cleaning process, a set of cleaning equipment can be used, including spray cleaning machine, bath cleaning machine, spray gun machine, comprehensive cleaning machine, circulation cleaning machine, special cleaning machine and so on. To decide what kind of equipment can be used should consider its use and production sites.

3.3 零件的检验（Checking of Parts or Components）

维修过程中检验工作包含的内容很广，在很大程度上，它是制定维修工艺措施的主要依据，决定零部件的弃取和装配质量，影响维修成本，是一项重要的工作。

In the maintenance process, the content of check is very wide. To a large extent, it is the main basis of establishing maintenance process measures. It is an important work to determine whether parts and components are scraped or kept. Check determines assembly quality and affects maintenance cost.

3.3.1 检验的原则（Principles of Checking）

零件检验一般遵循以下三项原则：

Parts checking generally follows the following three principles:

（1）在保证质量的前提下，尽量缩短维修时间，节约原材料、配件、工时，提高利用率，降低成本。

Under the premise of ensuring quality, try to shorten the maintenance time, save raw materials, accessories, working hours, improve the utilization rate, and reduce costs.

（2）严格掌握技术规范、修理规范，正确区分能用、需修、报废的界限，从技术条件和经济效果综合考虑。既不让不合格的零件继续使用，也不对不必维修或应报废的零件进行修理或报废。

Strictly grasp the technical specifications, repair specifications, correctly distinguish the boundaries of can be used, need to be repaired or must be scrapped. Comprehensive consideration should be done from the technical conditions and economic effects. The unqualified parts are forbidden to be used continuously. Parts that do not need to be repaired are not allowed to be repaired. Parts that should not be scrapped should not be scrapped.

（3）努力提高检验水平，尽可能消除或减少误差，建立健全合理的规章制度。按照检验对象的要求，特别是精度要求选用检验工具或设备，采用正确的检验方法。

Strive to improve the check level, as far as possible to eliminate or reduce errors, establish and improve reasonable rules and regulations. Use the correct check method and select of check tools or equipment according to the requirements of parts or components need to be checked, especially the requirements of accuracy.

检验一般分为修前检验、修后检验和装配检验。

Check is generally divided into check before repair, check after repair and check while assembling.

修前检验是在机械设备拆卸后进行。对已确定需要修复的零部件，可根据损坏情况及生产条件选择适当的修复工艺，并提出技术要求。对报废的零部件，要提出需补充的备件型号、规格和数量；不属备件的需要提出零件蓝图或测绘草图。

Check before repair is done after the machine has been disassembled. For the parts or components that have been determined to be repaired, the appropriate repair process can be selected according to the damage situation and production conditions, and the technical requirements shall be put forward. For scrapped parts or components, the model, specification and quantity of spare parts or components to be supplemented shall be put forward. The blueprints or drawings of parts which are not spare parts or components should be measured and finished.

修后检验是指零件加工或修理后检验其质量是否达到了规定的技术标准，确定是成品、废品或返修品。

Check after repair refers to check whether the quality of the parts or components after being processed or repaired has reached the specified technical standards or not. Then the parts or components are determined to be the finished products, the waste products or the parts need to be repaired again.

装配检验是指检验待装零部件质量是否合格、能否满足要求。在装配中，对每道工序或工步都要进行检验，以免产生中间工序不合格，影响装配质量。组装后，检验累积误差是否超过技术要求。总装后要进行调整，包括工作精度、几何精度及其他性能检验、试运转等，确保维修质量。

Check while assembling means to check whether the parts or components to be installed are qualified and meet the requirements or not. In the process of assembly, check should be carried out on every working procedure or working step to avoid disqualification of intermediate working procedure and influence on assembly quality. After the parts or components have been assembled, the accumulated error should be checked whether exceeds the technical requirements or not. After the final assembly, adjustment should be made to ensure the quality of maintenance, including working accuracy, geometric accuracy and other performance check and test run.

3.3.2 零件检验的内容（Contents of Checking）

零件检验的内容有很多，包括零件的几何精度、表面质量、力学性能、内疵、零部件的质量和静动平衡、零件的材料性质、零件表层材料与基体的结合强度、组件的配合情况、零件的磨损程度、密封性等。

There are many main items for parts or components to be checked, including geometric accuracy, surface quality of parts or components, mechanical properties, internal defects, quality and static and dynamic balance, material properties, surface material and matrix bonding strength, component matching, wear degree and sealing property of parts or components, etc.

（1）几何精度。零件的几何精度包括尺寸、形状和表面相互位置精度。经常检验的是尺寸、圆柱度、圆度、平面度、直线度、同轴度、平行度、垂直度、跳动等项目。根据维修特点，有时不是追求单个零件的几何尺寸精度，而是要求相对配合精度。

Geometric accuracy. The geometric accuracy of a part includes dimension, cylinder degree, roundness, flatness, straightness, coaxiality, parallelism, verticality, runout and other items. According to the characteristics of maintenance, sometimes the geometric dimension accuracy of a single part is not important, but the relative matching accuracy is required.

（2）表面质量。零件的表面质量包括表面粗糙度、表面有无擦伤、腐蚀、裂纹、剥落、烧损、拉毛等缺陷。

Surface quality. The surface quality of a part includes surface roughness, surface with or without abrasion, corrosion, cracking, peeling, burning, napping and other defects.

（3）物理力学性能。除硬度、硬化层深度外，对零件制造和修复过程中形成的性能，如应力状态、平衡状况、弹性、刚度、振动等也需根据情况适当进行校测。

Physical and mechanical properties. Physical and mechanical properties of a part refer to the performance formed in the process of the part is manufacturing and repair. Besides hardness and case depth, stress, balance, elasticity, rigidity, vibration also need to be checked and corrected appropriately.

（4）内疵。零件的内疵包括制造过程中的内部夹渣、气孔、疏松、空洞、焊缝等缺陷，还有使用过程中产生的微观裂纹。

Internal defects. Internal defects of a part include internal slag, porosity, loosen, empty, weld defects in the manufacturing process. There are micro-cracks generated during being used.

（5）质量和静动平衡。如活塞、连杆组之间的质量和曲轴、风扇、传动轴、车轮等高速转动的零部件的静动平衡都需要进行检验。

Mass and balance. For example, the mass between pistons and connecting rods and the static and dynamic balance of crankshafts, fans, driving shafts, wheels and other parts rotating at high speeds should be checked.

（6）材料性质。如零件合金成分、渗碳层含碳量、各部分材料的均匀性、铸铁中石墨的析出、橡胶材料的老化变质程度等都需要进行检验。

Material property. For example, the alloy composition of parts, the carbon content of carburized layer, the uniformity of each part of the material, the precipitation of graphite in cast iron, the aging and metamorphism of rubber materials, etc., all need to be tested.

（7）表层材料与基体的结合强度。如电镀层、喷涂层、堆焊层与基体金属的结合强度，机械固定连接件的连接强度，轴承合金和轴承座的结合强度等都需要进行检验。

Bonding strength. For example, the bonding strength of electroplating layer, spraying layer, surfacing layer and base metal, the bonding strength of mechanical fixed connectors, the bonding strength of bearing alloy and bearing housing, etc., all need to be tested.

（8）组件的配合情况。如组件的同轴度、平行度、啮合情况与配合的严密性等都需要进行检验。

Matching of the sub-assembly. For example, the coaxiality, parallelism, meshing condition and the tightness of matching all need to be tested.

（9）磨损程度。正确识别摩擦磨损零件的可行性，由磨损极限确定是否能继续使用。

Wear degree. Checking the wear degree of parts or components refers to identify the feasibility of the worn parts or components correctly and to determine whether to continue to use or not according to the wear limit.

（10）密封性能。如内燃机缸体、缸盖需进行密封试验，检查有无泄漏。

Sealing property. Checking sealing property of parts refers to doing checking test to identify whether there is leak or not. The sealing property of engine block and cylinder heads should be checked.

3.3.3 检验的方法（The methods of checking）

零件缺陷的检验方法主要有：感觉检验法、工具仪器检验法和物理检验法。

The main methods of check the defect of parts or components include feeling checking, tools checking, equipment checking and physical checking.

不用量具、仪器，仅凭检验人员的直观感觉和经验来鉴别零件的技术状况，统称感觉检验法。感觉检验法精度不高，只适于分辨缺陷明显的或精度要求不高的零件，要求检验人员有丰富的经验和技术。感觉检验法的具体方法有目测、耳听和触觉。

Measuring tools and instruments are not used is called feeling checking. Only the intuitive feeling and experience of the checking personnel can be used to identify the technical status of the parts or components. The precision of feeling checking is not high. It is only suitable for distinguishing parts or components with obvious defects or low precision requirements. Feeling checking requires the inspector have rich experience and technology. Feeling checking usually refers to visual (see) checking, aural (listen) checking, and tactile (touch) checking.

目测是指用眼睛或借助放大镜对零件进行观察和宏观检验，如倒角、圆角、裂纹、断裂、疲劳剥落、磨损、刮伤、蚀损、变形、老化等，做出可靠的判断。

Visual (see) checking refers to visual observation and macro check of parts or components with eyes or magnifying glass, such as chamfering, rounded corners, cracks, fractures, fatigue peeling, wear, scratch, corrosion, deformation, aging, etc. to make a reliable judgment.

耳听是根据机械设备运转时发出的声音，或敲击零件时的响声判断技术状态。零件无缺陷时声响清脆，内部有缩孔时声音相对低沉，若内部出现裂纹，则声音嘶哑。

Aural（listen）checking refers to listen the sound produced during the operation of mechanical equipment or the sound of knocking on parts or components to judge the technical state. The sound is clear when there is no defect in parts or components. The sound is relatively when there is low internal shrinkage cavity. The sound is hoarseness when there is internal crack.

触觉是用手与被检验的零件接触，可判断工作时温度的高低和表面状况；将配合件进行相对运动，可判断配合间隙的大小。

Tactile（touch）checking refers to touch the inspected parts or components with your hands to judge the temperature and surface conditions at work. The size of the fit clearance can be judged by the relative motion of the fitting part.

工具检验是指用检测工具和仪器进行检验。它能满足检测精度的要求，是目前应用最广泛的一种检测方法。

Tools checking refers to checking with testing tools and instruments. It is the most widely used because it can meet the requirement of checking accuracy.

用各种测量工具（如卡钳，钢直尺、游标卡尺、百分尺、千分尺或百分表、千分表、塞规、量块、齿轮规等）和仪器检验零件的尺寸、几何形状、相互位置精度。用专用仪器、设备对零件的应力、强度、硬度、冲击性、伸长率等力学性能进行检验。用静动平衡试验机对高速运转的零件做静动平衡检验。用弹簧检验仪或弹簧秤对各种弹簧的弹性和刚度进行检验。对承受内部介质压力并需防止泄漏的零部件，需在专用设备上进行密封性能检验。用金相显微镜检验金属组织、晶粒形状及尺寸、显微缺陷，分析化学成分。

A variety of measuring tools（such as calipers, steel ruler, vernier caliper, dial ruler, micrometer or dial gauge, micrometer, plug gauge, gauge block, gear gauge, etc.）and instruments can be used to check the size, geometric shape, mutual position accuracy of parts or components. Special instruments and equipment can be used to check stress, strength, hardness, impact, elongation and other mechanical properties of the parts or components. The static and dynamic balance testing machine can be used to check static and dynamic balance of high-speed running parts or components. Spring tester or spring balance can be used to check the elasticity and stiffness of various springs. Special equipment can be used to check sealing performance of the parts or components withstand the pressure of the internal medium. Metallographic microscope can be used to examine the metal structure, grain shape and size, microscopic defects and analyze chemical composition.

物理检验法是利用电、磁、光，声、热等物理量，通过零部件引起的变化来测定技术状况、发现内部缺陷。这种方法的实现是和仪器、工具检测相结合，它不会使零部件受伤、分离或损坏。目前普遍称无损检测。

Physical checking method can find internal defects to determine the technical status by the change caused the parts or components through the use of electricity, magnetic, light, sound, heat and other physical quantities. The implementation of this method is combined with instrument and tool detection, which will not cause parts or components to be injured, separated or damaged.

Currently, it is commonly called nondestructive testing.

对维修而言，这种检测主要是对零部件进行定期检查、维修检查、运转中检查，通过检查发现缺陷，根据缺陷的种类、形状、大小、产生部位、应力水平、应力方向等，预测缺陷发展的程度，确定采取修补或报废。目前在生产中广泛应用的物理检验法有：磁力法、渗透法、超声波法、射线法等。

For maintenance, this kind of checking is mainly to carry out regular checking, maintenance checking and in - operation checking of parts and components. Defects are found through checking. According to the type, shape, size, location, stress level and stress direction of defects, the development degree of defects is predicted and the repair or scrap is determined. At present, the physical checking methods widely used in production are: magnetic method, infiltration method, ultrasonic method, ray method and so on.

思考题（Questions）

3.1 机械零件常用的拆卸方法有哪些？

How many common methods of disassembling mechanical parts or components are there?

3.2 断头螺钉的拆卸方法有哪些？

How to disassembly the broken head screw?

3.3 打滑内六角螺钉的拆卸方法有哪些？

How to disassembly the sliding inner hexangular screw

3.4 锈死螺纹的拆卸方法有哪些？

How to disassembly the rusty thread?

3.5 说明成组螺纹连接件的拆卸顺序。

What is the order of disassembling the group thread connection?

3.6 拆卸后应该清除哪些污物？

What should be removed from parts or components after disassembly?

3.7 油污的清洗剂和清洗方法有哪些？

How many agents and cleaning methods for removing oil are there?

3.8 水垢的清洗方法有哪些？

How many methods for removing of scale are there?

3.9 积炭的方法有哪些？

How many methods for removing of carbon deposits are there?

3.10 去锈的主要方法有哪些？

How many methods for removing of rust are there?

3.11 清除漆层的方法有哪些？

How many methods for removing of lacquer are there?

3.12 零件检验的主要内容有哪些？

How many items should be checked for parts or components?

3.13 零件检验的主要方法有哪些？

How many methods can be used for parts or components checking?

4 机械零件的修复

Chapter 4 Repair Technologies of Mechanical Parts

4.1 修复前的准备工作（Preparation before Repair）

机械设备在维修前应进行有关准备工作，包括技术准备、拆卸、清洗、检验等。这些准备工作必须按照机械设备的结构特点、技术性能要求、需要修复或更换零部件的情况以及本单位的具体条件，依照一定的计划、方法和步骤，选用合适的工具及设备进行。

Preparation should be finished before repairing mechanical equipment including technical preparation, disassembly, cleaning, check, etc. These preparation must be done according to structural characteristics and technical performance requirements of mechanical equipment, the specific conditions of factory and situation of the parts or components need to repaired or need to be replaced, specified plans, methods and steps. The appropriate tools and equipment should be selected.

4.1.1 组织准备（Organizational Preparation）

根据需要，结合本单位的维修规模、机械设备情况、技术水平、承修机械设备的类型，以及材料供应等具体条件，全面考虑、分析比较，采用更合理更适用的组织形式和方法。

More reasonable and applicable organizational form and method should be adopted according to the needs, considering the specific conditions such as maintenance scale, mechanical equipment situation, technical level, type of mechanical equipment to be repaired, and material supply.

4.1.2 技术准备（Technical preparation）

技术准备主要是为维修提供技术依据。例如：准备现有的或需要编制的机械设备图册和备件图册，确定维修工作类别和年度维修计划，整理机械设备在使用过程中的故障及其处理记录，调查维修前机械设备的技术状况，明确维修内容和方案，提出维修后要保证的各项技术性能要求，提供必备的有关技术文件等。

Technical preparation provides technical basis for maintenance. For example, existing or needed mechanical equipment and spare parts or components catalogues should be prepared, the type of maintenance work and annual maintenance plan should be determined, the records of the failures and their treatment in the process of using the mechanical equipment should be sorted out, the technical status of mechanical equipment before maintenance should be investigated, the maintenance content and program should be defined, the technical performance requirements to be guaranteed after maintenance should be put forward, and necessary technical documents should be provided, etc.

4.2 修复技术（Repair Technology）

机械零件的修复技术主要有：机械加工修理技术、塑性变形修理技术、焊接维修技术、热喷涂技术、电镀修复技术、工件表面强化技术、金属扣合技术、粘接修复技术和研磨技术等。

Repair technology of mechanical parts or components mainly include machining repair technology, plastic deformation repair technology, welding repair technology, thermal spray technology, electroplating technology, surface hardening technology, metal stitch (metal buckle) technology, adhesive technology and grinding technology, etc.

4.2.1 机械加工修理技术（Machining Repair Technology）

机械加工修理技术是零件修理最主要、最基本的方法，可独立直接修理，也可成为其他工艺最后的加工工序。

Machining repair technology is the most important, and the most basic method of parts repair. It can be directly used as repair technology independently and can also become the final machining process of other repair technology.

机械加工修理技术简单易行，质量稳定可靠，修理成本低，应用广。

Machining repair technology is easy to operate. The quality of machining repair technology is stable and reliable. The repair cost of machining repair technology is low. Machining repair technology is widely used.

机械加工修理技术与制造新零件的区别是：加工对象不同，加工基准已被破坏，装夹定位难，加工余量小，要适应各种表面状态，数量少，品种杂。

The difference of machining a new part and machining an old part is that the machining object is different. The benchmark of the old part is broken, the clamping positioning is difficult to find, machining allowance is small. Machining an old part should adapt to all kinds of surface state. The quantity of parts need to be machined and repaired is small and the variety is mixed.

常用的机械加工修理方法有：修理尺寸法、尺寸选配法、附加零件法、局部更换法和成套换修法。

Commonly used machining repair methods include repairing dimension, choosing and matching the dimension, additional part repair method, local replacement repair method and replacement repair matched parts or components.

4.2.1.1 修理尺寸法（Repairing Dimension）

具有相对运动的配合件磨损后，配合间隙增大甚至超过极限间隙值，零件工作性能变坏。

The work performance of parts or components will be worse when fit clearance between the mating parts or components with relative movement increase even more than the limit value after wear.

修理尺寸法是将配合件中较重要的零件或较难加工的零件进行机械加工，消除其工作表面的损伤和几何形状误差，使之具有正确的几何形状和新的基本尺寸——修理尺寸。依此修理尺寸制造与之配合的另一零件，使二者具有原设计配合间隙值。

Repairing dimension method means machining the mating part which is more important or difficult to machine to eliminate the damage on the surface of the part and geometric error, so that, it has the correct geometry and the new basic dimension, that is repaired dimension. The other part will cooperate with it in accordance with the repaired dimension to make both have the original designed clearance value.

例如，曲轴主轴颈过度磨损后，在保证轴颈强度要求下，光车主轴颈，依光车后的尺寸重新配置主轴瓦使具有原有的轴承间隙。

For example, the spindle journal should be machined by turning after excessive wear of the crankshaft spindle neck in order to ensure the strength of the shaft journal. The spindle bushing with the original bearing clearance should be matched with the shaft according to the dimension of the shaft after turning.

修理尺寸法分为同心法和不同心法。同心法是指修复后的轴或孔的中心线与原轴或孔的中心线相同。不同心法是指修复后的轴或孔的中心线与原来的轴或孔的中心线不一样，稍有改变。

Repairing dimension method is divided into concentration method and decentration method. Concentration method refers to the centerline of shafts or holes after being repaired is the same with the centerline of the original shafts or holes. Decentration method refers to the centerline of shafts or holes after being repaired is not the same with the centerline of the original shafts or holes, is slightly changed.

修理尺寸法简单、经济，可延长零件的使用寿命。但使零件失去原有的互换性，给备件供应带来麻烦。

Repairing dimension method is simple and economical. The service life of parts or components can be prolonged. But the original interchangeability of the parts or components is losing. The supply of spare parts or components becomes difficult.

4.2.1.2 尺寸选配法（Choosing and Matching the Dimension）

集中一小批相同机型的已过度磨损的配合件，分别进行机械加工，消除配合表面的缺陷和几何形状误差，再按原配合间隙值重新配合成对。组成一些具有不同基本尺寸但具有相同配合间隙的新的配合件，此种方法称为尺寸选配法。

A small number of excessively worn mating parts or components with the same type were collected and machined to eliminate defects and geometric shape errors on the mating surface, and then matched in pairs according to the original mating clearance. Some new fitting parts or components with different basic dimensions but with the same clearance of distribution platform is formed. This method is called selection and match method of dimension.

例如，柴油机喷油泵和喷油器中的精密偶件就可采用此法修理。

For example, precision couplings in diesel fuel injection pumps and injectors can be repaired by this method.

此法简单、方便、经济，可使一部分已报废的配合件重新投入使用。缺点是必须有一小批配合件，数量太少，则不易组成新的配合件。

This method is simple, convenient, and economical, which can make part of the discarded parts or components be used again. The disadvantage is that there must be a small batch of pieces. If the quantity of parts is too small, it is not easy to form new pairs of matching part.

4.2.1.3 附加零件修理法（Additional Part Repair Method）

附加零件修理法是指有些零件个别工作表面磨损严重，当其结构和强度允许时，可以将磨损部位进行机械加工，再在这个部位镶上一个套或其他镶装件，以补偿磨损，最后将其加工到基本尺寸。附加零件修理法也称为镶套修理法。

Additional part repair method refers to a sleeve or other part is added to the work surface of some part badly worn when the structure and strength is permitted and wear part can be machined to compensate for wear and finally machined to basic dimension. Additional part repair method is also called insert sleeve repair method.

附加零件修理法特点是应用范围广，镶装件可更换。轴颈的镶套修理如图4.1所示，支承架的镶套修理如图4.2所示。

The characteristics of additional part repair method is being used widely and additional part is replaceable. Insert sleeve repair of shaft journal is shown in Figure 4.1, and insert sleeve repair of support frame is shown in Figure 4.2.

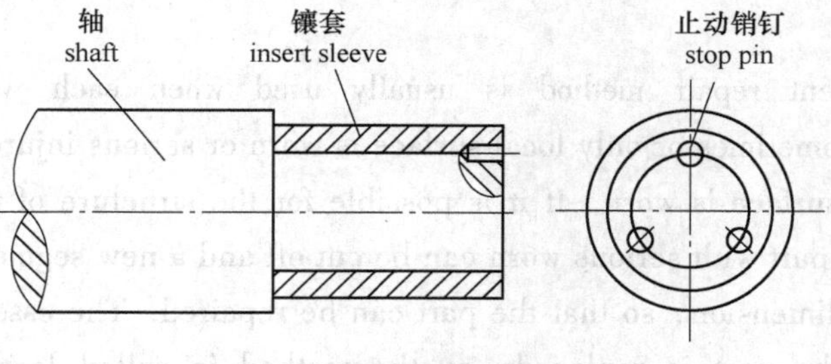

图4.1 轴颈的镶套修理

Figure 4.1 Insert sleeve for repair shaft journal

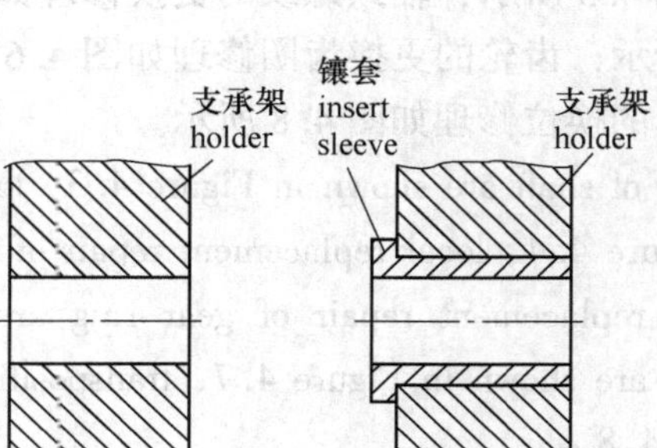

图4.2 支承架的镶套修理

Figure 4.2 Insert sleeve for repair holder

附加零件修理法的注意事项：镶装件的材质应根据零件所处的工况来选择。例如：高温下工作的镶装件应尽量选用与基体一致的材料，使两者的热膨胀系数相同，保证镶装件在工作中的稳固性；再如：若要求镶装件耐磨，则选用耐磨材料等。

Tips for additional part repair: Material of additional part should be chosen by work conditions of parts or components need to be repaired. For example, the material of additional part working at high temperature should be selected the same material of original part, so that the thermal expansion coefficients of them are the same to ensure the work stability of additional part. For another example, if the requirement inserts wear - resisting, then chooses wear - resisting material and so on.

镶套工艺往往受到零件结构和强度的限制，镶套壁厚一般只有 2~3mm，且镶装后为保证一定的紧固性，镶套和基体之间应采用过盈配合。

Additional part repair technology is often limited by the structure and strength of the parts or components because the wall thickness of additional part is generally only 2~3mm, and proper interference fit should be used between additional part and basal body to ensure a certain degree of tightness after repaired.

4.2.1.4 局部更换修理法（Local Replacement Repair Method）

局部更换修理法是指机械设备零件在使用过程中，各个表面的磨损程度往往不一致，有时只有个别表面磨损严重或损伤，其余表面尚好或只有轻微磨损，这时，如果零件结构允许，可把有严重缺陷的部分切除，更换新的部分，并把它加工到原有的形状和尺寸，使零件得以修复，这种方法的实质是对零件某一部分进行了更换，所以称之为局部更换修理法。

Local replacement repair method is usually used when each work surface is worn inconsistently and sometimes or only local surface is worn or serious injured and the rest is still good or only minor surface is worn. If it is possible for the structure of parts or components, the local area of the part with serious worn can be cut off and a new segment is machined to the original shape and dimension, so that the part can be repaired. The essence of this method is that a segment of the part is replaced, so the method is called local replacement repair method.

轴的局部更换修理如图 4.3 所示，轴颈螺纹的更换修理如图 4.4 所示，机床主轴锥孔的局部更换修理如图 4.5 所示，齿轮的更换齿圈修理如图 4.6 所示，键上键槽的换位修理如图 4.7 所示，轮盘上螺孔的换位修理如图 4.8 所示。

Local replacement repair of shaft are shown in Figure 4.3, local replacement repair of shaft neck thread are shown in Figure 4.4, local replacement repair of machine spindle cone hole are shown in Figure 4.5, local replacement repair of gear ring are shown in Figure 4.6, local replacement repair of keyway are shown in Figure 4.7, transposition repair of the screw hole on the disc are shown in Figure 4.8.

4.2.1.5 成套换修法（Replacement Repair Matched Parts or Components）

成套换修法是为了缩短修理时间，拆下有严重磨损或损伤零件的部件或设备，迅速换上备件继续运转，称为成套换修。设备或部件经修理后作为备件使用或供同类机型的船舶使用。例如柴油机运转中高压油泵柱塞——套筒偶件咬死，立即换上备用油泵，而损坏的油泵经修理后作为备件使用。

In order to shorten the repair time, the parts or components or equipment with serious wear or damage are removed and the spare parts or components are quickly replaced in order that the equipment can continue to operate. This method is known as replacement repair matched parts or components. After repaired, parts or components are used as spare parts or components. For example, when seizure occurs in the plunger of high-pressure oil pump, the spare oil pump is replaced immediately, and the damaged oil pump is used as spare parts or components after repair.

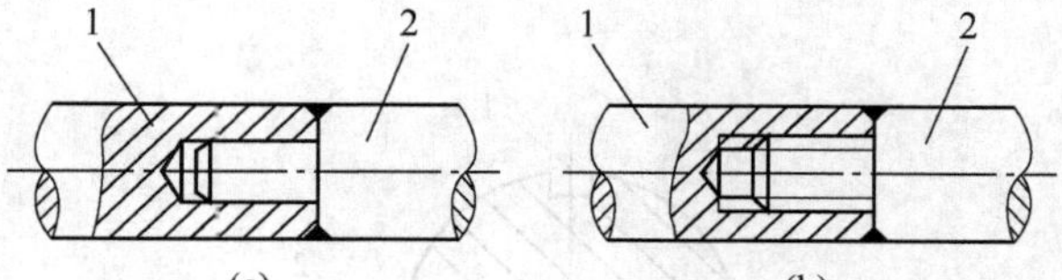

图 4.3 轴的局部更换修理

Figure 4.3 Local replacement and repair of shaft

1—轴的保留部分（reserved part of shaft）；2—轴的更换部分（replacement of shaft）

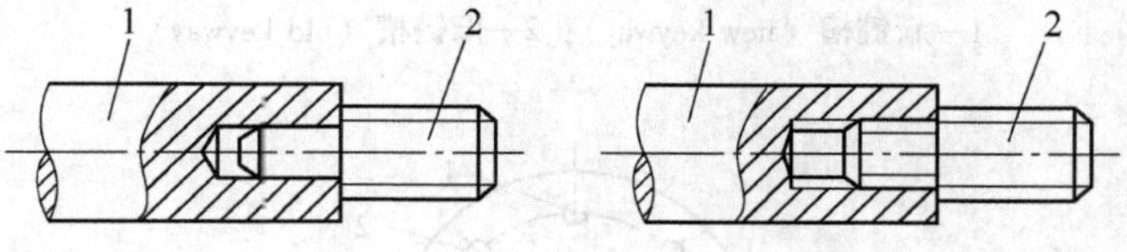

图 4.4 轴颈螺纹的更换修理

Figure 4.4 Local replacement and repair of journal thread

1—轴的保留部分（reserved part of shaft）；2—轴的更换部分（replacement of shaft）

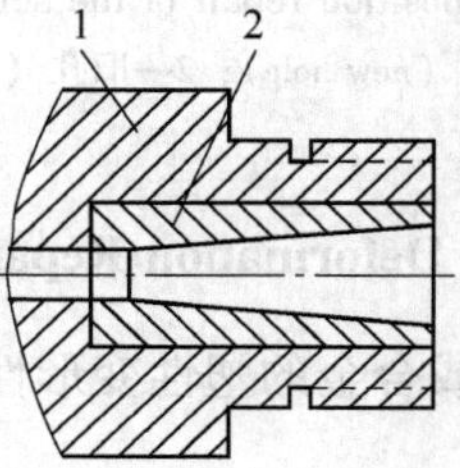

图 4.5 机床主轴锥孔的局部更换修理

Figure 4.5 Local replacement and repair of the spindle cone hole of the machine tool

1—轴（shaft）；2—镶套（sleeve）

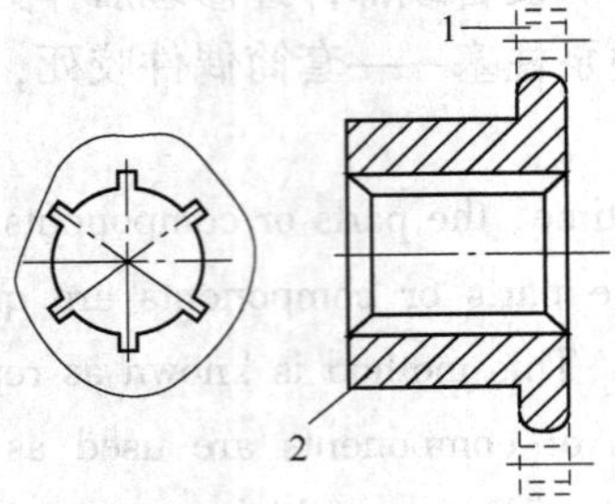

图 4.6　齿轮的更换齿圈修理

Figure 4.6　Local replacement repair of gear ring

1—齿圈（gear ring）；2—轮芯（gear core）

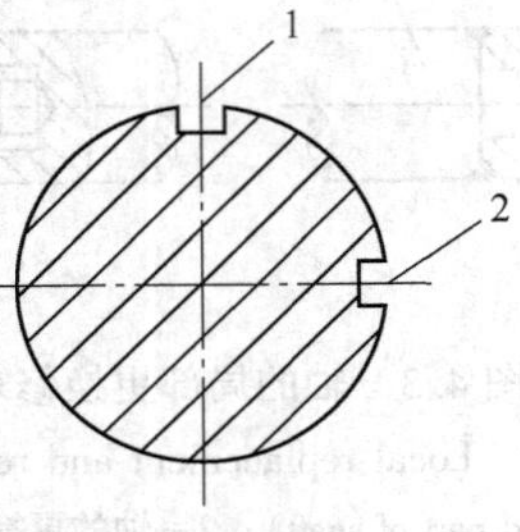

图 4.7　键上键槽的换位修理

Figure 4.7　Local replacement repair of keyway

1—新键槽（new keyway）；2—旧键槽（old keyway）

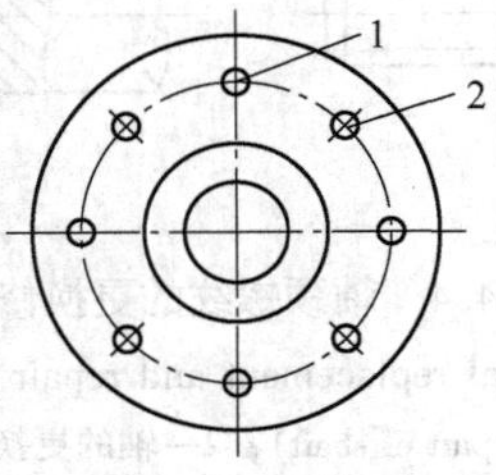

图 4.8　轮盘上螺孔的换位修理

Figure 4.8　Transposition repair of the screw hole on the disc

1—新孔（new hole）；2—旧孔（old hole）

4.2.2　塑性变形修复技术（Plastic Deformation Repair Technology）

塑性变形修复技术是利用金属或合金的塑性变形性能，使零件在一定外力作用的条件下改变其几何形状而不损坏。

Plastic deformation repair technology is that the plastic deformation properties of metal or alloy is used to change the geometric shape of parts or components under the condition of a certain external force without damage.

这种方法是将零件不工作部位的部分金属向磨损的工作部位移动，以补偿磨损掉的金属，恢复零件工作表面原来的尺寸和形状。它实际上也就是一般的压力加工方法，但其工作对象不是毛坯，而是具有一定尺寸和形状的磨损零件。因此，用这种方法不仅可以改变零件的外形，而且可改变金属的力学性能和组织结构。

In this method, the metal in the non - working part is moved to the worn part to compensate for the worn metal and to restore the original dimension and shape of the working surface of the part. It is actually the general pressure processing method, but its work object is not blank, but worn parts with a certain dimension and shape. Therefore, this method can not only change the shape of the parts, but also change the mechanical properties and microstructure of the metal.

塑性变形修复零件的方法一般有：镦粗法、挤压法、扩张法和校正法。

Plastic deformation methods generally include upsetting, extrusion, expansion and correction method.

4.2.2.1 镦粗法（Upsetting Method）

镦粗法是借助压力来减小零件的高度、增大零件的外径或缩小内径尺寸的一种方法，主要用来修复有色金属套筒和圆柱形零件。

The principle of upsetting method is reducing the height of the parts, increasing or reducing the outer diameter of the part by using pressure. Upsetting method is mainly used to repair the non-ferrous metal sleeve and cylindrical parts.

例如，当铜套的内径或外径磨损时，在室温下通过专用模具进行镦粗，设备一般可采用压床、手压床或用锤手工敲击，作用力的方向应与塑性变形的方向垂直，如图 4.9 所示。

For example, when the inner or outer diameter of the copper sleeve is worn, upsetting is carried out through a special die at room temperature. The equipment can be generally used, such as press machine, hand press or manually struck with a hammer. The direction of force should be perpendicular to the direction of plastic deformation, as shown in Figure 4.9.

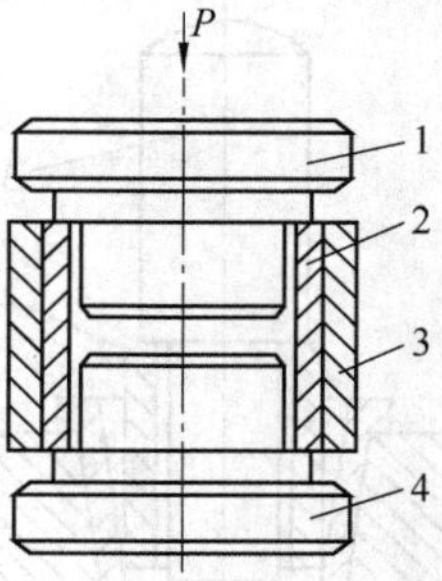

图 4.9　铜套的镦粗

Figure 4.9　Repair copper sleeve with upsetting method

1—上模（upper mold）；2—铜套（copper sleeve）；3—轴承（bearing）；4—下模（lower mold）

镦粗法可修复内径或外径磨损量小于 0.6mm 的零件，对必须保持内外径尺寸的零件，可采用镦粗法补偿其中一项磨损量后，再用其他的修复方法来保证另一项恢复到原来尺寸。

Upsetting method can repair the parts with inner or outer diameter wear less than 0.6mm. For the parts that must maintain the inner and outer diameter dimension, upsetting method can be used to compensate one of the wear amounts, and then other repair methods can be used to ensure that the other one returns to the original dimension.

用镦粗法修复零件，零件被压缩后的压缩量不应超过其原高度的15%，对于承载较大的则不应超过其原高度的8%。为保证镦粗均匀，其高度与直径之比不应大于2，否则不宜采用这种方法。

When a part is repaired by upsetting method, the compression of the part should not exceed 15% of its original height and 8% of its original height for the part with a large bearing capacity. To ensure even upsetting, the ratio of height to diameter should not be greater than 2.

4.2.2.2 挤压法（Extrusion Method）

挤压法是利用压力将零件不需严格控制尺寸部分的材料挤压到已磨损部位，主要用于筒形零件内径的修复。

Extrusion method is used to squeeze the material of the part whose dimension doesn't need to be strictly controlled by using pressure to the easy wear area of the part. Extrusion method is mainly used for repairing the inner diameter of cylindrical part.

一般都利用模具进行挤压，挤压零件的外径来缩小其内径尺寸，再进行加工以达到恢复原尺寸的目的。例如，修复轴套可用图4.10所示的模具进行。把轴套2放在外模1的锥形孔中，利用冲头3在压力的作用下使轴套2的内径缩小。

Generally, mold is used for extrusion. The outer diameter of part is extruded to reduce the inner diameter, then the part is processed to achieve the restoration of the original dimension. For example, the shaft bushing can be repaired using the mold shown in Figure 4.10. The shaft bushing 2 is placed in the tapered hole of the external mold 1, and the inner diameter of the shaft bushing 2 is reduced by the use of punch 3 under the action of pressure.

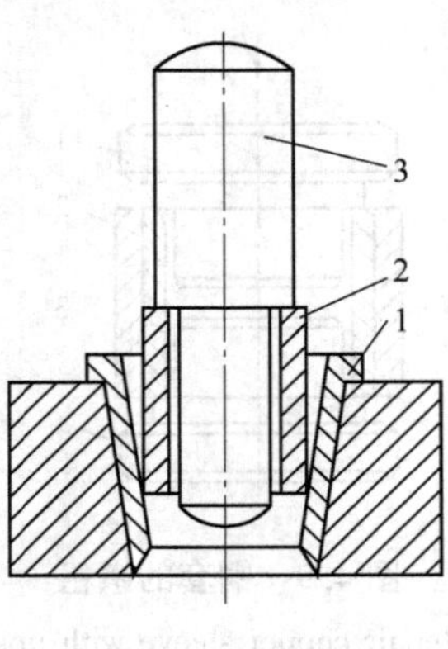

图4.10 挤压法修复轴套

Figure 4.10 Repair shaft bushing with extrusion method

1—外模（external mold）；2—轴套（shaft bushing）；3—冲头（punch）

4.2.2.3 扩张法（Expansion Method）

扩张法的原理与挤压法相同，所不同的是零件受压向外扩张，以增大外形尺寸，补偿磨损部分，主要应用于外径磨损的套筒形零件。

The principle of expansion method is the same as that of extrusion method. The difference is that the part expands outward after being pressed to increase the outer dimension to compensate for worn parts. Expansion method is mainly used in repairing the sleeve-type part with worn outer diameter.

例如，空心活塞销外圆磨损后，一般用镀铬法修复，但若没有镀铬设备时，可用扩张法进行修复，活塞销的扩张既可在热状态下进行，也可以在冷状态下进行。扩张后的活塞销，应按技术要求进行热处理，然后磨削其外圆，直到达到尺寸要求。扩张活塞销可用图 4.11 的方法进行。

For example, after the hollow piston pin is worn out, it is usually repaired by chromium plating method. However, if there is no chromium plating equipment, it can be repaired by expansion method. The expansion of piston pin can be carried out either in the hot state or in the cold state. The expanded piston pin shall be heat-treated according to the technical requirements, and its outer circle shall be ground until it reaches the dimension requirements. The expansion of piston pins may be performed as shown in Figure 4.11.

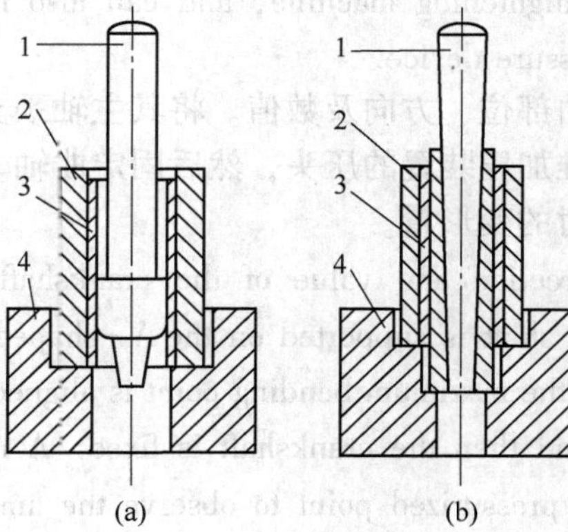

图 4.11 扩张活塞销

Figure 4.11 Repair piston pin with expansion method

1—冲头（punch）；2—活塞销（piston pin）；3—胀缩套（swell-shrink set）；4—模具（mold）

4.2.2.4 校正法（Correction Method）

零件在使用过程中，常会发生弯曲、扭曲等残余变形。利用外力或火焰使零件产生新的塑性变形，从而消除原有变形的方法称为校正。

Bent, twisted and residual strain often occur when parts are using. The original deformation can be eliminated by force or flame to make part produce new plastic deformation. This method is called correction.

校正分为冷校和热校。

Correction is divided into cold correction and hot correction.

A 冷校（Cold Correction）

冷校又分为冷压校正与冷作校正。

Cold correction is divided into cold press correction and cold working correction.

冷压校正是将变形零件放在压力机的V形铁中，使凸面朝上，施加压力使零件发生反方向变形，保持1~2min后去除压力，利用材料的弹性后效作用将变形抵消。检查校正情况，若一次不能校正，可进行多次，直到校正为止。

During cold press correction, the deformed parts are placed in the V-shaped iron block. The convex surface is placed upward, and the parts are deformed in the opposite direction by exerting pressure. The pressure is removed after being maintained for 1~2min, and the deformation is offset by the elastic after-effect of the material. Check the result of correction. If the correction cannot be done at one time, it can be done many times until the correction is completed.

对于弯曲变形不大的小型钢制曲轴，可采用此方法校直。曲轴的弯曲度，如小于0.05mm时，可结合磨修曲轴得以修整。如超过0.05mm时，则须加以校正。冷压校正一般在压力校直机上进行，也可用手动螺旋压力装置在地平台上进行。

This method can be used to straighten small steel crankshaft with little bending deformation. If bending of crankshaft is less than 0.05mm, the crankshaft can be repaired by grinding. If bending of crankshaft exceeds 0.05mm, correction is required. Cold pressure correction is usually carried out on the pressure straightening machine, and can also be carried out on the ground platform with manual spiral pressure device.

校正前应测出曲轴的弯曲部位、方向及数值。将其主轴颈支承在V形铁上，使弯曲凸面朝上，并使最大弯曲点对准加压装置的压头，然后固定曲轴。在加压点相对180°的位置架设百分表，借以观察加压时的变形量。

The bending position, direction and value of the crankshaft should be measured before correction. The journal of main shaft is supported on the V-shaped iron block, the convexity of bending is placed upward, and the maximum bending point is aligned with the pressing head of the pressure correction machine, and then the crankshaft is fixed. A dial indicator is set up at the position of 180° relative to the pressurized point to observe the amount of deformation when the crankshaft is pressed.

当曲轴的弯曲变形较大时，必须分次进行，以防压校时，反向弯曲变形量过大，而使曲轴折断。校正时的反向弹性变形量不宜超过原弯曲量的1~1.5倍。

When the bending deformation of the crankshaft is large, the crankshaft must be pressed several times to prevent the crankshaft from being broken due to the too large reverse bending deformation being made. The inverse elastic deformation should not exceed 1~1.5 times of the original amount of bending.

冷压校正简单易行，但校正的精度不容易控制，零件内留下较大的残余应力，效果不稳定，疲劳强度下降。

Cold pressure correction is simple and easy to operate, but the accuracy of correction is not easy to control. A large residual stress can be left in the parts. The effect is not stable. The fatigue strength decreases.

冷作校正是用手锤敲击零件的凹面，使其产生塑性变形。该部分的金属被挤压延展，在塑性变形层中产生压缩压力。弯曲零件在变形层应力的推动下被校正。

The aim of cold working correction is to make the part produce plastic deformation by hitting the concave surface of the part with a hand hammer. This part of metal is extruded and stretched to create compression pressure in the plastic deformation layer. Bending parts are corrected under the stress of the deformation layer.

利用冷作校正法来校正弯由的曲轴时，根据曲轴弯曲的方向和程度，使用球形手锤或气锤，沿曲柄臂的左右两侧进行敲击（锤击区应选在弯曲后曲柄臂受压应力的一侧），由于冷作而产生残余应力，曲柄臂敲击侧伸长变形，曲轴轴线产生位移，在各个曲柄臂变形的综合作用下，达到校直曲轴的目的。

Cold working correction method is used to correct bending of the crankshaft, according to the direction and level of crankshaft bending, spherical hammer or pneumatic hammer is used. Knock the crankshaft along the both side of the crank arm. Residual stress will produce during cold working. Lateral elongation deformation of the crank arm will occur. Axial displacement of the crankshaft will occur too. The crankshaft can be straightened under the combination deformation of each crank arm.

冷作校正的校正精度容易控制，效果稳定，且不降低零件的疲劳强度。但是，它不能校正弯曲量太大的零件，通常零件的弯曲量不能超过零件长度的 0.03%~0.05%。

The accuracy of cold working correction is easy to be controlled. The effect of cold working correction is stable. And the fatigue strength of parts is not reduced. However, it does not correct the parts with too much bending. The amount of bending of the part is usually less than 0.03%~0.05% of the length of the part.

B　热校（Hot Correction）

热校一般是将零件弯曲部分的最高点用气焊的中性焰迅速加热到 450℃ 以上。然后快速冷却，由于加热区受热膨胀，塑性随温度升高而增加，又因受周围冷金属的阻碍，不可能随温度增高而伸展。当冷却时，收缩量与温度降低幅度成正比，造成收缩量大于膨胀量，收缩力很大，靠它校正零件的变形。

Hot correction is a method that the highest point of the bending part is quickly heated to 450℃ or above with gas welding neutral flame. Then the highest point of the bending part is cooled rapidly. The heating zone expands while being heated. The plasticity of the heating zone increases with the temperature increase. But the expansion is limited by the surrounding cold metal. It is impossible to extend with the temperature increase. When the heating zone is cooled, the shrinkage is proportional to the amplitude of temperature decrease. As a result, the shrinkage is larger than the expansion. The deformation of parts is corrected by the shrinkage force which is very large.

热校适合于校正变形量较大，形状复杂的大尺寸零件，其校正保持性好，对疲劳强度影响较小，应用比较普遍。热校正的关键在于弯曲的位置及方向必须找正确，加热的火焰也要和弯曲的方向一致，否则会出现扭曲或更多的弯曲。轴类零件的热校正如图 4.12 所示。

Hot correction is suitable for the correction of large dimension parts with large deformation and complex shape, which has good correction retention and little impact on fatigue strength. The key to thermal correction is to find the correct position and direction of the bending, and to heat the flame in the same direction as the bending, otherwise distortion or more bending will occur. Hot correction of shaft parts is shown in Figure 4.12.

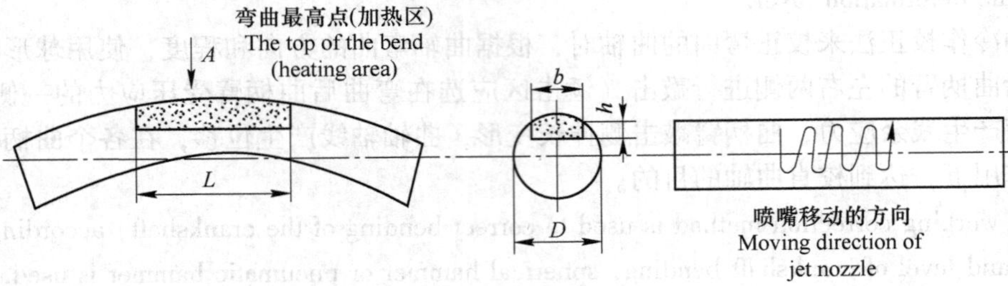

图 4.12 轴类零件的热校正

Figure 4.12 Repair shaft with hot correction method

4.2.3 焊接维修技术（Welding Repair Technology）

焊接是指两个或两个以上的零件（同种或异种材料），通过局部加热或加压达到原子间的结合，造成永久性连接的工艺过程。

Welding refers to two or more parts (the same or different materials) are permanent connected through local heated or pressured to achieve the combination of atoms.

焊接的具体措施包括加压和加热。加压用以破坏结合面上的氧化膜或其他吸附层，并使接触面发生塑性变形，以扩大接触面。在变形足够时，也可直接形成原子间结合，得到牢固接头。对连接处进行局部加热，使之达到塑性或熔化状态，激励并加强原子的能量，从而通过扩散、结晶和再结晶的形成与发展，以获得牢固接头。

Welding includes pressing and heating. The oxidation mode or other adsorption layer on the bonding surface is broken by pressing in order to produce plastic deformation on the contact surface to enlarge the contact surface. When the deformation is enough, it can also form the interatomic bond directly, and the solid connection is got. The connection is heated to a plastic or molten state and excites and strengthens the energy of the atoms. A strong connection through diffusion, crystallization and recrystallization is formed and developed.

焊接技术应用广泛，涉及电站、钢铁、水泥、纺织、建筑、采矿、石油、机械、造纸、化学和交通运输等许多行业。

Welding technology is widely used in power plants, steel, cement, textile, construction, mining, petroleum, mechanical equipment, paper, chemical and transportation industries.

4.2.3.1 焊接技术的优点和缺点（Advantages and Disadvantages of Welding Repair Technology）

焊接技术的优点如下：

Advantages of welding technology include the followings:

（1）能修理由各种原因引起损坏的零件，如磨损、断裂、腐蚀等。

Be able to repair parts damaged by various reasons, such as wear, fracture, corrosion, etc.

（2）能修理多种材料的零件，机械零件中常用的金属材料绝大部分是可焊的。

Can repair parts of various materials, most of the metal materials commonly used in mechanical parts are weldable.

（3）修理质量高，有的零件修后更为耐用。

Repair quality is high. Some parts are more durable after repair.

（4）生产率高，且成本较低。

High productivity and low cost.

（5）一般常用的焊修设备均较简单，操作容易，且便于野外抢修。

The commonly used welding repair equipment is simple, easy to operate and convenient for field repair.

焊接技术的缺点是局部加热使零件不可避免地产生应力，大的内应力常会使零件产生变形、裂隙纹等缺陷。零件在焊后进行机械加工时，内应力的释放会影响加工精度。

The disadvantage of welding technology is that local heating will inevitably produce stress on the parts, and large internal stress will often cause the parts to have deformation, cracks, and other defects. When parts are machined after welding, the release of internal stress will affect the machining accuracy.

残留的焊接内应力对零件的疲劳强度也是不利的。另外焊修容易使零件内产生气孔，从而导致焊缝区强度下降，以及焊修后焊层的加工难度增大，特别是耐磨堆焊层的加工难度更大，这些都是在焊修时应注意的问题。

Residual welding internal stress is also unfavorable to fatigue strength of parts. In addition, it is easy to produce pores in the parts during welding repair, which will lead to the decrease of strength in the weld zone, and the difficulty of processing the welding layer after welding repair increases, especially the difficulty of processing the wear-resistant surfacing layer.

在现代金属加工中，焊接技术发展晚于铸造和锻造技术，但发展迅速。焊接结构的重量占钢材产量的45%，铝及铝合金焊接结构的比重在不断增加。

In modern metal processing, welding technology developed later than casting and forging technology, but it developed very fast. The weight of welded structure accounts for 45% of the steel output, and the proportion of aluminum and aluminum alloy welded structure is increasing

4.2.3.2 焊接技术的分类（Classification of Welding Repair Technology）

金属焊接方法有40种以上，主要分为熔化焊、压力焊和钎焊三大类。

There are more than 40 kinds of metal welding methods, which are mainly divided into three categories: melting welding, pressure welding and brazing.

A 熔化焊（Melting Welding）

a 熔化焊的原理（Principle of Melting Welding）

熔化焊是在焊接过程中将工件接口加热至熔化状态，不加压力完成焊接的方法。

Melting welding refers to a welding process in which the work piece is heated to a molten state at the interface, the welding is completed without the pressure.

熔化焊时，热源将待焊两工件接口处迅速加热熔化，形成熔池。熔池随热源向前移动，冷却后形成连续焊缝而将两工件连接成为一体。在熔化焊过程中，如果大气与高温的熔池直接接触，大气中的氧就会氧化金属和各种合金元素。

During fusion welding, heat source rapidly heats and melts the interface of two work piece to be welded, forming a molten pool. The molten pool moves forward with the heat source, forming a continuous weld after cooling and connecting the two work pieces into one. During fusion welding, if the atmosphere is in direct contact with the hot pool, the oxygen in the atmosphere will oxidize the metal and various alloy elements.

大气中的氮、水蒸气等进入熔池，还会在随后冷却过程中在焊缝中形成气孔、夹渣、裂纹等缺陷，恶化焊缝的质量和性能。为了提高焊接质量，人们研究出了各种保护方法。

Nitrogen and water vapor in the atmosphere will enter the molten pool and form defects such as pores, slag inclusion and cracks in the welding seam during the subsequent cooling process, worsening the quality and performance of the welding seam. In order to improve welding quality, various protection methods have been developed.

例如，气体保护电弧焊就是用氩、二氧化碳等气体隔绝大气，以保护焊接时的电弧和熔池率；又如钢材焊接时，在焊条药皮中加入对氧亲和力大的钛铁粉进行脱氧，就可以保护焊条中有益元素锰、硅等免于氧化而进入熔池，冷却后获得优质焊缝。

For example, argon, carbon dioxide and other gases isolated from the atmosphere are used in gas shielded arc welding to protect welding arc and pool rate. For example, when steel is welded, titanium iron powder with high affinity for oxygen is added to the electrode coating to deoxidize titanium iron powder in order to protect the beneficial elements such as manganese and silicon in the electrode from oxidation and enter the molten pool to obtain high quality welding seam after cooling.

b 熔化焊的分类（Classification of Melting Welding）

熔化焊包括气焊、电弧焊、电渣焊、激光焊、电子束焊、堆焊和铝热焊等。

Melting welding includes gas welding, electrode arc welding, electroslag welding, laser welding, electron beam welding, surfacing welding and Aluminothermic welding etc.

（1）气焊。气焊是利用可燃气体与助燃气体混合燃烧生成的火焰为热源，熔化焊件和焊接材料使之达到原子间结合的一种焊接方法。图4.13是一台气焊机的实物图。

Gas welding. Gas welding is a way that combustion flame combined with combustible gas and combustion-supporting gas mixture is used as a heat source, weldment and welding materials are melted to make them reach the combination between atoms. Figure 4.13 is a picture of a gas welding machine.

助燃气体主要为氧气，可燃气体主要采用乙炔、液化石油气等。所使用的焊接材料主要包括可燃气体、助燃气体、焊丝、气焊熔剂等。特点设备简单不需用电。设备主要包括氧气瓶、乙炔瓶（如采用乙炔作为可燃气体）、减压器、焊枪、胶管等。由于所用储存气体的气瓶为压力容器、气体为易燃易爆气体，所以该方法是所有焊接方法中危险性最高的之一。

图 4.13 气焊机

Figure 4.13 Gas welding machine

Oxygen is mainly used as combustion gas. Acetylene, liquefied petroleum gas, etc. are mainly used as flammable gas. The welding materials used mainly include combustible gas, combustion-supporting gas, welding wire, gas flux and so on. Gas welding machine includes oxygen cylinders, acetylene cylinders, pressure reducer, welding gun, hose, etc. This method is one of the most dangerous of all welding methods because the cylinders used to store the gas are pressure vessels and the gas is flammable and explosive.

气焊的优点是：设备简单、使用灵活；对铸铁及某些有色金属的焊接有较好的适应性；在电力供应不足的地方需要焊接时，气焊可以发挥更大的作用。

The advantages of gas welding include simple and flexible equipment, good adaptability to welding of cast iron and certain non-ferrous metals. Where electricity supply is insufficient, gas welding can play a greater role.

气焊的缺点是：生产效率较低；焊接后工件变形和热影响区较大；较难实现自动化。

The disadvantages include low production efficiency, large deformation and heat affected zone of the work piece after welding, and difficulty to achieve automation.

（2）焊条电弧焊。焊条电弧焊是工业生产中应用最广泛的焊接方法，它的原理是利用电弧放电（俗称电弧燃烧）所产生的热量将焊条与工件互相熔化并在冷凝后形成焊缝，从而获得牢固接头的焊接过程。

Electrode arc welding. Electrode arc welding is the most widely used welding method in industrial production. Its principle is to use the heat generated by arc discharge (commonly known as arc combustion) to melt the welding rod and the work piece and form a welding seam after condensation, so as to obtain a solid connection welding process.

气体电离和阴极电子发射是电弧燃烧的必要条件。电弧焊是手工控制焊条进行焊接的工作。电弧焊可用于水平焊、垂直焊、水平焊和向上焊等位置焊。

Gas ionization and cathode electron emission are necessary conditions for arc combustion. Electrode arc welding is manual control of the electrode for welding work. Electrode arc welding can be used for horizontal welding, vertical welding, horizontal welding and upward welding and other position welding.

另外，由于焊条电弧焊设备轻便灵活，只要有电源就可以使用焊条电弧焊。电弧焊适用于各种金属材料、各种厚度、各种结构形状的焊接。

In addition, since the welding rod arc welding equipment is portable and can be carried out flexibly, the welding rod arc welding can be used wherever there is a power source. Electrode arc welding is suitable for welding all kinds of metal materials, all kinds of thickness, all kinds of structural shapes.

电焊条电弧焊不安全。焊条电弧焊设备的空载电压一般为50~90V，人体能承受的安全电压为30~45V。它会给人们带来风险。使用手工电弧焊设备的人员必须佩戴劳动保护用品。

Welding electrode arc welding is not safe. The no-load voltage of welding electrode arc welding equipment is generally 50~90V, and the safety voltage the human body can withstand is 30~45V. It will be risk to people. The people using manual arc welding equipment must wear labor protection supplies.

电弧焊可分为手工电弧焊、半自动（电弧）焊、自动（电弧）焊。自动（电弧）焊通常是指埋弧自动焊，在焊接部位覆有起保护作用的焊剂层，由填充金属制成的光焊丝插入焊剂层，与焊接金属产生电弧，电弧埋藏在焊剂层下，电弧产生的热量熔化焊丝、焊剂和母材金属形成焊缝，其焊接过程是自动化进行的。最普遍使用的是手工电弧焊。

Arc welding can be divided into manual arc welding, semi-automatic arc welding, automatic arc welding. Automatic arc welding usually refers to the automatic submerged arc welding. The protective flux layer is put on the welding parts. The bare welding wire made of filler metal is inserted into the flux layer. Arc is produced between the bare wire and weld metal. The heat generated by the arc buried under the flux layer melt welding wire, flux and base metal to form welding line. The welding process is automatic. The most commonly used is manual arc welding.

手工电弧焊的基本工艺如下：

The basic process of manual arc welding is as follows:

1）在焊接前清理焊接表面，以免影响电弧引燃和焊缝的质量。

Clean the welding surface before welding to avoid affecting the quality of arc ignition and welding seam.

2）准备好接头形式（坡口形式）。

Prepare the connection type (groove type).

坡口的作用是使焊条、焊丝或焊炬（气焊时喷射乙炔-氧气火焰的喷嘴）能直接伸入坡口底部以保证焊透，并有利于脱渣和便于焊条在坡口内作必要的摆动，以获得良好的熔合。

The function of groove is to make welding rod, wire or torch (acetylene-oxygen flame spray nozzle) can directly extend into the bottom of the groove to ensure welding penetration, and is conducive to slag removal and easy to make welding rod necessary swing in the groove in order to obtain a good fusion.

坡口的形状和尺寸主要取决于被焊材料及其规格（主要是厚度）以及采取的焊接方法、焊缝形式等。

The shape and dimension of groove mainly depend on the material to be welded and its specification（mainly thickness）as well as the welding method and welding form.

电弧焊的焊接规范中最主要的参数有焊条种类（取决于母材的材料）、焊条直径（取决于焊件厚度、焊缝位置、焊接层数、焊接速度、焊接电流等）、焊接电流、焊接层数、焊接速度等。

The main parameters in the welding specification of arc welding include the type of welding rod（depending on the material of base material）, the diameter of welding rod（depending on the thickness of welding piece, the position of welding seam, the number of welding layers, the welding speed, the welding current, etc.）, the welding current, the number of welding layers, the welding speed, etc.

除了上述的普通电弧焊外，为了进一步提高焊接质量，还采用气体保护电弧焊、钨极氩弧焊和等离子电弧焊等技术。

In addition to the above ordinary arc welding, gas shielded arc welding, tungsten argon arc welding, plasma arc welding and other technologies are adopted in order to further improve the welding quality.

气体保护电弧焊是例如利用氩气作为焊接区域保护气体的氩弧焊、利用二氧化碳作为焊接区域保护气体的二氧化碳保护焊等，其基本原理是在以电弧为热源进行焊接时，同时从喷枪的喷嘴中连续喷出保护气体把空气与焊接区域中的熔化金属隔离开来，以保护电弧和焊接熔池中的液态金属不受大气中的氧、氮、氢等污染，以达到提高焊接质量的目的。

Gas shielded arc welding includes arc welding using argon shielding gas in the welding region of argon, CO_2 shielded welding area using CO_2 as a protective gas. The basic principle of gas shielded arc welding is that the protective gas is continuously jetted from the spray gun nozzle to isolate air from the molten metal in the weld area during the welding with arc as heat source in order to protect the electric arc and welding pool of liquid metal are not affected by oxygen, nitrogen, hydrogen, and other pollution in atmosphere to improve the quality of welding.

钨极氩弧焊是以高熔点的金属钨棒作为焊接时产生电弧的一个电极，并处在氩气保护下的电弧焊，常用于不锈钢、高温合金等要求严格的焊接。

Argon tungsten arc welding is a kind of welding in which high-melting metal tungsten bar is used as an electrode and argon gas is used to protect welding. Argon tungsten arc welding is often used for stainless steel, high temperature alloy and other welding with strict requirements.

等离子电弧焊是由钨极氩弧焊发展起来的一种焊接方法，气体由电弧加热发生离解，在高速通过水冷喷嘴时受到压缩，增大能量密度和离解度，形成等离子弧。它的稳定性、发热量和温度都高于一般电弧，因而具有较大的熔透力和焊接速度。形成等离子弧的气体和它周围的保护气体一般用氩气。根据各种工件的材料性质，也有使用氦或氩氦、氩氢等混合气体的。

Plasma arc welding is a kind of welding method developed by argon tungsten arc welding. The gas is heated by arc to produce dissociation and compressed when it passes through the water-cooled nozzle at high speed. The energy density and dissociation degree are increased. Its stability, heat value and temperature are higher than general arc, so it has greater penetration force and welding speed. The gas that forms an arc of plasma and the protective gas around it are usually argon. According to the material properties of various work pieces, mixed gases such as helium or argon-helium and argon-hydrogen are also used.

等离子弧有两种工作方式。一种是“非转移弧”，电弧在钨极与喷嘴之间燃烧，主要用于等离子喷镀或加热非导电材料；另一种是“转移弧”，电弧由辅助电极高频引弧后，电弧燃烧在钨极与工件之间，用于焊接。

Plasma arcs work in two ways. One is the “non-transfer arc”. Non-transfer arc refers to the arc burns between the tungsten electrode and the nozzle. Non-transfer arc is mainly used for plasma spraying or heating non-conductive materials. The other is the “transfer arc”, in which an arc is triggered by an auxiliary electrode at high frequency, and the arc burns between the tungsten electrode and the work piece for welding.

形成焊缝的方式有熔透式和穿孔式两种。前一种形式的等离子弧只熔透母材，形成焊接熔池，多用于0.8~3mm厚的板材焊接；后一种形式的等离子弧只熔穿板材，形成钥匙孔形的熔池，多用于3~12mm厚的板材焊接。此外，还有小电流的微束等离子弧焊，特别适合于0.02~1.5mm的薄板焊接。等离子弧焊接属于高质量焊接方法。

There are two ways of forming weld line: penetration type and perforation type. The former only penetrates the base material and forms the welding pool, which is mostly used for welding plates with thickness of 0.8~3mm. The latter only melts through the plate to form a keyhole shaped molten pool, which is mostly used for welding 3~12mm thick plates. In addition, there is micro-beam plasma arc welding with small current. It is especially suitable for thin plate welding of 0.02~1.5mm. Plasma arc welding is one of high quality welding methods.

焊缝的深/宽比大，热影响区窄，工件变形小，可焊材料种类多。特别是脉冲电流等离子弧焊和熔化极等离子弧焊的发展，更扩大了等离子弧焊的使用范围。

The depth/width ratio of welding line is large. The heat affected area is narrow. The work piece deformation is small. There are many weldable materials. In particular, the development of pulsed current plasma arc welding (pulsed current plasma arc welding) and fusion plasma arc welding (plasma arc welding) has expanded the application range of plasma arc welding.

（3）电渣焊。电渣焊是利用电流通过熔渣所产生的电阻热作为热源，将填充金属和母材熔化，凝固后形成金属原子间牢固连接。在开始焊接时，使焊丝与起焊槽短路起弧，不断加入少量固体焊剂，利用电弧的热量使之熔化，形成液态熔渣，待熔渣达到一定深度时，增加焊丝的送进速度，并降低电压，使焊丝插入渣池，电弧熄灭，从而转入电渣焊焊接过程。

Electroslag welding. Electroslag welding the resistance heat generated by electric current through the slag as a source of heat. The filler metal and base metal are melted and formed a firm connection between metal atoms after solidified. At the start of welding, short-circuit arc starting occurs between welding wire and welding groove. A small amount of solid flux is constantly added and melt by arc heat to form liquid slag. When the depth of slag reaches a certain value, welding wire feeding speed increase and voltage decrease. The welding wire is inserted into the slag pool. The arc extinguishes and turns to electroslag welding process.

电渣焊主要有熔嘴电渣焊、非熔嘴电渣焊、丝极电渣焊、板极电渣焊等。电渣焊的缺点是输入的热量大，接头在高温下停留时间长，焊缝附近容易过热，焊缝金属呈粗大结晶的铸态组织，冲击韧性低，焊件在焊后一般需要进行正火和回火热处理。图 4.14 为电渣焊原理图。

Electroslag welding mainly includes melt nozzle electroslag welding, non-melt nozzle electroslag welding, silk electrode electroslag welding, plate electrode electroslag welding and so on. The disadvantages of electroslag welding are that the heat input is large. The welding seam is easy to overheat when the connection head stays at a high temperature for a long time. The metal in welding line shows the as-cast structure of coarse crystallization. The impact toughness is low. The welding parts generally need normalizing and tempering heat treatment after welding. Figure 4.14 is the principle of electroslag welding.

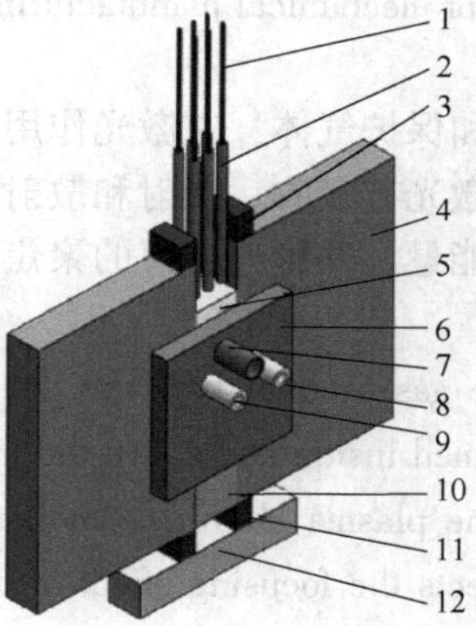

图 4.14 电渣焊原理图

Figure 4.14 Principle of electroslag welding

1—电极电线（electrode wires）; 2—熔嘴（consumable guide）; 3—熄弧板（run off tab）; 4—竖板（vertical plates）; 5—渣（slag）; 6—铜冷却板（copper cooling shocs）; 7—保护气体供应（shielding gas supply）; 8—进水（water in）; 9—出水（water out）; 10—焊接金属（weld metal）; 11—引弧板（run on tab）; 12—启动板（starting plate）

（4）激光焊接。激光焊接有两种基本模式：热导焊和深熔焊，前者所用激光功率密度较低（10^5~10^6W/cm^2），工件吸收激光后，仅达到表面熔化，然后依靠热传导向工件内部传递热量形成熔池。

Laser welding. There are two basic modes of laser welding: thermal conductivity welding and deep fusion welding. The laser power density of the former is low (10^5~10^6W/cm^2). After only the surface of the work piece meltsvafter absorbing the laser. Then the molten pool is formed by heat conduction to transfers heat tc the interior of the workpiece.

这种焊接模式熔深浅，深宽比较小。后者激光动车密度高（$10^6 \sim 10^7 W/cm^2$），工件吸收激光后迅速熔化乃至气化，熔化的金属在蒸汽压力作用下形成小孔。激光束可直照孔底，使小孔不断延伸，直至小孔内的蒸汽压力与液体金属的表面张力和重力平衡为止。

The melting pool is relatively small and the ratio of depth to width is small too. The laser power density of the latter is high（$10^6 \sim 10^7 W/cm^2$）. The workpiece melt rapidly and even gasification after absorbing laser. The molten metal forms a small hole laser beam under the effect of steam pressure and can shine directly on the bottom of the hole. The small hole continues to extend until there is a balance among the small hole vapor pressure and liquid metal surface tension and gravity.

小孔随着激光束沿焊接方向移动时，小孔前方熔化的金属绕过小孔流向后方，凝固后形成焊缝。这种焊接模式熔深大，深宽比也大。在机械制造领域，除了那些微薄零件之外，一般应选用深熔焊。

When the small hole moves along the direction of welding with the laser beam, the molten metal in front of the small hole bypasses the small hole and flows to the rear. The weld seam is formed after solidification. In this welding mode, the depth of welding pool is large and the ratio of depth to width is large. In the field of mechanical manufacturing, deep fusion welding can be used except those thin parts.

深熔焊过程产生的金属蒸汽和保护气体，在激光作用下发生电离，从而在小孔内部和上方形成等离子体。等离子体对激光有吸收、折射和散射作用，因此一般来说熔池上方的等离子体会削弱到达工件的激光能量，并影响光束的聚焦效果，对焊接不利。通常可辅加侧吹气驱除或削弱等离子体。

The metal vapor and protective gas produced by the deep welding process ionize under the action of the laser and plasma is formed inside and above the hole. The plasma can absorb, refract and scatter the laser. In general, the plasma above the molten pool can weaken the laser energy reaching the work piece, it also affects the focusing effect of beam and is unfavorable to welding. The plasma can usually be removed or weakened by side blowing.

小孔的形成和等离子体效应，使焊接过程中伴随着具有特征的声、光和电荷产生，研究它们与焊接规范及焊缝质量之间的关系，以及利用这些特征信号对激光焊接过程及质量进行监控，具有十分重要的理论意义和实用价值。

The formation of small holes and plasma effect make the welding process accompanied by the generation of characteristic sound, light and charge. It is of great theoretical significance and practical value to study the relationship between them and the welding specification and weld quality. These characteristic signals can be used to monitor the laser welding process and quality.

由于经聚焦后的激光束光斑小（0.1~0.3mm），功率密度高（$10^5 \sim 10^7 W/cm^2$），比电弧焊（$5\times10^2 \sim 5\times10^4 W/cm^2$）高几个数量级，因而激光焊接具有传统焊接方法无法比拟的显著优点：

Because the laser beams after focusing is small（0.1 ~ 0.3mm）and the power density（$10^5 \sim 10^7 W/cm^2$）is higher than the arc welding（$5\times10^2 \sim 5\times10^4 W/cm^2$）, so the laser welding has significant advantages than traditional welding methods：

1）加热范围小，焊缝和热影响区窄，接头性能优良。

Small heating range, narrow weld and heat affected area, excellent connection performance.

2）残余应力和焊接变形小，可以实现高精度焊接。

The residual stress and welding deformation are small, which can realize high precision welding.

3）可对高熔点、高热导率，热敏感材料及非金属进行焊接。

High melting point, high thermal conductivity, thermal sensitive materials and non-metal welding.

4）焊接速度快，生产率高。

Fast welding speed, high productivity.

5）具有高度柔性，易于实现自动化。

High flexibility, easy to achieve automation.

激光焊（如图 4.15 所示）与电子束焊有许多相似之处，但它不需要真空室，不产生 X 射线，更适合生产中推广应用。激光焊接实际上已取得了电子束焊接 20 年前的地位，成为高能束焊接技术发展的主流。

Laser welding (as shown in Figure 4.15) is similar to electron beam welding. But laser welding does not need vacuum chamber and does not produce X-ray. Laser welding is more suitable for production and application. In fact, laser welding has achieved the status of electron beam welding 20 years ago and become the mainstream of high-energy beam welding technology.

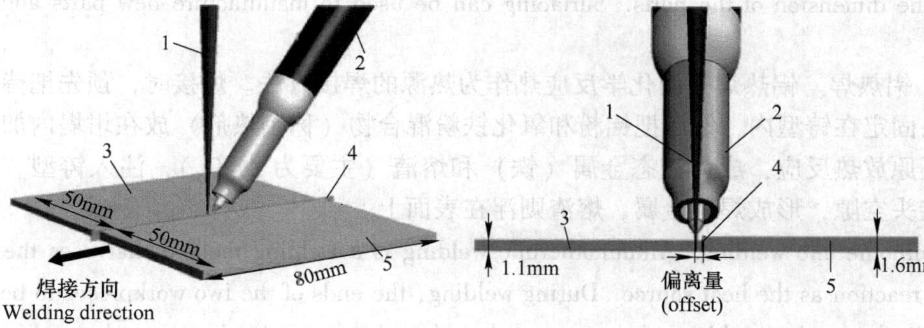

图 4.15　激光焊

Figure 4.15　Laser welding

1—激光（laser）；2—钨极（tungsten electrode）；3—钢（steel）；4—内层（interlayer）；5—镁合金（Mg alloy）

（5）电子束焊接。电子束焊接的基本原理是电子枪中的阴极由于直接或间接加热而发射电子，该电子在高压静电场的加速下再通过电磁场的聚焦就可以形成能量密度极高的电子束，用此电子束去轰击工件，巨大的动能转化为热能，焊接处工件熔化，形成熔池，从而实现对工件的焊接。

Electron beam welding. During electron beam welding, cathode of electron gun emits electrons due to directly or indirectly heated. The acceleration of the electron is finished in the high voltage electrostatic field under again through the electromagnetic field and form high energy density of electron beam. With the electron beam bombardment artifacts, huge kinetic energy is translated into heat energy make the welding workpiece melting. The molten pool is formed, so as to realize the welding of workpieces.

（6）堆焊。堆焊是用焊接方法在机械零件表面堆敷一层具有一定性能金属材料的工艺过程。堆焊作为材料表面改性的一种经济而快速的工艺方法，越来越广泛地应用于各个工业部门零件的制造修复中。为了最有效地发挥堆焊层的作用，希望采用的堆焊方法有较小的母材稀释、较高的熔敷速度和优良的堆焊层性能，即优质、高效、低稀释率的堆焊技术。

Surfacing welding. Surfacing welding is a process of metal materials with certain properties stacked on the surface of mechanical parts. As an economical and rapid process method for surface modification of materials, surfacing welding is more and more widely used in the manufacture and repair of various industrial parts. In order to give full play to the role of the surfacing layer, the surfacing technology should have smaller base metal dilution, higher deposition speed and excellent performance of the surfacing layer, that is, high quality, high efficiency and low dilution rate.

堆焊的目的不是连接零件，而是使零件获得具有耐磨、耐热、耐蚀等特殊性能的熔敷金属层，或是为了恢复或增加零件的尺寸。堆焊可既可应用于制造新零件又可应用于恢复旧零件。

The purpose of surfacing welding is not to connect the parts, but to make the parts have the special properties of wear resistance, heat resistance and corrosion resistance, or to restore or increase the dimension of the parts. Surfacing can be used to manufacture new parts and restore old parts.

（7）铝热焊。铝热焊是用化学反应热作为热源的焊接方法。焊接时，预先把待焊两工件的端头固定在铸型内，然后把铝粉和氧化铁粉混合物（称铝热剂）放在坩埚内加热，使之发生还原放热反应，成为液态金属（铁）和熔渣（主要为 Al_2O_3），注入铸型。液态金属流入接头空隙，形成焊缝金属，熔渣则浮在表面上。

Aluminothermic welding. Aluminothermic welding is a welding method that uses the heat of chemical reaction as the heat source. During welding, the ends of the two workpieces to be welded are fixed in the casting mold in advance, and then the mixture of aluminum powder and iron oxide powder (called aluminothermic agent) is heated in the crucible, so that reduction and exothermic reaction take place, and liquid metal (iron) and slag (mainly Al_2O_3) are injected into the casting mold. Liquid metal flows into the connection gap, forming weld metal, slag floating on the surface.

为了调整熔液温度和焊缝金属化学成分，常在铝热剂中加入适量的添加剂和合金。铝热焊具有设备简单、使用方便、不需要电源等特点，常用于钢轨、钢筋和其他大截面工件的焊接。

In order to adjust the melt temperature and weld metal chemical composition, an appropriate amount of additives and alloys are often added into the thermite. Thermal-aluminum welding has the characteristics of simple equipment, easy to use, no need for power supply, etc. It is often used for the welding of rails, rebar and other large section workpieces.

起重机钢轨焊接工艺往往比较复杂，需要熟练的工人进行有效的工艺操作，以保证钢轨在使用过程中的使用寿命。强烈推荐使用铝热焊接将钢轨连接在一起（如图 4.16 所示）。

Crane Rail welding is often complicated and requires skilled labor carry out the process effectively to ensure longevity of the rails during usage. It is highly recommending the use of aluminothermic welding to join rails together (as shown in Figure 4.16).

图 4.16 钢轨的铝热焊

Figure 4.16 Aluminothermic welding of the rails

B 压力焊（Pressure Welding）

压力焊是在加压条件下，使两工件在固态下实现原子间结合，又称固态焊接。

Pressure welding is finished under pressurized conditions so that the atoms of two parts can be combined in the solid state. It is also known as solid phase bonding.

常用的压力焊工艺是电阻对焊，当电流通过两工件的连接端时，该处因电阻很大而温度上升，当加热至塑性状态时，在轴向压力作用下连接成为一体。

The pressure welding process commonly used is upset butt welding. The resistance is high and the temperature rises when the current goes through the connection of the two workpieces end and two workpieces are connected into a whole when heated to the plastic state under the action of the axial pressure into one.

压力焊包括爆炸焊、冷压焊、摩擦焊（如图 4.17 所示）、扩散焊、超声波焊、高频焊和电阻焊等。

Pressure welding includes explosive welding, cold pressure welding, friction welding (as shown in Figure 4.17), diffusion welding, ultrasonic welding, high frequency welding and resistance welding.

各种压焊方法的共同特点是在焊接过程中施加压力而不加填充材料。多数压焊方法如扩散焊、高频焊、冷压焊等都没有熔化过程，因而没有像熔焊那样的有益合金元素烧损和有害元素侵入焊缝的问题，从而简化了焊接过程，也改善了焊接安全卫生条件。

The common feature of various pressure welding methods is that pressure is applied during welding without filling material. Most pressure welding methods, such as diffusion welding, high-frequency welding, cold pressure welding, etc., have no melting process, so there are no good

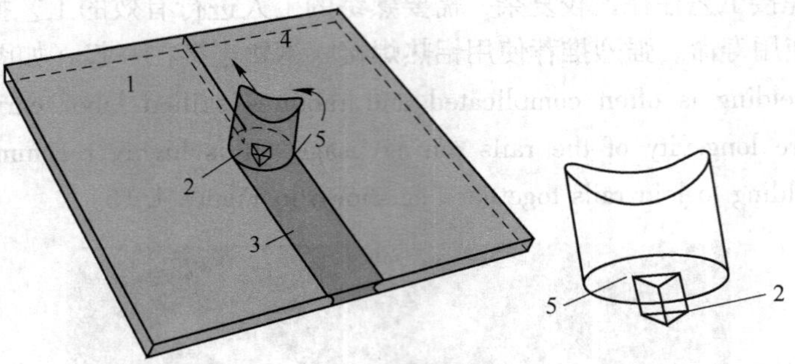

图 4.17 摩擦焊

Figure 4.17 Friction welding

1—焊缝回转侧（retreating side）；2—搅拌针（pin）；3—焊接区（welded zone）；
4—焊缝前进侧（advancing side）；5—搅拌肩（shoulder）

alloy elements are burning loss, and harmful elements enter into the welding seam, which simplifies the welding process and improves the welding safety and health conditions.

同时由于加热温度比熔焊低、加热时间短，因而热影响区小。许多难以用熔化焊焊接的材料，往往可以用压焊焊成与母材同等强度的优质接头。

At the same time, the heating temperature is lower than that of fusion welding, and the heating time is shorter, so the heat affected zone is smaller. Many materials difficult to use fusion welding, often can be pressure welding and get the same strength as the base.

C 钎焊（Brazing）

钎焊是采用比母材熔点低的金属材料作钎料，在加热温度高于钎料低于母材熔点的情况下，利用液态钎料润湿母材，填充接头间隙，并与母材相互扩散实现连接焊件的方法。它包括硬钎焊、软钎焊等。

Brazing is a way that a kind of metal material with lower melting point than the base material is used for the brazing filler to fill the connection space in the case of the heating temperature is higher than the solder melting point and lower than the base metal, liquid solder is used to wet the base material and the interdiffuse with base metal to connect weldments. It comprises hard brazing and soft brazing.

钎焊特点是只加热钎料，被焊零件处于固态。焊接零件加热温度低，组织性能变化小，变形也很小，接头光滑平整。

Brazing is characterized by only heating the solder and the welding parts are in solid state. The temperature of welding parts heating is low. The change of microstructure and properties is small. The deformation is small. The connection is smooth and even.

钎焊可焊异种金属。但接头强度低，工作温度不高，焊前工作要求严。激光焊-钎焊加工原理图如图 4.18 所示。

Brazing can be used to weld dissimilar metal, but the connection strength is low, the working temperature is not high, the work requirement before welding is strict. Schematic illustration of laser welding-brazing processing is shown in Figure 4.18.

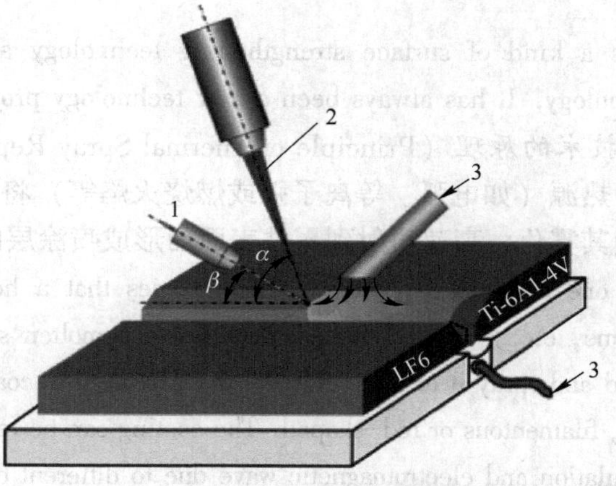

图 4.18 激光焊-钎焊加工原理图

Figure 4.18 Schematic illustration of laser welding-brazing processing

1—填充焊丝（filler wire）；2—激光束（laser beam）；3—氩气（Ar gas）

焊接产品比铆接件、铸件和锻件重量轻，对于交通运输工具来说可以减轻自重，节约能量。焊接的密封性好，适于制造各类容器。焊接已成为现代工业中一种不可缺少，而且日益重要的加工工艺方法。

Welding products are lighter than riveting parts, castings and forgings, which can reduce weight and save energy for transportation. It is suitable for making all kinds of containers due to good sealing. Welding has become an indispensable and increasingly important processing technology in modern industry.

未来的焊接工艺，一方面要研制新的焊接方法、焊接设备和焊接材料，以进一步提高焊接质量和安全可靠性，如改进现有电弧、等离子弧、电子束、激光等焊接能源；运用电子技术和控制技术，改善电弧的工艺性能，研制可靠轻巧的电弧跟踪方法。

In the future, on the one hand, new welding methods, welding equipment and welding materials should be developed to further improve welding quality and safety and reliability, such as improving existing welding energy sources such as arc, plasma arc, electron beam and laser. Electronic technology and control technology are used to improve the process performance of arc, and a reliable and lightweight arc tracking method is developed.

另一方面要提高焊接机械化和自动化水平，如焊机实现程序控制、数字控制；研制从准备工序、焊接到质量监控全部过程自动化的专用焊机；在自动焊接生产线上，推广、扩大数控的焊接机械手和焊接机器人，可以提高焊接生产水平，改善焊接卫生安全条件。

On the other hand, the level of mechanization and automation of welding will be improved. For example, the welding machine can be operated by program control, digital control. In automatic welding production line, popularizing and expanding CNC welding manipulator and welding robot can improve welding production level and improve welding health and safety conditions.

4.2.4 热喷涂技术（Thermal Spray Repair Technology）

热喷涂是一种表面强化技术，是表面工程技术的重要组成部分，一直是中国重点推广的新技术项目。

Thermal spraying is a kind of surface strengthening technology and an important part of surface engineering technology. It has always been a new technology project promoted in China.

4.2.4.1 热喷涂技术的原理（Principle of Thermal Spray Repair Technology）

热喷涂是利用某种热源（如电弧、等离子弧或燃烧火焰等）将喷涂材料加热到熔融状态，并通过气流吹动使其雾化，高速喷射到零件表面而形成喷涂层的表面加工技术。

Thermal spraying is one of surface processing technologies that a heat source (such as arc, plasma arc or burning flame, etc.) is used to heat material to a molten state and atomize it by air blowing with a high-speed and spray it to the surface of the part to form coating. The coating material may be powdery, banded, filamentous or rod-shaped. The coating can be resistant to high temperature corrosion, wear, heat insulation and electromagnetic wave due to different coating materials.

在热喷涂过程中，细微而分散的金属或非金属的涂层材料，以一种熔化或半熔化状态，沉积到一种经过制备的基体表面，形成某种喷涂沉积层。

In the process of thermal spraying, a fine and dispersed coating material of metal or non-metal is deposited on the surface of a prepared substrate in a molten or semi-molten state to form a coating layer.

热喷涂枪由燃料气、电弧或等离子弧提供必需的热量，将热喷涂材料加热到塑态或熔融态，再经受压缩空气的加速，使受约束的颗粒束流冲击到基体表面上。冲击到表面的颗粒，因受冲压而变形，形成叠层薄片，黏附在经过制备的基体表面，随之冷却并不断堆积，最终形成一种层状的涂层。

The thermal spray gun is heated to the plastic or molten state by fuel gas, arc or plasma arc, and then subjected to the acceleration of compressed air to impact the constrained particle beam onto the surface of the substrate. The particles that strike the surface are deformed by stamping, forming laminated flakes that adhere to the prepared substrate surface and then cool and accumulate, eventually forming a layered coating.

该涂层因涂层材料的不同可实现耐高温腐蚀、抗磨损、隔热、抗电磁波等功能。图 4.19 为热喷涂机械手操作装置。图 4.20 为热喷涂原理。

The coating can be resistant to high temperature corrosion, wear, heat insulation and electromagnetic wave due to different coating materials. Figure 4.19 is an anipulator for thermal spraying. Figure 4.20 shows the principle for thermal spraying.

图 4.19 热喷涂机械手操作装置

Figure 4.19 Manipulator for thermal spraying

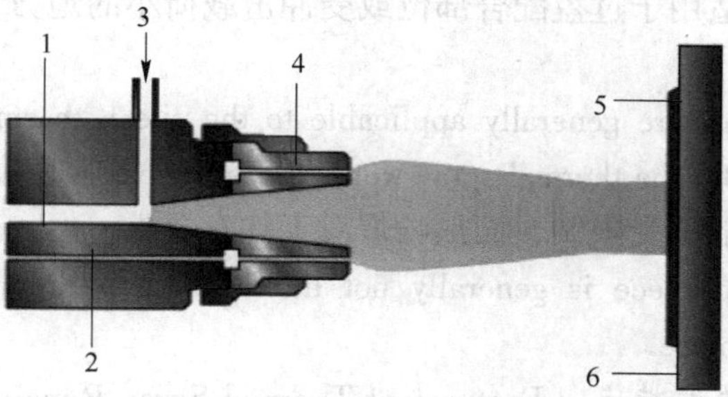

图 4.20 热喷涂原理图

Figure 4.20 Principle for thermal spraying

1—燃气（fuel gas）；2—氧气（oxygen）；3—粉末（powder）；
4—喷嘴（nozzle）；5—涂层（coating）；6—工件（workpiece）

4.2.4.2 热喷涂和热喷焊的区别（Differences of Thermal Spraying and Thermal Spray Welding）

热喷焊是在喷涂过程中或喷涂后，加热喷涂层，使其熔化并润湿工件表面，通过涂层与基体的互溶和扩散，形成与基体具有冶金结合的喷焊层的加工方法。

Thermal spray welding is in the process of spraying or after spraying, heating spray layer, make it melt and wet the surface of the workpiece, through the coating and matrix of mutual dissolution and diffusion, forming and matrix with metallurgical combination of spray welding layer processing method.

热喷涂和热喷焊不同，区别在于：

Thermal spraying and thermal spray welding are different, differences between them are followings:

（1）工件受热情况不同，喷涂无重熔过程，工件表面温度始终不超过 300℃，零件不会发生变形和组织变化，但热喷焊时零件温度达 900℃ 以上，某些工件可能会发生退火及变形。

Workpiece heating situation is different. There is not remelting in thermal spraying process. The surface temperature is always no more than 300℃ in thermal spraying process. There is not part deformation and organizational changes. However, the temperature of parts in thermal spray welding is more than 900℃. Annealing and deformation may occur in some workpieces.

（2）与基体结合的状态不同。

The state combine with base body is different.

（3）热喷焊的材料一般是自熔合金，而热喷涂的材料无限制。

The materials for thermal spray welding are generally self-fluxing alloy, but there is not limit for the material of thermal spraying.

（4）热喷焊层均匀致密，一般无孔隙，而热喷涂层有孔隙。

The density of thermal spray welding coating is uniform. There is not porous generally in thermal spray welding coating. But there is porous in thermal spraying coatings.

（5）热喷涂层一般适用于过盈配合部位或受冲击载荷小的地方，而热喷焊层则可承受较大的冲击荷载。

Thermal spray coatings are generally applicable to the site with small local impact loads or interference fit position. But the thermal spray welding coating can withstand greater impact loads.

（6）热喷涂工件一般不受限制，而热喷焊对工件有要求。

Thermal spraying workpiece is generally not limited, while thermal spraying welding has requirements on the workpiece.

4.2.4.3 热喷涂技术特点（Features of Thermal Spray Repair Technology）

热喷涂技术特点：

The features of thermal spraying technology are followings:

（1）基体材料不受限制，可以是金属和非金属，可以在各种基体材料上喷涂。

There is no limit for base material. It can be metal or non-metal and a variety of matrix material.

（2）可喷涂的涂层材料极为广泛，热喷涂技术可用来喷涂几乎所有的固体工程材料，如硬质合金、陶瓷、金属、石墨等。

Coating materials can be sprayed very wide. Thermal spraying technology can be used to spray almost all solid engineering materials, such as hard alloy, ceramics, metal, graphite, etc.

（3）喷涂过程中基体材料温升小，不产生应力和变形。

The temperature of the base material rises small and no stress or deformation is generated during spraying.

（4）操作工艺灵活方便，不受工件形状限制，施工方便。

Operation process is flexible and convenient. Not limited by the shape of the workpiece. Construction is convenient.

（5）涂层厚度可以从0.01mm至几毫米。

Coating thickness can be from 0.01mm to a few millimeters.

（6）涂层性能多种多样，可以形成耐磨、耐蚀、隔热、抗氧化、绝缘、导电、防辐射等具有各种特殊功能的涂层。

The coating has a variety of properties such as wear resistance, corrosion resistance, heat insulation, oxidation resistance, insulation, conductive, radiation protection and other special functions.

（7）具有适应性强及经济效益好等优点。

It has strong adaptability and good economic benefits.

（8）喷涂层与基体结合强度不太高而且存在孔隙。

The bonding strength of spraying layer and base material is not too high and there are porous.

（9）喷焊件存在变形问题。

Deformation of spray welding parts is big.

4.2.4.4 热喷涂的种类（Classification of Thermal Spray Repair Technology）

常用的热喷涂有火焰喷涂、电弧喷涂、等离子喷涂和爆炸喷涂。

The common thermal spraying include flame spraying, arc spraying, plasma spraying and explosive spraying.

4.2.4.5 热喷涂材料（Thermal Spraying Materials）

热喷涂材料按外形分为线材和粉末。线材喷涂材料包括丝材和棒材，如铝、锌、铜、钼、镍等有色合金，碳钢不锈钢等黑色金属，复合线材和陶瓷棒材等。

Thermal spraying materials are divided into wire and powder according to shape. Wire spraying materials include wire and bar, such as aluminum, zinc, copper, molybdenum, nickel and other non-ferrous alloys, carbon steel, stainless steel and other ferrous metals, composite wire and ceramic bar, etc.

粉末喷涂材料除可做成线材的材料，还有难熔金属，自熔合金、陶瓷、复合材料等。

In addition to wire materials, powder spraying materials includes refractory metals, self-melting gold, ceramics, composite materials.

热喷涂材料按成分分为金属及其合金、陶瓷、塑料、复合材料和自熔合金等五类。

Thermal spraying materials are divided into five categories according to composition: metal and its alloy, ceramics, plastics, composite materials and self-fusing alloy.

4.2.5 电镀修复技术（Electroplating Repair Technology）

电镀是利用电解作用使金属或其他材料制件的表面附着一层金属膜的工艺。

Electroplating technology is a process the surface of the metal or other material parts are attached to a layer of metal film by the use of electrolytic action.

电镀包括让电流通过一种叫做电解质的溶液。这是通过将两个被称为电极的端子浸入电解液中，并将它们连接到与电池或其他电源相连的电路中来实现的。

Electroplating involves passing an electric current through a solution called an electrolyte. This is done by dipping two terminals called electrodes into the electrolyte and connecting them into a circuit with a battery or other power supply.

电极和电解质由精心挑选的元素或化合物制成。当电流通过他们制造的电路时，电解液分裂，其中的一些金属原子被沉积在一个电极上的一层薄层中——它被电镀了。

The electrodes and electrolyte are made from carefully chosen elements or compounds. When the electricity flows through the circuit they make, the electrolyte splits up and some of the metal atoms it contains are deposited in a thin layer on top of one of the electrodes, it becomes electroplated.

各种金属都可以用这种方法镀，包括金、银、锡、锌、铜、镉、铬、镍、铂和铅。

All kinds of metals can be plated in this way, including gold, silver, tin, zinc, copper, cadmium, chromium, nickel, platinum, and lead.

4.2.5.1 电镀修复技术的原理（Principle of Electroplating Repair Technology）

电镀非常类似于电解（用电来分解化学溶液），这是电池产生电流的相反过程。所有这些都是电化学的例子：由电引起或产生的化学反应，提供科学或工业上有用的最终产品。

Electroplating is very similar to electrolysis (using electricity to split up a chemical solution), which is the reverse of the process by which batteries produce electric currents. All these things are examples of electrochemistry: chemical reactions caused by or producing electricity that give scientifically or industrially useful end-products.

电镀过程中，镀层金属作为阳极，氧化成阳离子进入电镀液；将被镀的金属制品做阴极，在金属表面涂敷金属阳离子，使之还原形成涂层。为了消除其他阳离子的干扰，使镀层均匀牢固，电镀液中应使用含有镀层金属阳离子的溶液，以保持镀层金属阳离子的浓度不变。图 4.21 给出了电镀原理。

In electroplating, coating metal as an anode, is oxidized into cation into the bath. To be plated metal products do cathode, coating metal cation on the metal surface is reduced to form a coating. In order to eliminate the interference of other cations and make the coating uniform and firm, the solution containing metal cations of the coating should be used for electroplating solution to keep the concentration of metal cations of the coating unchanged. Figure 4.21 gives the principle of electroplating.

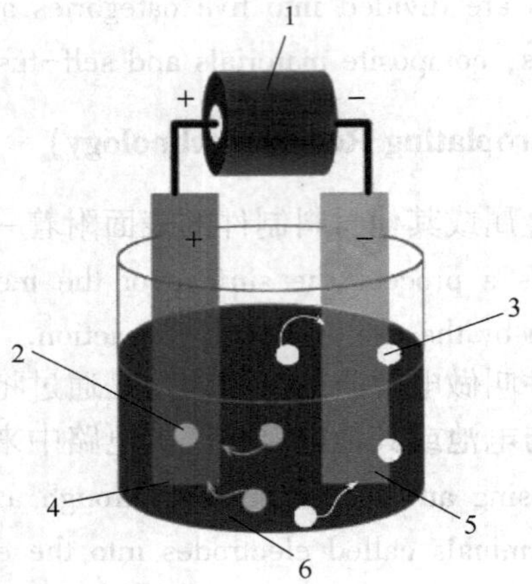

图 4.21　电镀的原理
Figure 4.21　Principle of electroplating
1—电池（battery）；2—硫酸盐离子（sulfate ions）；3—铜离子（copper ions）；
4—铜电极（copper electrode）；5—黄铜电极（brass electrode）；6—电解液（electrolyte）

4.2.5.2　电镀修复技术的目的（Purpose of Electroplating Repair Technology）

电镀的目的是通过在基材或零件上电镀金属涂层来改变基材的表面性质或尺寸。电镀可以提高金属的耐腐蚀性（所涂金属通常是耐腐蚀金属），增加硬度，防止磨损，提高导电性、润滑性、耐热性，以及表面的外观。

The purpose of electroplating is to change the surface property or dimension of the substrate by plating the metal coating on the base material or part. Electroplating can enhance the corrosion resistance of the metal (the coating metal is usually corrosion resistant metal), increase hardness, prevent abrasion, improve electrical conductivity, lubricity, heat resistance, and the appearance of the surface.

4.2.5.3　电镀修复技术的分类（Classification of Electroplating Repair Technology）

电镀分为挂镀、滚镀、连续镀和刷镀等方式，主要与待镀件的尺寸和批量有关。

Electroplating can be divided into hanging plating, rolling plating, continuous plating and brush plating. They can be chosen according to the dimension and batch of the parts to be plated.

挂镀适用于一般尺寸的制品，如汽车的保险杠，自行车的车把等。滚镀适用于小件，如紧固件、垫圈、销子等。连续镀适用于成批生产的线材和带材。刷镀适用于局部镀或修复。刷镀是在电镀基础上发展起来的修复新技术，是涂镀、快速电镀、无槽电镀、擦镀等的统称。

Hanging plating is suitable for general dimension products, such as car bumpers, bicycle handlebars, etc. Rolling plating is suitable for small parts, such as fasteners, gaskets, pins, etc. Continuous plating is suitable for batch production of wire and strip. Brush plating is suitable for partial plating or repair. Brush plating is a new technology developed on the basis of electroplating. It is a general term of coating plating, rapid electroplating, slotless electroplating, rubbing plating, etc.

常见的金属电镀修复技术有：低温镀铁修复技术和镀铬修复技术以及镀铜技术。其他电镀方法还有复合电镀、化学镀镍、脉冲电镀、周期转向镀、摩擦电喷镀等。

Common repair technologies of metal plating include low-temperature iron plating, chromium plating and copper plating. Other electroplating methods include composite electroplating, electroless nickel plating, pulse electroplating, periodic turning electroplating, friction electroplating, etc.

电镀液有酸性的、碱性的和加有铬合剂的酸性及中性溶液。无论采用何种镀覆方式，与待镀制品和镀液接触的镀槽、吊挂具等应具有一定程度的通用性。

Electroplating solution has acid, alkaline and add chromium mixture acid and neutral solution. No matter what kind of plating method is adopted, there should be a certain degree of universality among products to be plated, plating tank contact with plating solution and tools for hanging parts.

4.2.5.4 镀层（Coating）

镀层分为装饰保护性镀层和功能性镀层两类。装饰保护性镀层，主要是在铁金属、非铁金属及塑料上的镀铬层，特别是钢的铜-镍-铬层，锌及钢上的镍-铬层。

Coating is divided into decorative protective coating and functional coating. Decorative protective coating mainly refers to the chromium layer electroplated on iron metals, non-ferrous metals and plastics on, especially the copper-nickel-chromium layer of steel, zinc and steel on the nickel-chromium layer.

为了节约镍，在钢上镀铜-镍、铁-高硫镍-镍、铁-低固分镍-铬层。与镀铬层相似的锡、镍镀层，可用于分析天平、化学泵、阀和流量测量仪表上。

In order to save nickel, copper-nickel, iron-high sulfur-nickel, iron-low solid-nickel-chromium coating has been applied to steel. A tin, nickel coating similar to the chrome-plated coating used in analytical balances, chemical pumps, valves and flow gauges.

功能性镀层种类很多，如：

There are many kinds of functional coatings, such as:

（1）提高与轴颈的相容性和嵌入性的滑动轴承罩镀层，铅-锡，铅-铜-锡，铅-铟等复合镀层。

Lead－tin，lead－copper－tin，lead－indium composite coating are used to improve the compatibility and embeddability with the journal and bearing bushing of the sliding bearing.

（2）用于耐磨的中、高速柴油机活塞环上的硬铬镀层，这种镀层也可用在塑料模具上，具有不粘模具和使用寿命长的特点。

The hard chromium coating is used for wear resistant，high－speed diesel engine piston ring. This coating can also be used in plastic mold，with non－stick mold and long service life characteristics.

（3）在大型人字齿轮的滑动面上镀铜，可防止滑动面早期拉毛。

Copper plating is used on the sliding surface of large herringbone gear to prevent the sliding surface early hair.

（4）用于防止钢铁基体遭受大气腐蚀的镀锌。

Galvanized is used to prevent steel base body from atmospheric corrosion.

（5）防止渗氮的铜锡镀层。

Copper tin coating is used to prevent from nitriding.

4.2.5.5 电镀的过程（Electroplating Process）

电镀的过程基本如下：

The electroplating process is basically as follows：

（1）把镀上去的金属接在正极。

The plated metal is connected to the positive electrode.

（2）要被电镀的物件接在负极。

The part to be electroplated is connected to the negative electrode.

（3）正负极以镀上去的金属的正离子组成的电解质溶液相连。

The positive and negative electrodes are connected by an electrolyte solution composed of positive ions of the plated metal.

（4）通以直流电的电源后，正极的金属会进行氧化（失去电子），溶液中的正离子则在负极还原（得到电子）成原子并积聚在负极表层。

When a direct current is applied，the metal at the positive electrode oxidizes（loses electrons），and the positive ions in the solution are reduced（gains electrons）to atoms and accumulate on the negative surface.

电镀后被电镀物件的美观性和电流大小有关系，电流越小，被电镀的物件便会越美观；反之则会出现一些不平整的形状。

After electroplating，the beauty of the electroplated object is related to the current. The smaller the current，the more beautiful the electroplated object will be. On the contrary，some uneven shapes will appear.

4.2.5.6 电镀修复技术的特点（Characteristics of Electroplating Repair Technology）

电镀层比热浸层均匀，一般都较薄，从几微米到几十微米不等。通过电镀，可以在机械制品上获得装饰保护性和各种功能性的表面层，还可以修复磨损和加工失误的工件。镀层大多是单一金属或合金，如锌、镉、金或黄铜、青铜等。

Coating of electroplating is uniform，generally thin，from several microns to dozens of microns.

Through electroplating, the surface layer of decorative protection and various functions can be obtained on mechanical products, and the worn and mis-processed workpiece can be repaired. The coating is mostly single metal or alloy, such as zinc, cadmium, gold or brass, bronze, etc.

电镀层也有弥散层，如镍-碳化硅、镍-氟化石墨等；还有覆合层，如钢上的铜-镍-铬层、钢上的银-铟层等。电镀的基体材料除铁基的铸铁、钢和不锈钢外，还有非铁金属，如ABS塑料、聚丙烯、聚砜和酚醛塑料，但塑料电镀前，必须经过特殊的活化和敏化处理。

There is also dispersion layer, such as nickel-silicon carbide, nickel-fluorine fossil ink, etc. There are also cladding layers, such as copper-nickel-chromium layer on steel, steel-silver-indium layer. In addition to iron-based cast iron, steel and stainless steel, the substrate materials for electroplating include non-ferrous metals such as ABS plastics, polypropylene, polysulfone and phenolic plastics. However, before electroplating plastics, they must be specially activated and sensitized.

电镀因在低温下进行（一般远低于100℃），基体金属的性质几乎不受影响，原来的热处理状况不会改变，零件也不会受热变形，镀层的结合强度高，这是常规的热喷涂（焊）、焊接维修所不能比拟的。但镀层的力学性能随厚度的增加而变化，镀层沉积速度慢，修复层尺寸厚时更应重点考虑这些问题。

Electroplating is finished at low temperature (generally far below 100℃). The property of the base metal is almost unaffected. The original heat treatment condition will not change. The parts will not be thermal deformation. Coating bonding strength is high. However, the mechanical properties of the coating change with the increase of thickness, and the deposition rate of the coating is slow. These problems should be considered when the repair layer is thick.

4.2.6 工件表面强化技术（Surface Hardening Repair Technology）

零件的修复，有时不仅仅是补偿尺寸，恢复配合关系，还要赋予零件表面更好的性能，如耐磨性、耐高温性等。采用表面强化技术可以使零件表面获得更好的性能。

Parts being repaired, refers to sometimes not only compensation dimension, restore the relationship, but also giving parts surface better performance, such as wear resistance, high temperature resistance and so on. Using surface strengthening technology can make the surface of parts get better performance.

工件表面强化技术是指采用某种工艺手段，通过材料表层的相变、改变表层的化学成分、改变表层的应力状态以及提高工件表面的冶金质量等途径来赋予基体材料本身所不具备的特殊力学、物理和化学性能，从而满足工程上对材料及其制品提出的要求的一种技术。

Surface hardening technology refers to the technology of giving base material a special mechanical, physical and chemical properties through changing the phase change of the surface of materials, the chemical composition of the surface layer, surface stress state to meet the requirements of engineering materials and products.

表面强化技术作为表面工程学的一项重要技术，对于改善材料的表面性能，提高零件表面的耐磨性，抗疲劳性，延长其使用寿命等具有重要意义。它可以节约稀有、昂贵材料，对各种高新技术发展具有重要作用。

Surface hardening technology is used to improve the surface properties of materials, improve the wear resistance, fatigue resistance and prolong the service life of parts. It can save rare and expensive materials and play an important role in the development of various high and new technologies.

工件表面强化技术包括：表面形变强化、表面热处理强化、表面化学热处理强化和三束表面改性技术。

Workpiece surface hardening technology including surface deformation hardening, surface heat treatment hardening, surface chemical heat treatment hardening, and three-beam surface modification technology.

4.2.6.1 表面形变强化（Surface Deformation Hardening）

表面形变强化的基本原理是通过喷丸、滚压、挤压等手段使工件表面产生压缩变形，表面形成强化层，其深度可达0.5~1.5mm，从而有效地提高工件表面强度和疲劳强度。

The basic principle of surface deformation hardening is to produce compression deformation on the surface of the workpiece by means of shot peening, rolling, extrusion, etc., and a hardening layer on the surface is formed with a depth of 0.5~1.5mm, so as to the surface strength and fatigue strength of the workpiece can be effectively improved.

表面形变强化成本低廉，强化效果显著，在机械设备维修中常用，其中喷丸强化应用最为广泛。

The cost of surface deformation hardening is low. Hardening effect is significant. Surface deformation hardening is commonly used in mechanical equipment maintenance. Shot peening hardening is the most widely used among them.

4.2.6.2 表面热处理（Surface Heat Treatment）

表面热处理是仅对零件表层进行热处理，使表层发生相变，从而改变表层组织和性能的工艺，是最基本、应用最广泛的表面强化技术之一。它可使零件表层具有高强度、硬度、耐磨性及疲劳极限，而心部仍保留原组织状态。表面热处理的淬火炉如图4.22所示。

Surface heat treatment is a process of heat treatment only act on the surface of parts to change the surface structure and performance. It is one of the most basic and widely used surface hardening technologies. It can make the surface of parts have high strength, hardness, wear resistance and fatigue limit, and the heart still retains the original tissue state. A quenching furnace used for surface heat treatment is shown in Figure 4.22.

4.2.6.3 表面化学热处理（Surface Chemical Heat Treatment）

表面化学热处理强化是将工件加热到一定温度，使活性介质通过扩散和释放活性原子渗透元素，渗透到工件的表面，从而改变成分、组织和性能的表面。

Surface chemical heat treatment intensification is to place the workpiece in a certain active medium, heating to a certain temperature, so that the active medium through diffusion and release the active atoms to infiltrate into the elements, infiltrated into the surface of the workpiece, so as to change the composition, organization and performance of the surface.

图 4.22 表面热处理的淬火炉

Figure 4.22 Quenching furnace used for surface heat treatment

表面化学热处理强化可以提高工件的表面强度、硬度和耐磨性，提高表面疲劳强度，提高表面耐腐蚀性，使工件表面具有良好的附着力和低摩擦系数。

Surface chemical heat treatment intensification can improve the surface strength, hardness and wear resistance of the workpiece, improve the surface fatigue strength, improve the corrosion resistance of the surface, and make the workpiece surface with good adhesion resistance and low friction coefficient.

表面化学热处理有很多种，一般以元素的渗入来命名。常用的表面化学热处理方法有渗碳、渗氮、渗碳、渗硼和金属化（通常是 W、V、Cr 等）。

There are many kinds of surface chemical heat treatment, generally named after the infiltration of elements. The commonly used methods of surface chemical heat treatment include carburizing, nitriding, carburizing, boriding, and metalizing (usually W, V, Cr, etc.).

根据加热方式不同，常用的表面热处理强化包括：感应加热表面淬火、火焰加热表面淬火、电接触加热表面淬火、浴炉加热表面淬火等。

According to different heating methods, commonly used surface heat treatment hardening includes: induction heating surface quenching, flame heating surface quenching, electrical contact heating surface quenching, furnace heating surface quenching, etc.

4.2.6.4 三束表面改性技术（Three Beam Surface Modification Technology）

近年来，随着激光束、离子束、电子束和出现与发展，采用激光束、离子束、电子束对材料表面进行改性已成为材料表面增强新技术，通常称为“三束表面改性技术”。

In recent years, with the emergence and development of laser beam, ion beam and electron beam, it has become a new technology of material surface hardening to use laser beam, ion beam and electron beam to modify the material surface, which is usually called “three beam surface modification technology”.

A 激光表面处理技术（Laser Surface Treatment Technology）

激光表面处理技术是应用光学透镜将激光束聚集到很高的功率密度与很高的温度，照射到材料表面，借助于材料的自身传导冷却，改变表面层的成分和显微结构，从而提高表面性能的方法。

Laser surface treatment technology refers to the laser beam optical lens is used to gather high power density and very high temperatures and irradiated to the material surface. The composition and microstructure of the surface layer material are changed by means of the conduction cooling of material itself in order to improve performance.

激光表面处理技术可以解决其他表面处理方法无法解决或解决不好的材料强化问题，可大幅提高材料或零部件抗磨损、抗疲劳、耐腐蚀、防氧化等性能，延长其使用寿命。它被广泛应用于汽车、冶金、机床以及刀具、模具等的生产和修复中。

Laser surface treatment technology can solve the problems other surface treatment methods cannot solve and can greatly improve the properties of material or parts, such as anti-wear, anti-fatigue, corrosion resistance, anti-oxidation, and other properties. The service life can be extended. It is widely used in the production and repair of automobile, metallurgy, machine tools, cutting tools and molds.

与其他光源相比，激光光源发射激光束的功率密度较大，经过光学透镜聚焦后，功率密度进一步增强，可达到 $10^{14}W/cm^2$，焦斑中心温度可达几千度到几万度，比太阳的表面亮度高 10^{10} 倍。

The power density of the laser beam emitted by the laser source is large compared with other light sources. After the optical lens gathered, the power density is further enhanced, which can reach $10^{14}W/cm^2$. The temperature at the center of the focal spot can reach several thousand degrees to tens of thousands of degrees, 10^{10} times higher than the surface brightness of the sun.

激光光束的发射角很小，小到一至几毫弧度，所以可以认为光束基本上是平行的。

High directivity. The emission angle of highly directional laser beam is very small, ranging from one to a few milliradians, so it can be considered that the beam is basically parallel.

激光具有相同的位相和波长，单色性好。

Laser has the same phase, wavelength, and good monochromaticity.

激光表面处理技术与其他表面处理相比，无需使用外加材料，仅改变被处理材料表面层的组织结构，处理后的改性层具有足够的厚度，可根据需要调整深浅一般可达 0.1～0.8mm。处理层和基体结合强度高，激光表面处理的改性层和基体材料之间是致密的冶金结合，而且处理层本身是致密的冶金组织，具有较高的硬度和耐磨性。被处理件变形极小。由于激光功率密度高，与零件的作用时间极短，故零件的热影响区和整体变化极小。适合于高精度零件处理，作为材料和零件的最后处理工序。加工柔性好，适用面广，利用灵活的导光系统可随意将激光导向处理部位，从而可方便地处理深孔、内孔、盲孔和凹槽等，可进行选择性的局部处理，工艺简单优越。

Laser surface treatment technology has the following characteristics compared to other surface treatment. Without additional material, only the structure of the surface layer of the material to be treated is changed. The processed and modified layer has a sufficient thickness which can be adjusted according to the depth, generally up to 0.1～0.8mm. Strength of processing layer and the substrate is high. There is a compact metallurgical bond between the laser surface treated modified layer and the substrate material. Treatment layer is a dense metallurgical structure with high hardness and wear resistance. Deformation of the processed part is very small. The time of interaction with the parts is very short due to the high laser power density. So the thermal influence zone and the overall change of the parts is very small. It is suitable for high precision parts processing and taken as the final processing of materials and parts. Processing flexibility is good, and application is wide. The laser light guide system can be used to guide the processing part optionally. It can easily handle deep hole, inner hole, blind hole and groove, etc. Process is simple and superior.

激光表面处理均在大气环境中进行，免除了镀膜工艺中漫长的抽真空时间，没有明显的机械作用力和工具损耗，无噪声、无污染、无公害，劳动条件好。再加上激光器配以微机控制系统，很容易实现自动化生产，易于批量生产。产品成品率极高，几乎达到100%，效率很高、经济效益显著。

Laser surface treatment is carried out in atmospheric environment. The long vacuum time in the coating process is eliminated. There is no significant mechanical force or tool loss. There is no noise, no pollution, no pollution, and working conditions is good. It is easy to achieve automatic production with microcomputer laser control system, easy to mass production. The rate of finished products is very high, almost 100%. Efficiency and economic benefits are all very high.

常见的激光表面处理技术有激光表面淬火、激光表面涂敷、激光表面合金化、激光表面非晶态处理、激光气相沉积等。

Common laser surface treatment technology includes laser surface hardening, laser surface coating, laser surface alloying, laser surface amorphous processing, laser surface amorphous processing.

激光表面淬火也称为激光相变强化，指用激光向零件表面加热，在极短的时间内，零件表面被迅速加热到奥氏体温度以上，在激光停止辐照后，快速自冷淬火得到马氏体组织的一种工艺方法。

Laser surface quenching is also known as laser phase change hardening, which refers to a process that uses laser to heat the surface of parts to the temperature of austenite within a very short period of time. After the laser stops irradiation, the surface of parts is rapidly heated to the temperature of austenite.

目前激光表面淬火强化已广泛应用于工业生产中，如采用球墨铸铁制造的汽车曲轴，其圆弧处以表面淬火后，硬度可升高到55~62HRC，耐磨性与疲劳强度也大为提高；采用钢或铸铁制造的凸轮、齿轮、活塞环、缸套、模具等，以激光表面淬火后，大大提高其表面硬度和耐磨性。

At present, laser surface quenching has been widely used in industrial production. For example, automobile crankshaft made of ductile iron, the hardness of its arc section can be increased to 55~62HRC after surface quenching, and the wear resistance and fatigue strength are also greatly improved. The surface hardness and wear resistance of cam, gear, piston ring, cylinder liner and die made of steel or cast iron are greatly improved by laser surface quenching.

激光表面淬火对材料的性能有如下影响：硬度提高、提高耐磨性能、改善疲劳强度、降低残余应力。

The influence of laser surface hardening on the performance of the material includes increasing hardness, improving wear resistance, improving the fatigue performance, reducing residual stress.

激光表面涂敷的原理与堆焊相似，将预先配好的合金粉末预涂到基材表面。在激光的辐射下，混合粉末熔化形成熔池，直到基材表面微熔。

The principle of laser surface coating is similar to that of surfacing. Pre-prepared alloy powder is applied to the surface of substrate. Under the laser radiation, the mixed powder is melted to form a molten pool until the substrate surface is slightly melted.

激光停止辐照后，熔化池凝固，并在界面处与基材达到冶金结合。它可避免热喷涂方法使涂层内有过多的气孔、熔渣夹杂、微观裂纹和涂层结合强度低等缺点。

After the laser stops irradiating, the melting pool solidifies and metallurgical bonding is achieved with the substrate at the interface. It can avoid the defects such as too many pores, slag inclusions, micro cracks and low bonding strength of the coating by thermal spraying.

基材一般选择廉价的钢铁材料，有时也可选择铝合金、铜合金、镍合金、钛合金。涂敷材料一般为 Co 基、Ni 基、Fe 基自熔合金粉末。

The cheap steel material is generally chosen as base material. Sometimes the aluminum alloy, the copper alloy, the nickel alloy, the titanium alloy also may be chosen. Co-based, Ni-based and Fe-based self-fusing alloy powder can be the coating materials.

也可将此三种自熔合金作为陶瓷增强涂层的黏结材料，耐增强陶瓷颗粒选择为 WC、SiC、TiC、TiN 等。

The Co-based, Ni-based and Fe-based self-fusing alloy can also be used as bonding materials for ceramic reinforced coating. WC, SiC, TiC, TiN, and so on are selected as ceramic reinforced particles.

激光表面涂敷的目的是提高零部件的耐磨、耐热与耐腐蚀性能。例如，汽轮机和水轮机叶片表面涂敷 Co-Cr-Mo 合金，提高耐磨与耐腐蚀性能。

The purpose of laser surface coating is to improve the wear resistance, heat resistance and corrosion resistance of components. For example, gas and water turbine blades are laser surface coated with Co-Cr-Mo alloy to improve wear resistance and corrosion resistance.

激光表面合金化是一种既改变表面的物理状态，又改变其化学成分的激光表面处理技术。它是预先用电镀或喷涂等技术把所需合金元素涂敷在金属表面，再用激光照射该表面。

Laser surface alloying is a laser surface treatment technology that both the physical state of the surface and its chemical composition are changed. Alloying elements are pre-coated to the surface of the metal by electroplating or spraying, and then the surface is irradiated with a laser.

也可以涂敷与激光照射同时进行。由于激光照射，涂敷层合金元素和基体表面薄层熔化、混合而形成物理状态、组织结构和化学成分不同的新的表层，从而提高表层的耐磨性、耐腐蚀性和高温抗氧化性等。

Laser irradiation can also be applied simultaneously with laser. As a result of laser irradiation, thin layers of alloy elements and substrate surface melt and mix, new surface layer with different physical state, organizational structure and chemical composition is formed, so as to improve the wear resistance, corrosion resistance and high temperature oxidation resistance of the surface layer.

如在碳钢表面预涂一定配比的 W、V、Cr、C 元素混合粉末，经激光表面合金化后，可在钢表面获得类似 W18Cr4V 高速钢的成分、组织与性能，从而大大提高表面硬度与耐磨性。它特别适用于工件的重要部位的表面处理。

For example, the carbon steel surface pre-coating a certain ratio of W, V, Cr, C elements mixed powder, after laser surface alloying, the composition, structure and performance similar to W18Cr4V high-speed steel can be got, so as to greatly improve the surface hardness and wear resistance. It is especially suitable for surface treatment of important parts of workpiece.

激光表面非晶态处理是指金属表面在激光束辐照下熔化并快速冷却，熔化的合金在快速凝固过程中来不及结晶，从而在表层形成厚度为 1~10μm 的非晶相。

Laser surface amorphous processing refers to the metal surface under the laser irradiation is melt and rapid cooled. There is not enough time for molten alloy crystallization. The surface layer thickness of 1~10μm of amorphous phase is formed in the process of rapid solidification.

这种非晶相薄层不仅具有高强度、高韧度、高耐磨性和高耐腐蚀性，而且具有独特的电磁性和氧化性。

The thin layer of amorphous phase not only has high strength, high tenacity, high abrasion resistance and high corrosion resistance, but also has a unique electric magnetic and oxidizing.

激光气相沉积是以激光束作为热源在金属表面形成金属膜，通过控制激光的工艺参数可精确控制膜的形成。用这种方法可以在普通材料上涂敷与基体完全不同的具有各种功能的金属或陶瓷，节省资源效果明显。

Laser vapor deposition uses laser beam as heat source to form a metal film on the metal surface. The formation of the film can be precisely controlled by controlling the technological parameters of laser. By this method, metal, or ceramic with various functions completely different from the substrate can be coated on ordinary materials, which can save resources obviously.

B 离子束表面处理技术（Ion Beam Surface Treatment Technology）

离子束表面处理是把所需要元素的原子电离成离子，并使其在几十至几百千伏的电压下进行加速，进而轰击零件表面，使离子注入表层一定深度的真空处理工艺技术，从而改变材料表面层的物理化学和力学性能。

Ion beam surface treatment refers to a kind of vacuum treatment technology in which the atoms of the required elements are ionized into ions. The beam is accelerated under the voltage of dozens to hundreds of kilovolts to bomb the surface of parts, so that the ion is injected to surface with a certain depth, so as to change the physical, chemical, and mechanical properties of the surface layer of materials.

离子束注入技术的优缺点：

The advantages of ion implantation surface treatment are as follows:

(1) 离子注入是一个非热力学平衡过程，注入离子能量很高，可以高出热平衡能量的 2~3 个数量级。因此，原则上讲，周期表上的任何元素，都可注入任何基体材料内。

Ion beam surface treatment is a non-thermodynamic equilibrium process. The energy of ion implantation is very high and higher than the heat balance energy 2 to 3 orders of magnitude.

Therefore, in principle, any element on the periodic table can be injected into any base material.

（2）离子注入表层与基体材料无明显界面，使力学性能在流入层至基材为连续过渡，保证了流入层与基材之间具有良好的动力学匹配性，与基体结合牢固，避免了表面层的破裂与剥落。

There is no obvious interface between the ion - implanted surface layer and the substrate material, so that the mechanical properties are continuously transitive between the inflow layer and the substrate. A good dynamic matching between the inflow layer and the substrate can be ensured, and the injection layer is firmly bound to the base body to avoid the surface layer of fracture and peeling.

（3）注入元素的种类、能量、剂量均可选择，用这种方法形成的表面合金，不受扩散溶解度的经典力学参数的影响。

The type, energy and dose of injected elements can be selected. The surface alloy formed by this method is not subject to the classical mechanical parameters of diffusion solubility.

（4）离子注入为常温真空表面处理技术，零部件经表面处理后，无形变、无氧化，能保持原有尺寸精度和表面状态，特别适合于高精密部件的最后工艺。

Ion beam surface treatment is a kind of normal temperature vacuum surface treatment technology. There is no deformation, invisible change, no oxidation, can maintain the original dimension accuracy and surface state after surface treatment, especially suitable for the final process of high precision parts.

与其他表面处理技术相比，离子束注入技术也存在一些缺点：设备昂贵，成本较高。故目前主要用于重要的精密关键部件。

There are some disadvantages of ion beam surface treatment compared with other surface treatment technologies. The equipment of ion beam injection technology is expensive. Therefore, it is mainly used for important precision key components at present.

另外，离子注入层较薄，如 10 万电子伏的氮离子注入 GCr15 轴承钢中的平均深度仅为 0.1μm，这就限制了它的应用范围，受到真空室尺寸的限制。

In addition, the ion implantation layer is very thin, for example, the average depth of nitrogen ion implantation in GCr15 bearing steel of 100000 electron volts is only 0.1μm, which limits its application range and limits the dimension of vacuum chamber.

C 电子束表面处理技术（Electron Beam Surface Treatment Technology）

电子束由电子枪阴极发射后，在加速电压的作用下，速度高达光束的 2/3。高速电子束经电磁透镜聚焦后辐照在待处理工件的表面。

After the beam is fired from the cathode of the gun, the speed of the beam is up to 2/3 of that of the beam under the action of the accelerating voltage. The high - speed electron beam is irradiated on the surface of the workpiece after being focused by electromagnetic lens.

当高速电子束照射到金属表面时，电子能达到金属表面一定深度，与基体金属的原子核及电子发生相互作用。电子与原子核碰撞可看作为弹性碰撞，因此能量传递主要通过电子与金属表层碰撞而完成的。

When the high-speed electron beam hits the metal surface, the electron can reach a certain depth on the metal surface and interact with the nucleus and electrons of the base metal. The collision of electrons with the nucleus can be seen as an elastic collision, so the energy transfer is mainly completed by the collision of electrons with the metal surface.

所传递的能量立即以热能形式传给金属表层电子，从而使金属表层温度迅速升高。

The transfer of energy is immediately transferred to the surface of the metal in the form of heat, which causes the surface temperature of the metal to rise rapidly.

电子束加热与激光加热不同，激光加热时金属表面吸收光子能量，激光并未穿过金属表面。目前电子束加速电压达125kV，输出功率达150kW，能量密度达10MW/m，这是激光器无法比拟的。因此，电子束加热的深度和尺寸比激光大。

The heating of electron beam differs from that of laser. The laser does not pass through the metal surface in the heating of laser beam and the metal surface absorbs photon energy. At present, the accelerating voltage of electron beam is up to 125kV, the output power is up to 150kW, and the energy density is up to 10MW/m, which is incomparable by laser. As a result, the electron beam is heated to a greater depth and dimension than a laser.

电子束表面处理的主要特点有：加热和冷却速度快；零件变形小；与激光表面处理相比，使用成本低；能量利用率高；处理在真空中进行减少了氧化、氮化的影响，可得到纯净的表面处理层；不论形状多么复杂，凡是能观察到的地方就可用电子束处理。

Main features of electron beam surface treatment are as followings. Heating and cooling speed are fast. The deformation of parts is small. The cost is low compared with laser surface treatment. The energy utilization rate is high. Treatment is finished in the vacuum. The impact of oxidation and nitriding can be reduced. A pure surface treatment layer can be got. Wherever the shape can be observed, however complex, it can be treated with an electron beam.

常用的电子束表面处理技术有：电子束表面淬火、电子束表面重熔、电子束表面合金化和电子束表面非晶态处理。

Common electron beam surface treatment technology include electron beam surface quenching, electron beam surface remelting, electron beam surface alloying and electron beam surface amorphous treatment.

4.2.7 金属扣合技术（Metal Stitch Repair Technology）

金属扣合技术是利用扣合件的塑性变形或热胀冷缩的性质将损坏的零件连接起来，以达到修复零件裂纹或断裂的目的。这种技术可用于不易焊补的钢件、不允许有较大变形的铸件，以及有色金属的修复，对于大型铸件如机床床身、轧机机架等基础件的修复效果更为突出。

In terms of the technique of metal stitch, also called metal buckle, the plastic deformation or thermal expansion and cold contraction of the buckle are used to connect the damaged parts in order to repair the cracks or fracture of the parts. This technology can be used for repairing steel parts that are not easy to be repaired by welding, casting parts that are not allowed to have large deformation, and non-ferrous metal repair, especially for large castings, such as machine bed, mill stand and other basic parts.

扣合件一般由高强度合金材料或 10 号、15 号、20 号钢制成。如修复气缸盖、机座、螺旋桨等的裂纹。尤其是铸铁零件不易用焊补方法修复，用此法更为合适。金属扣合工艺与黏结剂配合使用，不仅提高了连接效果，而且可增强密封性。

The buckle is usually made of high strength alloy material or steel of 10#, 15# and 20# to repair cylinder head, frame, propeller cracks, etc. Especially, it is suitable to repair cast iron parts which are not easy repaired by welding. The use of the metal stitch technology in combination with the binder not only improves the connection effect, but also enhances the tightness.

4.2.7.1 金属扣合法的优点（Advantages of Metal Stitch Repair Technology）

金属扣合法的优点如下：

The advantages of metal buckle are as follows:

（1）修复工艺简单，成本低。所需设备和工具极为简单，可完全采用手工作业，而且修复工作一般不受场地限制，可就地施工。不需拆卸机件进车间，因此可以节省大量费用和时间。

The repair process is simple and cost is low. The required equipment and tools are very simple, and can be finished completely by manual operation. The repair work is generally not limited by the site. It can be finished on-site. There is no need to remove the parts into the workshop, which can save a lot of money and time.

（2）不破坏原有精度。修复是在常温下进行的，不会引起机件变形。

The original accuracy is not damaged. The repair work is carried out at normal temperature without causing deformation of the parts.

（3）修复质量可靠。用这种方法修复的机件有足够的强度和良好的密封性能。

Reliable repair quality can be got. The parts repaired by this method have sufficient strength and good sealing performance.

（4）波形槽分散排列，扣合件（波形槽）分层装入，逐片铆击，避免了应力集中。

The situation of scattered arrangement of corrugated grooves, layered loading of buckle parts (corrugated grooves), riveting piece by piece can avoid stress concentration.

4.2.7.2 金属扣合技术的分类（Classification of Metal Stitch Repair Technology）

金属扣合技术可分为强固扣合法、强密扣合法、优级扣合法和热扣合法四种。

Metal fastening technology can be divided into strong buckle method, strong dense buckle method, superior buckle method and heat buckle method.

在实际应用中，可根据具体情况和技术要求，选择其中一种或多种联合使用，以达到最佳效果。

In practical application, one or more combinations can be selected according to specific conditions and technical requirements to achieve the best effect.

A 强固扣合法（Strong Buckle Method）

强固扣合法或称波浪键扣合法。强固扣合法是先在垂直于损坏零件的裂纹或折断面上，加工出若干个一定形状和尺寸的凹槽（波形槽），然后把形状与波形槽相吻合的高强度材料制成的扣合件（波形键）镶入槽中，并在常温下铆击波形键，使其产生塑性变形而充满波形槽腔，甚至使其嵌入零件基体之内。这样由于波形键的凸缘和波形槽相互扣合，便将损坏的零件重新牢固地连接成一体（如图 4.23 所示）。

Strong buckle method refers to the process that a number of a certain shape and dimension of the waveform grooves are machined and perpendicular to the crack or broken the surface of the damaged parts. Then the waveform keys with the same shape with waveform grooves made of high strength materials are inserted into the grooves, and waveform keys are riveting hammered at room temperature to produce plastic deformation and embedded within the base body. In this way, the waveform keys and waveform grooves mutual buckle to connect the damaged parts into a whole (as shown in Figure 4.23).

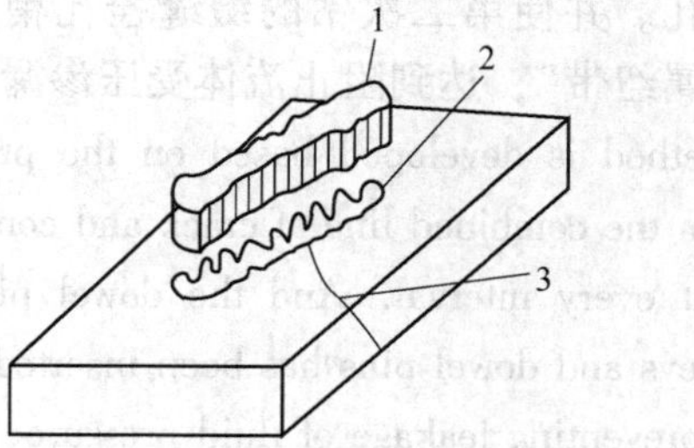

图 4.23　强固扣合法

Figure 4.23　Strong buckle

1—波形键（waveform key）；2—波形槽（waveform key groove）；3—裂纹（crack）

强固扣合法适用于修理裂纹处壁厚在 8~45mm，有一般强度要求的零件。采用频率高、冲击力小（控制压缩空气压力）的铆钉枪来铆击波形键。由两端向中间轮换对称铆击，最后铆击靠近裂纹凸缘时不宜过紧，以免撑开裂纹。

This method is suitable for repairing the parts with the wall thickness of 8~45mm with general strength requirement. The riveting gun with high frequency and small impact is used to rivet waveform keys. The waveform keys are symmetrically riveted from both ends to the middle. And finally the connection near to the edge should not be too tight, so as to avoid cracking.

先铆凸缘，后铆凸缘连接部分。先用圆弧冲头铆击中心部，再用平冲头铆击各部边缘。控制波形键的铆紧度，达到充分冷作硬化，从而提高抗拉强度。

The flange should be riveted first. The flange connection part is riveted later. The center should be riveted with arc punch first. Then the edge is riveted with flat punch. The waveform keys riveting degree should be controlled. The full cold hardening should be achieved. Thus the tensile strength can be enhanced.

强固扣合工艺过程如下：

The process of strong fastening is as follows:

（1）在零件上裂纹两端钻止裂孔，防止裂纹扩展。

Drill stop-crack holes at both ends of cracks in parts to prevent crack growth.

（2）设计并确定在零件裂纹处的波形槽位置。

Design and determine the position of waveform grooves at the crack of the part.

（3）利用专用钻模板和手电钻加工出波形槽。

Use special drill template and hand electric drill to drill the waveform grooves.

(4) 将波浪键镶嵌入波形槽中（可预先在槽中涂抹胶黏剂），用手锤敲击波浪键使之充满槽腔，将裂纹拉紧。

Insert waveform keys into the waveform grooves. Adhesive can be applied on the waveform keys in advance. Waveform keys are hit with a hand hammer to fill the waveform grooves and tighten the crack.

B 强密扣合法（Strong Dense Buckle Method）

强密扣合法基于强固扣合工艺原理。在两波形键之间、裂纹或折断面的结合线上，每间隔一定距离加工缀缝栓孔。并使第二次钻的缀缝栓孔稍微切入已装好的波形键和缀缝栓，形成一条密封的“金属纽带”，达到阻止流体受压渗漏的目的，如图 4.24 所示。

Strong dense buckle method is developed based on the principle of strong fastening. Many dowel pin holes are drilled on the combined line of crack and connection line between two waveform keys at a certain distance at every interval. And the dowel pin holes second drilled should be inserted in those waveform keys and dowel pins has been inserted to form a sealed “metal bond” in order to achieve the goal of preventing leakage of fluid pressure, as shown in Figure 4.24.

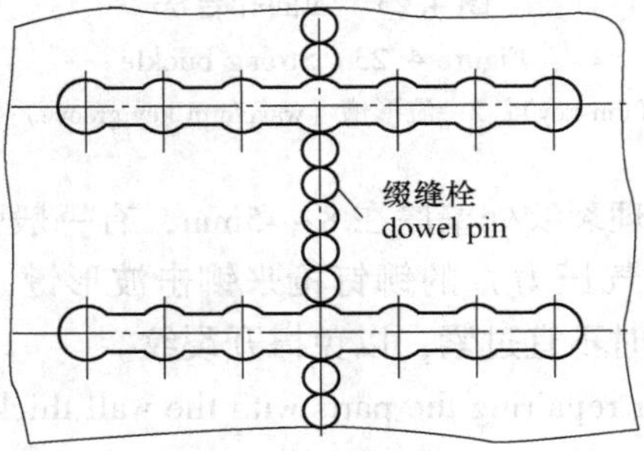

图 4.24 强密扣合

Figure 4.24 Strong dense buckle

采用强密扣合法修复有裂纹的零件时，在裂纹上可安装密封螺钉，也可安装密封圆柱销。前者适用于承受低压的有裂纹的零件，后者则适用于承受高压的有裂纹零件。密封螺钉可选用 M3 ~ M8 规格；圆柱销直径可选用 3 ~ 8mm，长度均与波浪键的厚度相同。此外，它们的材料亦同于波浪键，但不重要的零件也可选用低碳钢或紫铜等较软材料。

Sealing screw or sealing cylinder pin can be installed on the crack when strong dense buckle method is used to repair the parts with crack. The former is suitable for crack parts under low pressure, while the latter is suitable for crack parts under high pressure. M3 ~ M8 screw can be chosen as sealing screw. The diameter of cylinder pin is 3 ~ 8mm, and the length is the same as the thickness of waveform key. In addition, their material is the same as the waveform key, but low carbon steel or copper and other soft materials can be used for the parts not important.

强密扣合法不仅满足零件的强度要求，而且满足零件的密封性要求，所以可用来修理有密封要求的裂纹零件，例如承受高压的柴油机气缸套和气缸盖、压力容器等。

Strong dense buckle method not only meets the strength requirements of the parts, but also meets the sealing requirements of the parts, so it can be used to repair cracked parts with sealing requirements, such as the diesel engine cylinder liner and cylinder head, pressure vessel, etc.

C 优级扣合法（Superior Buckle Method）

优级扣合法也称加强扣合法或加强块扣合法，是在垂直于裂纹或断裂面的修复区上加工出一定形状的空穴，然后将形状、尺寸相同的钢制加强件镶入空穴中，在零件与加强件再加缀缝栓，使其一半嵌在加强件上，另一半在零件基体上。

Superior buckle method refers to the process that a certain shape of hole is machined perpendicular to the crack on the repair area, then the steel reinforcement part with the same shape, dimension of the hole is inserted into the hole. Dowel pins are added between the parts and the steel reinforcement parts to make half of dowel pin locates in the steel reinforcement part, the other half in the parts.

必要时还可再加入波形键，主要用于承受高载荷的厚壁机件，通常壁厚超过45mm。加强块的形状各异，有矩形、十字形和X形等，依机件和裂纹的情况选用，如图4.25所示。

Waveform keys still can be added when it is necessary. Waveform keys are mainly used for thick wall parts under high load, usually the wall thickness is more than 45mm. There are many shapes of reinforcement parts, including rectangular, cruciform and X-shape and can be chosen depending on the mechanism and crack, as shown in Figure 4.25.

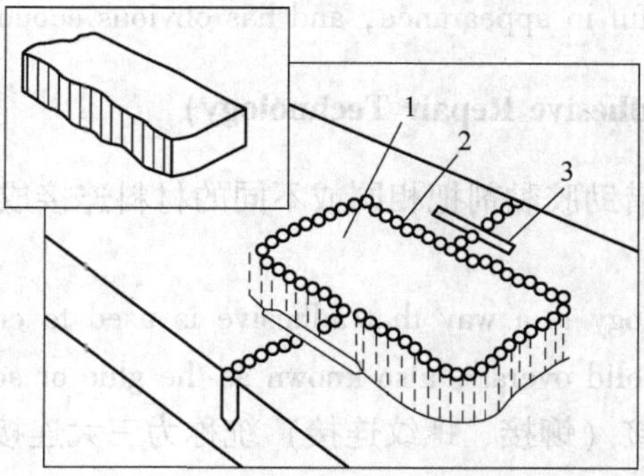

图4.25 优级扣合法

Figure 4.25 Superior grade buckle

1—加强件（strengthen part）；2—缀缝栓（dowel pin）；3—波形键（waveform key）

D 热扣合法（Heat Buckle Method）

热扣合法是利用金属热胀冷缩的原理，将选定的具有一定形状的扣合件经加热后放入机件损坏处已加工好的与扣合件形状相同的凹槽中，扣合件在冷却过程中产生收缩，将破裂的机件重新密合，如图4.26所示。这种方法比其他扣合法更为简便实用，多用来修复大型飞轮、齿轮和重型设备的机身等。

The heat buckle method is based on the principle of thermal expansion and contraction for metal. The buckle part with a certain shape is heated and put into the groove with the same shape of the buckle part. The buckle part contract during the cooling process and re-connect the broken parts, as shown in Figure 4.26. This method is simpler and more practical than other buckle methods, and is often used to repair the large flywheels, gears and house of heavy equipment.

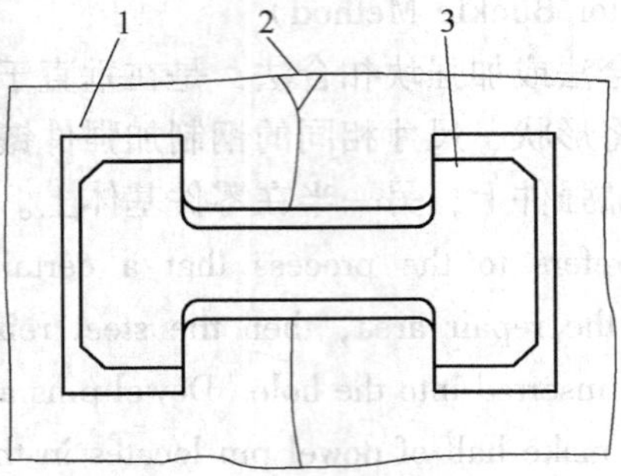

图 4.26 热扣合法

Figure 4.26 Heat buckle

1—零件（part）；2—裂纹（crack）；3—扣合件（buckle part）

扣合法目前已在各行各业越来越普遍地应用。实践证明，采用扣合法修复，质量可靠，精度能够得到保证，工艺成熟简便，成本低廉，外形美观，具有明显的经济效益。

Metal stitch/buckles technologies are already widely used in all walks of life. It has been proved by practice that the method of buckle repair is reliable in quality and accuracy, simple in process, low in cost, beautiful in appearance, and has obvious economic benefits.

4.2.8 粘接修复技术（Adhesive Repair Technology）

粘接修复技术是一种借助胶黏剂把相同或不同的材料连接成为一个连续牢固整体的方法，也称为胶接或粘合。

Adhesive repair technology is a way that adhesive is used to connect the same or different materials into a continuous solid overall, also known as the glue or adhesive.

粘接同焊接、机械连接（铆接、螺纹连接）统称为三大连接技术。

Adhesive , welding, mechanical connection (riveting, threaded connections) are collectively referred to as the three major connection technologies.

粘接的优点如下：

Advantages of adhesive are followings:

（1）不受材质的限制，相同材料或不同材料、软的或硬的、脆性的或韧性的各种材料均可粘接，且可达到较高的强度。

Not limited by the material, the same material, or different materials, soft or hard, brittle or toughness of all kinds of materials can be adhesive, and can achieve higher strength.

（2）粘接时的温度低，不会引起基体（或称母材）金相组织发生变化或产生热变形，不易出现裂纹等缺陷。因而可以修复铸铁件、有色金属及其合金零件、薄件及微小件等。

The adhesive temperature is low. Adhesive will not cause the microstructure changes or thermal deformation of the base part. There are not crack and other defects. Therefore, adhesive can repair cast iron, nonferrous metal and its alloy parts, thin parts, and mini parts.

（3）粘接工艺简便易行，不需要复杂设备，节省能源，成本低廉，生产率高，便于现场修复。

Adhesive process is simple and easy to operate. There is no need for complex equipment. It is a method of energy saving, low cost, high productivity.

（4）与焊接、铆接、螺纹连接相比，减轻结构质量20%~25%，表面光滑美观。

The mass of the structure can be reduced to 20%~25% compare to welding, riveting, thread connection. The surface of parts after adhesive is smooth and beautiful.

（5）粘接还可赋予接头密封、隔热、绝缘、防腐、防振，以及导电、导磁等性能。两种金属间的胶层还可防止电化学腐蚀。

Adhesive can also give connection seal, heat insulation, insulation, corrosion protection, anti-vibration, and electrical, magnetic and other properties. The adhesive layer between the two metals also prevents electrochemical corrosion.

粘接的缺点如下：

Disadvantages of adhesive are followings:

（1）不耐高温。

Adhesive is not able to withstand high temperature.

（2）粘接强度不高（与焊接、铆接相比）。

The strength of adhesive is not high (compared with welding, riveting).

（3）使用有机胶黏剂尤其是溶剂型胶黏剂，存在易燃、有毒等安全问题。

The organic adhesives, especially solvent-based adhesives are flammable, toxic and not safe.

（4）有机胶受环境条件影响易变质，抗老化性能差。其寿命由于使用条件不同而差异较大。

Organic glue is perishable by environmental conditions. Aging performance is poor. Its life varies greatly due to different service conditions.

（5）胶接质量尚无可行的无损检测方法，靠严格执行工艺来保证质量，因此应用受到一定的限制。

There is no feasible nondestructive testing method for adhesive bonding quality check. Its application is limited and the quality is ensured by strict execution of technology.

粘接的应用有下列几个方面：

Adhesive applications are used following areas:

（1）用结构胶粘接修复断裂件。

Broken parts are repaired with structural glue.

（2）用于补偿零件的尺寸磨损。

Used to compensate dimension wear of parts.

（3）用于零件的密封堵漏。

Used for leaking stoppage and sealing of parts.

（4）以粘代焊、代铆、代螺、代固等。

Welding, riveting, snail, fixing, etc. are replaced by adhesive.

（5）用于零件的防松紧固。

For the locking and fastening of parts.

（6）用粘接代替离心浇铸，制作滑动轴承的双金属轴瓦。

Centrifugal casting was replaced by bonding to produce bimetal bearing bushing for sliding bearing.

4.2.9 研磨技术（Grinding Repair Technology）

研磨是精密和超精密零件精加工的主要方法之一，是在精加工，如精车、精磨或精洗加工后的超精加工。研磨加工可使零件获得极高的尺寸精度、几何形状和位置精度，最高的表面粗糙度等级以及提高配合精度。

Grinding is the main method of precision and ultra-precision parts finishing. It is a superfinishing process after fine turning, fine grinding, or fine milling. The high dimensional accuracy and geometrical shape and position accuracy, and the surface roughness with highest precision of the parts can be got by grinding. The fit precision can be improved.

零件的内外圆表面、平面、圆锥面、斜面、螺纹面、齿轮的齿面及其他特殊形状的表面均可以采用此种方法进行加工。柴油机燃油系统中的三对精密偶件：柱塞-套筒偶件、针阀-针阀体偶件、出油阀-出油阀座偶件的内外圆表面、圆锥面、平面在制造时都需要采用研磨进行精加工。

The inner and outer circular surface, plane, conical surface, bevel surface, screw surface, tooth surface of gear and other special shape surface can be processed by this method. Three pairs of precision coupling parts in diesel engine fuel system: plunger-sleeve coupling parts, needle valve-needle valve body coupling parts, oil valve-oil valve seat coupling parts inside and outside circular surface, conical surface and plane need to be processed can be machined by grinding.

在针阀-针阀体配合锥面磨损和柴油机的进排气阀配合键面磨损后均需采用研磨技术进行修复，使配合面恢复密封性能。

After needle valve-needle valve body matching taper and diesel engine inlet and exhaust valve matching key are worn, the grinding technology can be used for repairing to restore the sealing performance of the mating surface.

进行研磨的零件材料可以是经淬火或未经淬火的碳钢、合金钢、硬质合金，也可以是铸铁、铜及其合金等有色金属材料，或玻璃、水晶和塑料等非金属材料。

The hardened or unhardened carbon steel, alloy steel, hard alloy, or non-ferrous metal materials such as cast iron, copper and its alloy, or non-metallic materials such as glass, crystal and plastic can be repaired by grinding.

灵活的研磨技术是进行精密零件修理的有效方法，尤其是在备件缺乏、时间紧迫的情况下此法尤为重要。

Flexible grinding technology is an effective method for precision parts, especially in the case of lack of spare parts and the repair is pressed for time.

4.3 零件修复技术的选择（The Choice of Part Repair Technology）

零件维修技术的选择包括维修技术的选择原则、制定零件维修程序、维修费用和零件维修更换原则。

The choice of part repair technology includes principles for the selection of repair technology, work out parts repair procedure, maintenance costs and principle of parts maintenance and replacement.

4.3.1 修复技术的选择原则（Principles for the Selection of Repair Technology）

对于一个损坏的零件，可能有几种修复工艺可供选择，究竟选用哪一种较为合适，这是修理前必须慎重考虑的问题。

For a damaged part, there may be several repair processes available for selection, which is more appropriate to choose, it is a matter of careful consideration before repair.

一般来说，对于一个具体零件的修复过程应遵守以下基本原则：工艺合理、经济性好、生产可行。

Generally, a specific part of the repair process should comply with the following basic principles: the process should be reasonable, the economy should be good, the production should be feasible.

工艺合理是指该修复工艺能满足待修机械零件的技术要求。

The process should be reasonable means that the repair process can meet the technical requirements of mechanical parts to be repaired.

经济性好是在保证机械零件修复工艺合理的前提下，应考虑到所选择修复工艺的经济性。

The economy should be good means that the economy of the selected repair process should be taken into account under the premise of ensuring the reasonable repair process of mechanical parts.

生产可行是指选择修复工艺要结合企业现有的修复用装备状况和修复水平进行。

The production should be feasible means that the selection of repair technology should be carried out in combination with the existing repair equipment status and repair level of the enterprise。

4.3.2 零件修复工艺规程的制订（Work Out the Repair Procedure of Parts）

机械修复工艺规程是规定零件完成修复工艺过程和操作方法等的工艺文件。它是在具体的生产条件下，将最合理或较合理的修复工艺过程和操作方法，按规定的形式制成工艺文本，经审批后用来指导生产并严格贯彻执行的指导性文件。

Mechanical repair process rules are the process documents that specify the parts to complete the repair process and operation methods. It is under specific production conditions, the most reasonable or more reasonable repair process and operation method, according to the prescribed form into the process text, after the examination and approval used to guide production and strictly implement the implementation of the guidance document.

工艺规程有以下几方面的作用：

The process rules have the following functions:

（1）工艺规程是指导生产的主要技术文件。合理的工艺规程是在总结生产实践的基础上，依据工艺理论和工艺实验制定的。它体现了一个企业或部门的集体智慧。因此，严格按工艺规程组织生产是保证产品质量、提高生产效率的前提。实践证明，不按科学的工艺进行生产，往往会引起产品质量的严重下降，生产效率显著降低，甚至使生产陷入混乱。

The technological procedure is the main technical document to guide repair. The reasonable technological procedure is formulated on the basis of summarizing the repair practice, according to the technological theory and technological experiment. Therefore, to organize repair strictly according to the process rules is the premise to ensure the quality of repair and improve repair efficiency. Practice has proved that repair without scientific technology often leads to serious decline in product quality, significant reduction in repair efficiency, and even chaos in repair.

（2）工艺规程是生产组织管理工作、计划工作的依据。由工艺规程所涉及的内容可以看出，在生产管理中，产品投产前原材料及毛坯的供应、通用工艺装备、机械负荷的调整、专用工艺装备的设计与制造、作业计划的编排、劳动力的组织以及生产成本的核算等，都是以工艺规程作为依据的。

Process planning is the basis of repair organization management, planning by the procedure, it can be seen that the content of the involved in the repair management, the supply of raw material and blank before a product is released to production, general process equipment, the adjustment of the mechanical load, special process equipment design and manufacturing, the work plan arrangement, the organization of labor and production cost accounting, etc., are based on process planning.

（3）工艺规程是新建或改建工厂或车间的基本资料。在新建、扩建或改造工厂或车间时，只有依据工艺规程和生产纲领，才能正确地确定生产所需要的机床和其他设备的种类、规格和数量，确定车间面积、机床布置、生产工人的工种、等级和数量及辅助部门的安排等。

Technological procedure refers to the basic data of a newly built or rebuilt factory or workshop. The types, specifications and quantities of machine tools and other equipment needed for production, determine workshop area, machine tool layout, job type, grade and quantity of production workers and arrangement of auxiliary departments can be correctly finished in the newly built, expanded or renovated factory or workshop according to the technological procedure and production program.

制订零件修复工艺规程的目的是保证修理质量及提高生产率和降低修理成本。

The purpose of drawing up the parts repair procedure is to ensure the repair quality and improve productivity and reduce the repair cost.

制订零件修复工艺规程的步骤：

The steps to work out the repair procedure of parts:

（1）调查研究（零件、设备和工艺）（Investigation and study（parts, equipment and process））。

（2）确定修复方案（择优）（Determine the repair plan（select the best one））。

（3）制订修复工艺规程（Drawing up technical rules for repair）。

4.3.3 维修费用（Maintenance Costs）

费用是机械设备设计中十分重要且比较复杂的因素。虽然可靠性、维修性与费用之间不存在总的函数关系，但是，为提高可靠性，应使用优质材料、故障率低的高级零部件，而这又需要更多的时间和资金。通常，应发展一种具有相当高的可靠性和维修性且总费用最低的机械设备。

Cost is a very important and complicated factor in the maintenance and repair of mechanical equipment. Although there is no general functional relationship between reliability, maintainability and cost, advanced parts with high quality materials and low failure rate should be used to improve reliability, which requires more time and money. In general, a machine with high reliability and maintainability and the lowest total cost should be developed.

一方面由于研究和设计费用将随可靠性的提高而增加，制造时因需要较精密复杂的生产加工设施和较新的加工工艺，生产费用也随之增加；另一方面因可靠性提高，维修和备件费用却相应减少。因此，机械设备总费用曲线就会有一个最低点，该点所代表的就是寿命周期费用的最佳化。

On the one hand, the cost of research and design will increase with the improvement of reliability. On the other hand, the cost of maintenance and spare parts is correspondingly reduced due to the improvement of reliability. Therefore, the total cost curve of mechanical equipment will have a minimum point, which represents the optimization of life cycle cost.

维修性在机械设备设计中很重要，如果把可靠性保持在一个适当的水准，提高维修性将会带来实质性的效果。可靠性和维修性均要立足于经济性才能得以存在，对这三项要素应进行综合平衡。

Maintainability is very important in the maintenance and repair of mechanical equipment. If the reliability is kept at an appropriate level, the improvement of maintainability will bring substantial effect. Reliability and maintainability must exist based on economy.

维修的直接费用包括：日常维修、检查、修理（指修复突然故障和劣化）等费用。

The direct cost of maintenance and repair includes: daily maintenance, check, repair（refers to the repair of sudden failure and deterioration）and other costs.

维修的间接费用是指停产、准备等损失的费用。

The indirect cost of maintenance and repair refers to the cost of loss such as suspension of production and preparation.

机械设备发生故障和损坏，如果用较长时间去维修，则直接费用较低。若短时间投入大量的维修手段进行总检修，则花费多。至于间接费用，显然与维修时间成正比，停机时间越长，损失越大。

If the mechanical equipment with failure and damage takes a longer time to repair, the direct cost is lower. If a large number of maintenance and repair means are used to carry out the general overhaul within a short time, the cost is higher. As for indirect costs, they are clearly proportional to maintenance and repair time, and the longer the downtime, the greater the damage.

采用何种维修方式都要考虑到直接费用和间接费用，并使它们保持平衡。对于定期维修而言，不论有无故障，定期地将全部可换的零部件进行更换。未到更换期发生故障的零部件，则坏一个更换一个。这时的费用是定期维修费用加上维修周期内因发生故障而进行修理的费用。

All maintenance methods should take into account direct and indirect costs and keep them in balance. For regular maintenance, all parts and components that can be replaced regularly, no matter there is no failure. Parts that is fail before the replacement period will be replaced one by one. At this time the cost is the cost of regular maintenance plus maintenance cycle due to the failure of the cost of repair.

如果对于发生故障的零部件，损坏一个修复一个，即采用个别更换式的修理费用应包括零部件的费用、更换的人工费，以及维修期间因停工而造成的损失等。

If the damaged parts are repaired one by one, that is to say, the cost of individual replacement shall include the cost of parts, replacement labor cost, and the loss caused by shutdown during maintenance and repair.

可见，不同的维修方式，其维修费用的计算是不相同的。

It can be seen that the calculation of maintenance cost is not the same with different maintenance methods.

4.3.4 零部件维修更换的原则（Principle of Parts Maintenance and Replacement）

在维修中，经常遇到零部件是更换还是修复的问题。它对机械设备的可靠性、维修内容、工作量、计划性和经济性都有重要影响。

In the maintenance, parts are often replaced or repaired. It has an important impact on the reliability, maintenance content, workload, planning and economy of mechanical equipment.

确定零部件修换应考虑的因素：

The factors to be considered in the repair or replacement of parts are followings:

（1）对机械设备精度的影响。应按工作精度决定修换零部件。例如机床床身导轨、主轴及轴承等基础零件磨损严重，对工作精度影响较大，应进行修换；若磨损不严重，估计能满足下一个维修周期使用，则可不修换。

Influence on the precision of mechanical equipment. The parts should be repaired or replaced according to the working accuracy. For example, the machine bed guide rail, spindle and bearing and other basic parts wear seriously, which has a greater impact on the working accuracy, should be repaired or replaced; If the wear is not serious, it is estimated to meet the next maintenance cycle, then do not repair.

（2）对完成预定使用功能的影响。当零部件不能完成预定的使用功能时，应考虑修换。零部件虽然还能完成预定的功能，但降低了机械设备的性能，应考虑修换。

Influence on the completion of specified functions. When the parts cannot complete the specified functions, they should be replaced or repaired. Although the parts can still complete the specified function, the performance of the mechanical equipment reduces, so it should be considered to repair or replace.

（3）对生产率的影响。由于机械设备本身的原因增加了工作时间和劳动强度，自动装置失常、废品率上升、生产率明显下降，应考虑修换有关零部件。

Influence on productivity. If the working time and labor intensity increase due to the mechanical equipment, and if the automatic device is out of order, the rejection rate rises, and the productivity decreases obviously, the parts should be replaced or repaired.

（4）对零部件强度和刚度的影响。有些零部件强度和刚度已达到最小值，因磨损严重而可能破坏时，必须修换；零件刚度下降引起精度降低或破坏了正常工作条件时，应修换；对安全性要求高的零件，当强度下降，出现裂纹等，必须修换。

Influence on the strength and stiffness of parts. The strength and stiffness of some parts have reached the minimum value. If the parts may be damaged because of serious wear, the parts must be replaced or repaired. When the accuracy is reduced or the normal working condition is damaged due to the stiffness decrease of the part, it should be repaired or replaced. For parts with high safety requirements, when the strength drops and cracks appear, they must be repaired or replaced.

（5）磨损条件。当零部件磨损急剧上升，破坏了正常配合、啮合和传动，使效率下降、发热量大增、润滑失常、表面拉伤、咬住或断裂，应修换；如果零件表面硬化层被磨掉或脱皮、配合间隙过大、表面拉伤、疲劳剥蚀等，应及时修换有关零件，保护较贵重的配对件。

Wear condition. When the wear and tear of parts increases sharply, which damages the normal coordination, engagement and transmission, and causes the efficiency decrease, heat increase, lubrication disorder, surface strain, seizure or fracture, it shall be repaired or replaced. If the surface hardening layer of the parts is worn off or peeling, fit gap is too large, surface strain, fatigue denudation, etc., the relevant parts should be timely repaired or replaced to protect the more expensive parts.

（6）经济分析。一些零件接近失效极限但未达到，要考虑其剩余寿命对今后使用维修的影响，从经济上分析是否应修换和什么时候修换最适宜。

Economic analysis. Some parts are close to but have not reached the failure limit. The influence of its remaining life on future service and maintenance should be taken into account. The economic analysis is on whether and when the repair or replacement is the most appropriate.

对允许事后维修的零件可使用到其寿命极限；对维持不了一个修理间隔期，而修换拆装劳动量大、停机损失大、对生产影响大，又无替代时，应修换；对安全要求高而又无可靠的监测手段，应定期修换，不过多考虑其剩余寿命。

Parts that allow after-service maintenance may be used to the limit of their life. If the parts cannot maintain a repair interval, and repair or disassembly labor volume is large, shutdown loss is large, the production impact is large, and no replacement, the parts should be replaced or repaired. If there are high safety requirements and no reliable means of monitoring, the parts should be regularly replaced or repaired, without over considering its remaining life.

对已决定修换的零部件，要进一步确定是更换还是修复再用。一般情况下，对失效零件进行修复可节约材料、减少配件的加工和备件的储备量，从而降低成本。

For the parts that have been decided to be replaced or repaired, the replacement or repair shall be further determined. In general, the repair of failed parts can save materials, reduce the processing of parts and the storage of spare parts, thus the cost can be reduced.

修复零部件应满足以下要求：

The repaired parts shall meet the following requirements:

（1）零部件修复后必须恢复和保持原有的各项技术要求，包括尺寸、形位公差、表面质量、足够的强度和刚度等。

The original technical requirements, including dimensions, shape and position tolerance, surface quality, sufficient strength and stiffness, etc., must be restored and maintained after the parts are repaired.

（2）修复零部件要求有足够的使用寿命，其耐用度至少应维持一个修理间隔期，防止因故障和事故而停机。

Repair parts shall have sufficient service life, and their durability shall be maintained at least one repair interval to prevent downtime due to failures and accidents.

（3）修复零部件要考虑经济效益，在保证前两项要求的前提下降低维修成本。在比较修复、更换成本和使用寿命时，若相对维修成本低于相对新制作件成本，应考虑修复。

Economic benefits should be considered when parts and components are repaired, and maintenance costs should be reduced under the premise of guaranteeing the first two requirements. In the comparison of repair, replacement cost and service life, if the relative repair cost is lower than the relative new production cost, the parts should be repaired.

零部件是修复还是更换也受到其他一些因素的影响，如本单位的制造和修复工艺水平、备件的储备、采购条件的影响、计划停机时间的限制。

Whether the parts are repaired or replaced is also affected by some other factors, such as the manufacturing and repair process level of the unit, the reserve of spare parts, the impact of procurement conditions, planned downtime restrictions.

4.3.5 典型零件修复技术的选择（Repair Technology of Typical Parts）

典型零件包括轴类件、孔类件和齿轮类零件等。

Shaft parts, parts with hole and gears are typical parts to be repaired.

4.3.5.1 轴的修复技术选择（Repair Technology of Shaft）

轴类零件是机械设备中比较常见的主要零件之一，轴类零件常见的失效形式是磨损，比如：滑动轴承的轴颈及外圆柱面、装滚动轴承的轴颈及静配合面、轴上键槽、轴上螺纹，以及圆锥孔、销孔等部位的磨损。常用的轴的修复技术选择见表 4.1。

Shaft parts are one of the most common main parts in mechanical equipment. The common failure mode of shaft parts is worn, such as journal of sliding bearing and external cylinder, journal of mounting rolling bearing and static mating surface, keyway on shaft, thread on shaft and wear of tapered hole, pin hole and other parts. The repair technology of shaft is shown in Table 4.1.

表 4.1 轴类件的修复技术

Table 4.1 Repair technology of shaft parts

零件磨损部分（The wear position of shaft）	修理方法（Repair method）	
	达到公称尺寸（Reach nominal dimension）	达到修配尺寸（Reach repair dimension）
滑动轴承的轴颈及外圆柱面（Journal and external cylindrical surface of sliding bearing）	镀铬、镀铁、金属喷涂、堆焊（Chrome plating，iron plating，metal spraying，surfacing）	达到修配尺寸（Reach repair dimension）
安装滚动轴承的轴颈及静配合面（Journal and static mating surface where installing rolling bearing）	镀铬、镀铁、堆焊、滚花、化学镀铜（0.05mm 以下）（Chrome plating，iron plating，surfacing，knurling，electroless copper plating（below 0.05mm））	
轴上键槽（Shaft keyway）	堆焊修理键槽，转位新铣键槽（Surfacing repair keyway，mill a new keyway for change）	键槽加宽，不大于原宽度的 1/7，重配键（The keyway is widened to no more than 1/7 of the original width）
花键（Spline）	堆焊重铣或镀铁后磨（最好用振动焊）（Surfacing for heavy milling or post-milling iron plating（preferably vibration welding））	
轴上螺纹（Screw thread on shaft）	堆焊，重车螺纹（Surfacing，rethreading）	车成小一级螺纹（Make a small thread）
外圆锥面（Outside the cone surface）	刷镀、喷涂、加工（Brush plating，spraying，machining）	磨到较小尺寸恢复几何精度（Grind to smaller dimensions to regain geometric accuracy）
圆锥孔（Cone hole）	刷镀、加工（Brush plating，machining）	磨到较大尺寸恢复几何精度（Grinding to a larger dimension regain geometric accuracy）
轴上销孔（Pin hole on the shaft）		重新铰孔（Back reaming）
扁头、方头及球面（Flat head，square head and sphere）	堆焊（Surfacing）	加工修整几何形状（Machining and trimming geometry shape）
一端损坏（One end of a shaft is broken）	切去损坏的一段，焊接一段，加工至标称尺寸（Cut off the damaged section，weld one section，process to nominal dimension）	
弯曲（Bending）	校正（Correction）	

4.3.5.2 孔类零件的修复技术（Repair Technology of Parts With Hole）

孔类零件的失效形式也主要表现为磨损失效，主要有孔径的磨损，键槽的磨损，螺纹孔、圆锥孔、销孔、凹坑、球面窝和小槽以及平面组成的导槽等的磨损。孔的修复技术选择见表 4.2。

The failure mode of parts with hole is also mainly manifested as wear failure, mainly including the wear of aperture, the wear of keyway, screw hole, taper hole, pin hole, pit, spherical socket and small groove, and the guide groove composed of plane, etc. The repair technology selection of parts with hole is shown in Table 4.2.

表 4.2 孔类件的修复技术

Table 4.2 Repair technology of parts with hole

零件磨损部分（The wear position of hole part）	修理方法（Repair methods）	
	达到公称尺寸（Reach nominal dimension）	达到修配尺寸（Reach repair dimension）
孔径（The inner surface of the hole）	镗大镶套、堆焊、刷镀、粘补（Boring to a large dimension，add a sleeve，surfacing，brush plating，bonding）	镗孔或磨孔，恢复几何精度（To bore or grind a hole，regain geometric accuracy）
键槽（Keyway）	堆焊修理，转位另插键槽（Surfacing repair，mill another keyway）	加宽键槽、另配键（Widen the keyway，match the other key）
螺纹孔（Screw hole）	镶螺塞，可改变位置的零件转位重钻孔（Add a screw plug，drill another hole at other position）	加大螺纹孔至大一级的标准螺纹（Increase screw hole to a larger degree of standard thread）
圆锥孔（Taper hole）	镗孔后镶套（Insert a sleeve after boring）	刮研或磨削恢复几何精度（Scraping or grinding to regain geometric accuracy）
销孔（Pin hole）	移位重钻，铰销孔（Drill another pin hole，ream pin hole）	铰孔、另配销子（Reaming，match the other pin）
凹坑、球面窝及小槽（Pit，Spherical socket and small groove）	铣掉重镶（Mill off and drill another）	扩大修整形状（Enlarge the dimension）
平面组成的导槽（Guide groove composed of plane）	镶垫板、堆焊、粘补（Set plate，surfacing，bonding）	加大槽形（Increase the dimension of guide groove）

4.3.5.3 齿轮类零件的修复技术选择（Repair Technology of Gears）

齿轮类零件是机械传动中的主要零件，也是比较容易产生失效的零件。齿轮类零件的失效主要发生在轮齿、齿角、孔径、键槽及离合器爪等部位。齿轮类零件的修复技术选择见表4.3。

Gear is the main part in mechanical transmission, and it is also easy to produce failure parts. The failure of gear mainly occurs in gear tooth, tooth angle, aperture, keyway and clutch claw. The selection of gear repair technology is shown in Table 4.3.

表4.3 齿轮类零件的修复技术选择

Table 4.3 Gear repair technology selection

零件磨损部分（Wear part）	修理方法（Repair methods）	
	达到公称尺寸（Up to nominal dimension）	达到修配尺寸（Reach repair dimension）
轮齿（Tooth of gear）	（1）利用花键孔，镶新轮圈插齿（Use spline holes, insert a new gear ring）； （2）齿轮局部断裂，堆焊加工成形（Partial fracture of gear surfacing machining）； （3）内孔镀铁后磨（Grinding after iron plating inner hole）	大齿轮加工成负修正齿轮（硬度低，可加工者）（Large gear is machined into negative correction gear（low hardness, machinable））
齿角（The tooth angle）	（1）对称形状的齿轮调头倒角使用（Change the head of gear with symmetrical shape）； （2）堆焊齿角后加工（Machining after surfacing welding tooth angle）	锉磨齿角（File grinding angle）
孔径（The inner surface of the hole）	镶套、镀铬、镀镍、刷镀、堆焊，然后加工（Insert set, chrome plating, nickel plating, brush plating, surfacing, and then the machining）	磨孔配角（Grinding hole）
键槽（Keyway）	堆焊加工或转位另开键槽（Surfacing processing or transposing another keyway）	磨孔配轴（Grinding hole, another shaft is matched）
离合器爪（Clutch）	堆焊后加工（Surfacing after processing）	加宽键槽、另配键（Widen the keyway, another key is matched）

思考题（Questions）

4.1 名词解释（Definition of noun）：

修理尺寸法（Repairing dimension），附加零件修理法（additional part repair method），局部更换修理法（local replacement repair method），焊接维修技术（welding maintenance technology），钎焊（brazing, surfacing），热喷涂（thermal spraying），热喷焊（thermal spray welding）

4.2 机械零件的修复技术主要有哪些？

How many main repair technologies of mechanical parts are there?

4.3 机械加工修理技术有哪些主要方法？

How many machining repair technologies are there?

4. 4 机械加工修理技术与制造新零件的区别有哪些?

What are the difference of machining a new part and machining an old part?

4. 5 修理尺寸法分为哪两种方法?

What are the two methods of repairing dimension?

4. 6 画图说明轴颈的镶套修理方法。

Illustrate the repair method of the insert sleeve on the shaft journal with a drawing.

4. 7 画图说明支架孔的镶套修理方法。

Illustrate the repair method of bracket holes with a drawing.

4. 8 应用附加零件修理法时应注意哪些问题?

What problems should be paid attention to when applying the additional part repair method?

4. 9 画图说明轴类零件的局部更换修理法。

Illustrate the method of local replacement repair method of shaft parts with a drawing.

4. 10 画图说明齿轮类零件的局部更换修理法。

Illustrate the local replacement repair method of gear parts with a drawing.

4. 11 画图说明键槽的局部更换修理法。

Illustrate the local replacement repair method of keyway with a drawing.

4. 12 画图说明螺孔的局部更换修理法。

Illustrate the local replacement repair method of screw hole with a drawing.

4. 13 局部更换零件法与附加零件修理法有何区别?

What is the difference between local replacement repair method and additional part repair method?

4. 14 塑性变形修复零件的方法一般有哪些?

How many methods of plastic deformation can be used to repair parts?

4. 15 冷压校正和冷作校正有什么区别?

What is the difference between cold pressure correction and cold working correction?

4. 16 焊接维修技术主要有哪些?

How many welding maintenance technologies are there?

4. 17 热喷涂和热喷焊的区别有哪些?

What are the differences between thermal spraying and thermal spray welding?

4. 18 常见的金属电镀修复技术有哪些?

How many electroplating technologies are there?

4. 19 工件表面强化技术的原理是什么?

What are the principle of surface hardening technology?

4. 20 激光表面处理技术与其他表面处理相比，具有哪些特点?

What are the characteristics of laser surface treatment compared with other surface treatments?

4. 21 激光表面淬火对材料的性能有哪些影响?

What are the effects of laser surface quenching on the properties of materials?

4. 22 金属扣合技术可分为哪四种方法?

What are the four methods of metal buckle technology?

4. 23 三大连接技术是指哪三项技术?

What are the three major connectivity technologies?

4. 24 如何修复轴类件?

How to repair the shaft parts when they are worn?

4. 25　如何修复孔类件？

How to repair the hole part when they are worn?

4. 26　如何修复齿轮类件？

How to repair the gear part when they are worn?

4. 27　零件修复工艺的选择原则是什么？

What is the selection principle of parts repair process?

5 机械设备的装配与安装

Chapter 5 Assembly and Installation of Mechanical Equipment

机器是由许多零件组合起来的，它们之间有一定的联系，必须把零件按一定的技术要求装配，才能组成一台机器而进行工作。机器在使用过程中难免产生磨损，必须进行检修和重新装配。研究零件的装配，特别是标准件的装配是非常重要的。

Machine is composed of many parts. There are some connections between the parts. The parts must be assembled according to certain technical requirements. In addition, wear is inevitable while machine is using. So machine must be repaired and reassembled. The assembly of parts is very important, especially the assembly of standard parts.

组成机器的零部件可以分为两大类，一类是标准零部件，如轴承、齿轮、联轴器、键销、螺栓等，它们是机器配置的主要组成部分，并且数量很多。另一类是非标准件，在机器上数量不多。在研究零部件的装配时，主要讨论标准件的装配问题。

The parts that make up the machine can be divided into two major categories. One is the standard parts, such as bearings, gears, couplings, key pins, bolts, etc., which are the main components of the machine configuration. There are a large number of standard parts. The other kind of non-standard parts are not many in the machine. When studying the assembly of parts, the assembly of standard parts is mainly discussed.

零件的连接分为固定连接和活动连接两种，固定连接是使零件或部件固定在一起而没有任何相对运动的连接。如螺栓连接、键连接、焊接及过盈配合等。活动连接是用以连接零件和部件，使它们保持一定性质的相对运动。例如，滑动轴承和轴颈的连接，保证轴的正常转动，齿轮和齿轮间的连接，保证动力的正常传递等。

The connection of parts can be divided into two kinds: fixed connection and moving connection. A fixed connection is one that holds parts or parts together without any relative motion, such as bolt connection, key connection, welding, and interference fit, etc. Movable connection is used to connection parts and components to maintain a certain property of relative motion. For example, the connection between sliding bearing and shaft journal is a kind of movable connection to ensure the normal rotation of shaft, the connection between gears to ensure the normal transmission of power, etc.

5.1 机械设备的装配（Assembly of Mechanical Equipment）

机器的性能和精度是在机械零件加工合格的基础上，通过良好的装配工艺实现的。机器装配的质量和效率在很大程度上取决于零件加工的质量。机械装配又对机器的性能有直接的影响，如果装配不正确，即使零件加工的质量很高，机器也达不到设计的使用要求。

The performance and precision of the machine are achieved by good assembly technology on the basis of qualified machining of mechanical parts. The quality and efficiency of machine assembly largely depend on the quality of parts processing. Mechanical assembly has a direct impact on the performance of the machine, if the assembly is not correct, even if the quality of parts machining is very high, the machine cannot meet the design requirements.

不同的机器的机械装配要求与注意事项各有特色，但机械装配需注意的共性问题通常有以下两个方面：保证装配精度和重视装配工作的密封性。

Different machines have their own characteristics of mechanical assembly requirements and points for attention, but the common problems of mechanical assembly usually have the following aspects: ensuring assembly accuracy and ensuring the sealing property of assembly work.

保证装配精度是机械装配工作的根本任务。装配精度包括配合精度和尺寸链精度。

Ensuring assembly precision is the fundamental task of mechanical assembly. Assembly accuracy includes fitting accuracy and dimension chain accuracy.

在机械装配过程中，大部分工作是保证零部件之间的正常配合。为了保证配合精度，装配时要严格按公差要求。目前常采用的保证配合精度的装配方法有完全互换法、分组选配法、调整法和修配法。

In the process of mechanical assembly, the task of most of the work is to ensure the normal fit between parts. In order to ensure the fitting accuracy, the assembly should be strictly in accordance with the tolerance requirements. At present, the commonly used assembly methods to ensure the matching accuracy include complete interchange method, group selection method, adjustment method and repairing before matching method.

（1）完全互换法。相互配合零件公差之和小于或等于装配允许公差，零件可以完全互换。对零件不需挑选、调整或修配就能达到装配精度要求。该方法操作方便，易于掌握，生产效率高，便于组织流水作业，但对零件的加工精度要求较高。适用于配合零件较少、批量较大的场合。

Complete interchange method. The sum of tolerance of mutual matching parts is less than or equal to the allowable tolerance of assembly, and parts are completely interchangeable. This method is called complete interchange method. The assembly precision can be achieved without selecting, adjusting or repairing. This method is easy to operate, easy to master, production efficiency is high and easy to organize flow operation, but it requires high machining accuracy of parts. It is suitable for small and large batch parts.

（2）分组选配法。分组选配法是将零件的加工公差按装配精度要求的允许偏差放大若干倍，对加工后的零件测量分组，对应的组进行装配，同组可以互换。零件能按经济加工精度制造，配合精度高，但增加了测量分组工作。适用于成批或大量生产，配合零件少，装配精度较高的场合。

Group selection method. In group selection method, the machining tolerance of parts is enlarged by several times according to the allowable deviation required by assembly accuracy. The machined parts after being measured are grouped and assembled by the corresponding groups. The parts in the same group can be interchanged. The parts can be manufactured according to economic machining accuracy. The accuracy of assembly is high. But the measurement and group is needed. It is suitable for the assembly of parts with large batch or mass production, few mating parts and high assembly accuracy.

（3）调整法。选定配合副中一个零件制造成多种尺寸作为调整件，装配时利用它来调整到装配允许的偏差；或采用可调装置如斜面、螺纹等改变有关零件的相互位置来达到装配允许偏差，这种方法被称为调整法。零件可按经济加工精度制造，能获得较高的装配精度。但装配质量在一定程度上依赖操作者的技术水平。调整法可用于多种场合。

Adjustment method. One part of the matching pair is selected to be machined into a variety of sizes as adjustment parts, which can be used to adjust the allowable deviation during assembly. Or adjustable device such as bevel, thread, etc., are used to change the relative position of the parts in order to achieve the allowable assembly deviation. This method is called adjustment method. The parts can be manufactured with economic machining accuracy, and can obtain higher assembly accuracy. But the assembly quality depends on the operator's technical level to some extent. Adjustment method can be used in a variety of occasions.

（4）修配法。在某零件上预留修配量，在装配时通过修去其多余部分达到要求的配合精度。这种方法零件可按经济加工精度加工，并能获得较高的装配精度。但增加了装配过程中的手工修配和机械加工工作量，延长了装配时间，且装配质量在很大程度上依赖工人的技术水平。适用于单件小批量生产，或装配精度要求高的场合。

Repairing before matching method. The amount of repairing is reserved on a part to achieve the required matching precision by repairing the excess parts during assembly. This method is called repair and match method. Parts can be machined with economical precision and higher assembly precision can be obtained. However, it increases the workload of manual repairing and mechanical machining in the assembly process. The assembly time is prolonged. And, assembly quality largely depends on the technical level of workers. Suitable for single small batch production, or assembly precision requirements of the occasion.

上述四种装配方法中，分组选配法、调整法、修配法过去采用得比较多，采用完全互换法比较少。但随着科学技术的进步，生产的机械化、自动化程度不断提高，零件较高的加工精度已不难实现。由于现代化生产的大型、连续、高速和自动化的特点，完全互换法已在机械装配中日益广泛采用，成为发展方向。

Above all, the group selection method, adjustment method and repairing before matching method are mostly used but the complete interchange method is seldom used. However, with the progress of science and technology, the degree of mechanization and automation of production continues to be improved, and the higher machining accuracy of parts is not difficult to achieve.

The complete interchange method has been widely used in mechanical assembly and has become the development direction due to the large, continuous, high speed and automatic characteristics of modern production.

机械装配过程中，有时虽然各配合件的配合精度满足了要求，但是累积误差所造成的尺寸链误差可能超出设计范围，影响机器的使用性能。因此装配后必须进行检验，当不符合设计要求时，重新进行选配或更换某些零部件。

In the process of mechanical assembly, the dimensional chain errors caused by cumulative errors may exceed the design scope and affect the machine's performance although the matching accuracy of each matching part meets the requirements. Therefore, checking must be carried out after assembly, and when it does not meet the design requirements, reselect or replace some parts.

在机械装配过程中，如果密封装置位置不当、选用密封材料和预紧程度不合适、密封装置的装配工艺不符合要求，都可能产生机械设备漏油、漏水、漏气等现象。轻则损失能源，造成环境污染，使机械设备降低或丧失工作能力；重则可能发生严重事故。

In the process of mechanical assembly, if the location of the sealing device is not appropriate, the selection of sealing materials and pre-tightening degree is not appropriate, the sealing device assembly process does not meet the requirements. This may produce mechanical equipment oil leakage, water leakage, air leakage and other phenomena, light loss of energy, environmental pollution, reduce or lose the working ability of mechanical equipment. Serious accidents may occur.

因此在装配工作中，对密封性必须给予足够重视。要恰当地选用密封材料，严格按照正确的工艺过程合理装配，要有合理的装配紧度，并且压紧要均匀。

Therefore, adequate attention must be paid to tightness in assembly work. Appropriate sealing materials should be chosen. The correct and reasonable process of assembly should be strictly executed. Reasonable assembly tightness of sealing should be reached and pressure on sealing should be uniform.

装配的工艺过程一般是：机械装配前的准备工作、装配、检验和调整。

Assembly process is generally the process of mechanical assembly before preparation, assembly, checking and adjustment.

5.1.1 过盈配合的装配（Assembly of Interference Fit Parts or Components）

采用过盈配合的目的主要是使配合零件的连接能承受大的轴向力、扭矩及动载荷，故零件的材料应能承受最大过盈所引起的应力。而配合零件的连接强度在最大过盈时应得到保证。

The purpose of using interference fit is to make the connection of the matching part able to bear large axial force, torque and dynamic load, so the material of the part should be able to bear the stress caused by maximum interference. And the connection strength of the matching parts or components should be guaranteed within the maximum interference.

常用的过盈配合的装配方法有常温下装配、热装配、冷装配、液压无键连接装配。

Common methods for interference fit assembly include room temperature assembly, heat assembly, cold assembly, and hydraulic keyless connection assembly.

5.1.1.1 常温装配（Room Temperature Assembly）

在室温下，采用打入法和压入法对过盈配合零件进行装配。通过敲打和按压来克服摩擦。打入力和压入力的确定应根据实验公式进行计算，如式（5.1）和式（5.2）所示。

At the room temperature, the parts or components with interference fit are assembled with hit method and press method. Friction is overcomed by beating and pressing. The force should be determined by calculation with experimental Formula (5.1) and Formula (5.2).

当轴、孔均为钢时（When the shaft and hole are steel）：

$$P = \frac{28\left[\left(\frac{D}{d}\right)^2 - 1\right] iL}{\left(\frac{D}{d}\right)^2} \tag{5.1}$$

当轴为钢、孔为铸铁时（When the shaft is steel and the hole is cast iron）：

$$P = \frac{42\left(\frac{D}{d} + 0.3\right) iL}{\frac{D}{d} + 6.35} \tag{5.2}$$

式中，P 为压入力，kN；i 为实测过盈量，mm；L 为配合面的长度，mm；D 为孔件外径，mm；d 为孔件内径，mm。

Where, P is pressure, kN, i is the measured interference, mm, L is length of fitting surface, mm, D is outer diameter of part with a hole, mm, d is inner diameter of part with a hole, mm.

为利于压入装配，可将计算的压入力再增大20%~30%，在压入配合件的孔和轴均涂以润滑油。表面粗糙度不低于$\sqrt{1.6}$，压入速度一般为2~4mm/s。

In order to make press-in assembly easy, the calculated press-in force can be increased by 20%~30%, and the holes and shafts of press-in fitting parts or components are coated with lubricating oil. The surface roughness is no less than *Ra* 1.6. The speed of pressing is 2~4mm/s.

5.1.1.2 热装配（Heat Assembly）

通过加热，使孔类件的孔径膨胀增大一定数值，再装配轴类件，使轴类件自由地送入孔中，待孔类件的孔冷却后，孔类件将轴类件紧紧地抱住。

The part with hole is heated in order to increase the diameter of hole so that the shaft can be easily assembled to the part with hole. The diameter of hole will decrease after cooled. The two parts or components can be assembled tightly together.

孔膨胀后的实际尺寸比轴大，否则无法装配。切忌出现装配没有结束，轴和孔就抱住的现象。

The actual size of the hole after expansion must be larger than that of the shaft, otherwise, it cannot be assembled. The phenomenon that the shaft and hole are held before the assembly is finished should be avoided.

热装主要用于没有压床或直径大和过盈量大的零件。

Heat assembly is mainly used for parts without press machine or the diameter and interference of part are all large.

A 加热温度的确定 (Temperature to be Heated)

常用加热温度可用式 (5.3) 计算。

The temperature to be heated can be determined by Formula (5.3).

$$t = \frac{3i}{K_d d} + t_0 \tag{5.3}$$

式中，t 为加热温度,℃；i 为实测过盈量，mm；K_d 为加热时孔材料线膨胀系数,℃$^{-1}$，$K_d = 12\times10^{-6}$℃$^{-1}$ (金属)；d 为未加热前孔的内径，mm；t_0 为室温,℃。

Where, t is the temperature to be heated, ℃, i measured interference, mm, K_d is coefficient of linear expansion of the part with hole, ℃$^{-1}$, for steel, $K_d = 12 \times 10^{-6}$ ℃$^{-1}$, d is inner diameter of hole before heated, mm, t_0 is room temperature, ℃.

B 加热方法和温度的测量 (Methods to Heat the Part and to Measure the Temperature)

对零件进行加热的方法有很多，如热浸加热法、氧-乙炔火焰加热法、木材或焦炭加热法、煤气加热法及电加热和电感应加热法。

There are many methods can be used to heat the part, such as hot dip heating method, oxygen-acetylene flame heating method, wood or coke heating method, gas heating method and electric heating and electric induction heating method.

测温材料可选择油类或有色金属。如：机油的闪点是 200~220℃，锡的熔点是 232℃，纯铅的熔点是 327℃；也可以用测温蜡笔、测温纸片及半导体点接触测温计测温。

Oil or non-ferrous can be chosen as temperature measuring materials. For example, oil flash point is 200~220℃, tin melting point is 232℃, pure lead melting point is 327℃. Temperature can also be measured with temperature crayon and temperature paper and semiconductor point contact thermometer.

C 零件尺寸的检测方法 (Methods to Check the Diameter of Part)

通常采用样杆检查加热后的孔径。样杆尺寸按实测过盈量大三倍制作，当样杆刚好能放入孔时，则加热温度正合适。样杆直径为 5mm，两头用砂轮打磨成锥形，操作手柄以 700mm 为宜，样杆和手柄焊在一起，如图 5.1 所示。

Sample rod usually used to check the inner diameter of the hole. The size of sample rod is made by three times of the measured interference. When the sample rod can just be put into the hole, the heating temperature is just right. The diameter of the sample bar is 5mm, and both ends are polished into a cone by grinding wheel. It is advisable to operate the handle 700mm. The sample bar and handle are welded together, as shown in Figure 5.1.

D 装配过程 (The Procedure of Assembly)

孔类件平放，水平仪找平，轴类件竖直吊起，挂线找正，迅速下落装配。

The part with hole is put horizontally, gradienter is used to find out if it is horizontal. The shaft is hung up vertically. A line with a weight is hung up to find out if it is vertical. The herringbone gear shaft is put down rapidly to finish assembly.

例 5.1 人字齿轮轴轴头热装齿形联轴器的外齿套公称尺寸为 $\phi800$，人字齿轮轴轴头公称尺寸为 $\phi500$，齿形联轴器的外齿套材质是 45 号钢，$k_d = 12 \times 10^{-6}$℃$^{-1}$，装配室温零

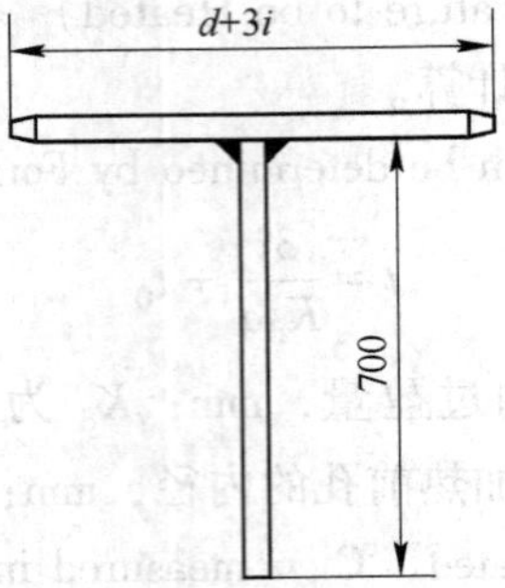

图 5.1 检测样杆

Figure 5.1 Sample rod

下 3℃，实测最大过盈量 i 为 0.42mm，计算加热温度，说明加热办法、测温检查方法和装配过程。

Example 5.1 An outer gear sleeve of gear coupling will be assembled with a $\phi800$ herringbone gear shaft with $\phi500$ shaft journal, the material of outer gear sleeve is steel, $k_d = 12 \times 10^{-6}℃^{-1}$, room temperature is 3℃ below zero, measured interference fit value i is 0.42mm, please calculate the temperature to be heated, describe the method of heating, the method of measuring temperature, the method of checking, and the procedure of assembly.

解（Solution）：

（1）计算加热温度（Calculate the temperature to be heated）：

$$t = \frac{3i}{k_d d} + t_0 = \frac{3 \times 0.42}{12 \times 10^{-6} \times 500}℃ + (-3)℃ = 207℃$$

（2）加热方法（The method of heating）：

外齿套用耐火砖垫好，下面均匀布置四个煤气烧嘴加热，外齿套上面盖大铁板，以便加热均匀。

The outer gear sleeve is put on the firebricks, some gas burner noses are put under the outer gear sleeve, a large iron plate is put on the outer gear sleeve to heat evenly.

（3）测温检查方法（The method of measuring temperature and the method of checking）：

半导体点接触测温计测温，采用样杆检查，样杆尺寸 $D = d + 3i = 500\text{mm} + 3 \times 0.42\text{mm} = 501.26\text{mm}$。

Semiconductor point contact thermometer is adopted to measuring temperature. Sample rod is made for checking. The diameter of sample rod is: $D = d + 3i = 500\text{mm} + 3 \times 0.42\text{mm} = 501.26\text{mm}$.

（4）装配过程（The procedure of assembly）：

外齿套平放，激光找平，人字齿轮轴竖直吊起，激光找正，迅速下落装配。

The outer gear sleeve is put horizontally, leveling is checked by laser. The herringbone-gear shaft is hung up vertically. laser is used to check whether it is vertical. The herringbone gear shaft is put down rapidly to finish assembly.

5.1.1.3 冷装配（Cold Assembly）

当带孔零件较大而压入零件较小时，采用加热孔零件法既不方便也不经济，则可采用把被压入零件低温冷却的方法使其尺寸缩小，然后迅速将此零件装入到带孔零件中去。

Cold assembly is adopted when part with a hole is big and the shaft is small. The shaft can be cooled to decrease its size to assemble with the part with a hole.

冷却温度可用式（5.4）计算。

The cooling temperature can be calculated by Formula (5.4).

$$t = \frac{(2 \sim 3)i}{K_a d} - t_0 \tag{5.4}$$

式中，t 为冷却温度，℃；i 为实测过盈量，mm；K_a 为冷却时孔材料线收缩系数，℃$^{-1}$；d 为被冷却件的公称尺寸，mm；t_0 为室温，℃。

Where, t is the temperature to be cooled, ℃, i is interference fit value by measured, mm, K_a is coefficient of linear shrink of the part with hole, ℃$^{-1}$, d is diameter of shaft before cooled, mm; t_0 is room temperature, ℃。

常用冷却剂及冷却温度如下：

Common coolants and their cooling temperature are following:

固体二氧化碳加酒精或丙酮：−75℃；

Solid carbon dioxide and alcohol or acetone: −75℃;

液氨：−120℃；

Liquid ammonia: −120℃;

液氧：−180℃；

Liquid oxygen: −180℃;

液氮：−190℃。

Liquid nitrogen: −190℃.

5.1.1.4 液压无键连接（Hydraulic Keyless Connection Assembly）

液压无键连接的原理是利用钢的弹性膨胀和收缩使套件紧箍在轴上所产生的摩擦力来传递动力。

The performance of elastic expansion and shrink of steel can be used to generate friction to transmit the power to shaft.

这种方法通常用于速度快、负荷大、需要经常拆卸的零部件。液压无键连接总成原理图如图5.2所示。

This method is usually used for parts or components with high speed, heavy load, and need to be disassembled frequently. The schematic diagram of hydraulic keyless connection assembly is shown in Figure 5.2.

如图5.2所示，当高压油液进入被连接件的连接面之间时，在油压作用下，孔件会产生弹性膨胀，轴件则会产生弹性压缩。通过轴向力的作用，将被连接件顺利安装到位。此时，渐渐给膨胀油缸卸压，在材料弹性作用下，孔与轴在恢复变形的过程中紧紧压合在一起，从而使连接面产生很大的弹性接触压力。为使装配顺利，常将连接面设计成圆锥面。

As shown in Figure 5.2, when high-pressure oil enters between the connecting surfaces of the connected parts, under the action of oil pressure, the hole parts will produce elastic expansion, while the shaft parts will produce projetiles compression. Through the action of axial force, the connecting piece will be successfully installed in place. At this point, gradually to the expansion cylinder pressure relief, under the action of material elasticity, the hole and shaft in the process of recovery of deformation tightly pressed together, so that the joint surface produces a lot of elastic contact pressure. In order to make assembly smooth, the connecting surface is often designed as a conical surface.

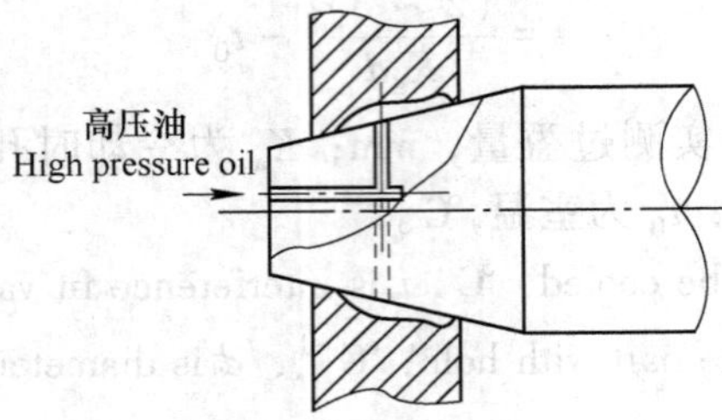

图 5.2　液压无键连接

Figure 5.2　The schematic diagram of hydraulic keyless connection assembly

5.1.2　联轴器的装配（Assembly of Coupling）

各种设备的旋转零件或部件都是靠轴来带动的，所以轴的装配质量对确保设备正常运行有很大的影响。

The rotating parts or components of various equipment are driven by the shaft. One shaft is connected to another shaft by a coupling. So the assembly quality of the coupling has a great impact on the normal operation of the equipment.

正确装配轴的基本要求是：

The basic requirements for the correct assembly of the shaft are as follows:

（1）轴与配合件间的组装位置正确，水平度、垂直度及同心度均应符合技术要求。

The assembly position between the shaft and the matching parts is correct. The levelness, verticality and concentricity should all meet the technical requirements.

（2）轴应均匀地支承在轴承上，转运轻松和平稳，并且保持位置的正确性。

The shaft should be evenly supported on the bearing, runs easily and smoothly, and maintain the correct position.

（3）轴上的轴承除一端定位外，其余轴承应有移动的余地，以适应轴的伸缩。

Only the bearing on one end of the shaft should be located. The rest of the bearings shall have space for movement to adapt to the expansion and contraction of the shaft.

（4）在装配轴前应对轴的轴颈部位进行清洗，对轴进行弯曲检查，对一般的机器，当轴的转速小于 500r/min 时，最大允许挠度为 0.3mm/m，大于 500r/min 时，则为 0.2mm/m。

Before assembling the shaft, the neck of the shaft should be cleaned. Bending checking of the shaft should be conducted. For general machines, the maximum allowable deflection is 0.3mm/m when the rotation speed of the shaft is less than 500r/min, and 0.2mm/m when the rotation speed is greater than 500r/min.

（5）在装配轴时，必须做好轴承的同心度、两根轴的垂直度或平等度以及轴与联轴器的同心度检查。

When assembling the shaft, it is necessary to check the concentricity of the bearing, the verticality or equality of the two shafts and the concentricity of the shaft and the coupling.

（6）轴承的同心度常用激光的方法来检查，两根轴的垂直度或平行度常用角尺、内径千分尺、弯规、激光和摇臂的测量方法。

The concentricity of the bearing is usually checked by the method of laser. The verticality or parallelism of the two shafts is usually measured by angle ruler, inside micrometer, bending gauge, laser and rocker arm.

联轴器用于连接不同机器或部件，将主动轴的运动及动力传递给从动轴。在装配各种联轴器时，总的要求是使连接的两根轴符合规定的同心度，保证它们的几何中心线相互重合。

Coupling is used to connect different machines or components to transfer the motion and power from the drive shaft to the driven shaft. When assembling various couplings, the general requirement is to make the two connected shafts conform to the specified concentricity and ensure that their geometric centerlines coincide with each other.

通常，从动轴应首先调整到正确的位置。因此，在装配联轴器时，应调整传动轴上的半联轴器。

Usually, the driven shaft should be adjusted to the correct position firstly. Therefore, the half-coupling on the drive shaft should be adjusted when the coupling is assembled.

联轴节的装配包括两方面：一是将轮毂装配到轴上，另一个是联轴节的找正和调整。

Assembly of coupling includes two steps: assembling the hub to the shaft and adjusting the alignment of the coupling.

5.1.2.1 联轴器的装配间隙（Coupling Assembly Clearances）

安装联轴器时，两个半联轴器应同轴。联轴器的装配间隙包括轴向间隙 S 和径向间隙 a（如图 5.3 所示）。

When coupling is installed, two half-couplings should be coaxial. The assembly clearances of coupling include the axial clearance S and radial clearance a (as shown in Figure 5.3).

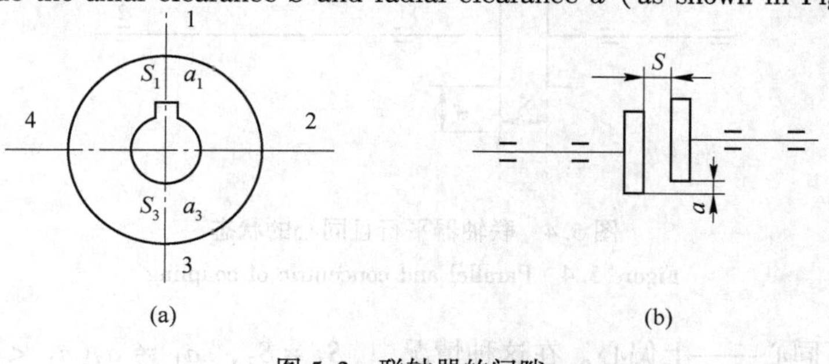

图 5.3 联轴器的间隙

Figure 5.3 Deviation of coupling

（a）半联轴节的端面示意图（Schematic diagram of the end face of the half coupling）；

（b）联轴器正视图（Front view of the coupling）

在图 5.3（b）中，左轴为从动轴（减速机输入轴、齿轮座输入轴或主机输入轴），右轴为待调整的驱动轴（减速机输出轴或电机输出轴）。根据轴向间隙 S 和径向间隙 a，可以将两个半联轴器组装在以下位置。

In Figure 5.3（b）, left shaft is the driven shaft, maybe the input shaft of reduction gears, gear box, or main machine. Right shaft is the drive shaft, maybe the output shaft of motor or reduction gears, to be adjusted later. Two half-couplings may be assembled on the following positions according to the axial clearance S and radial clearance a.

联轴器端面沿顺时针方向与垂直线、水平线相交为四个点，顶点记为点 1，其余三个点沿记为点 2、点 3、点 4（如图 5.3（a）所示）。

The end face of the coupling intersects the vertical line and the horizontal line as four points, and the top point is denoted as point 1, and the other three points is denoted as point 2, point 3 and point 4 along the clockwise direction（as shown in Figure 5.3（a））.

点 1 和点 3 的轴向间隙 S 和径向间隙 a 反映了联轴器的垂直偏差。点 2 和点 4 的轴向间隙 S 和径向间隙 a 反映了联轴器的水平偏差。

The axial clearance S and radial clearance a of point 1 and point 3 reflect the vertical deviation of coupling. The axial clearance S and radial clearance a of point 2 and point 4 reflect the horizontal deviation of coupling.

5.1.2.2 联轴器的偏斜（Deflections of Coupling）

下面根据点 1 和点 3 的轴向间隙 S 和径向间隙 a 对联轴器的垂直偏差进行分析。

The followings are an analysis of the vertical deviation of the coupling according to the axial clearance S and radial clearance a of point 1 and point 3.

（1）平行且同心。在这种情况中，$S_1 = S_3$，$a_1 = a_3$。图 5.4 给出了两个轴的位置。这是联轴器安装的正确位置。

Parallel and concentric. In this case, $S_1 = S_3$, $a_1 = a_3$. Figure 5.4 gives the position of two shafts. This is the correct position of coupling.

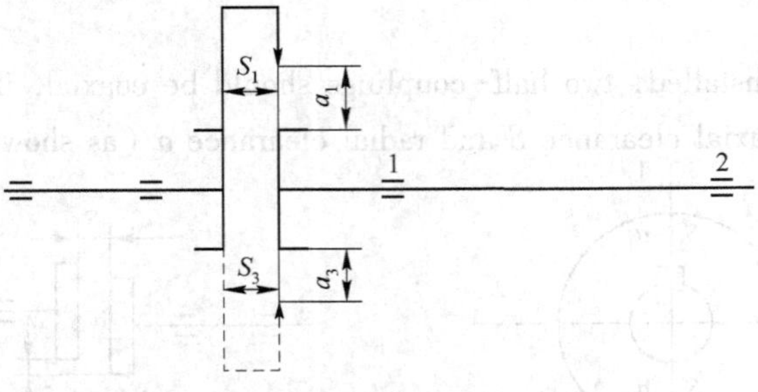

图 5.4 联轴器平行且同心的状态

Figure 5.4 Parallel and concentric of coupling

（2）平行不同心——上偏心。在这种情况中，$S_1 = S_3$，$a_1 \neq a_3(a_1 < a_3)$。图 5.5 给出了两个轴的位置。联轴器右侧的轴与左侧轴平行，但高于左侧轴。

Parallel and up decentration. In this case, $S_1 = S_3$, $a_1 \neq a_3(a_1 < a_3)$. Figure 5.5 gives the position of two shafts. The right shaft is parallel with the left shaft but higher than the left shaft.

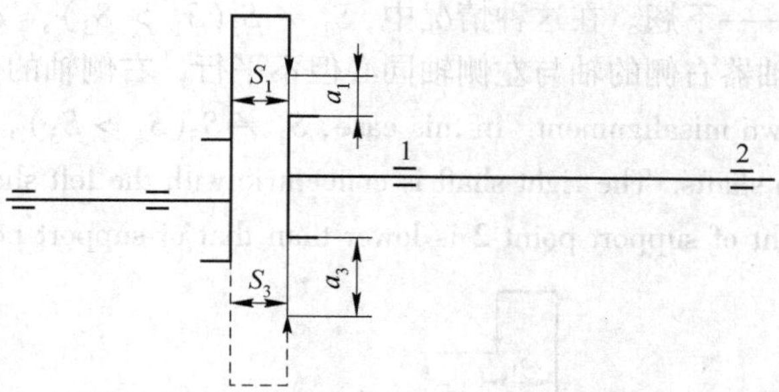

图 5.5　联轴器平行不同心——上偏心的状态

Figure 5.5　Parallel and up decentration of coupling

(3) 平行不同心——下偏心。在这种情况中，$S_1 = S_3$，$a_1 \neq a_3(a_1 > a_3)$。图 5.6 给出了两个轴的位置。联轴器右侧的轴与左侧轴平行，但低于左侧轴。

Parallel and down decentration. In this case, $S_1 = S_3$, $a_1 \neq a_3(a_1 > a_3)$. Figure 5.6 gives the position of two shafts. The right shaft is parallel with the left shaft but lower than the left shaft.

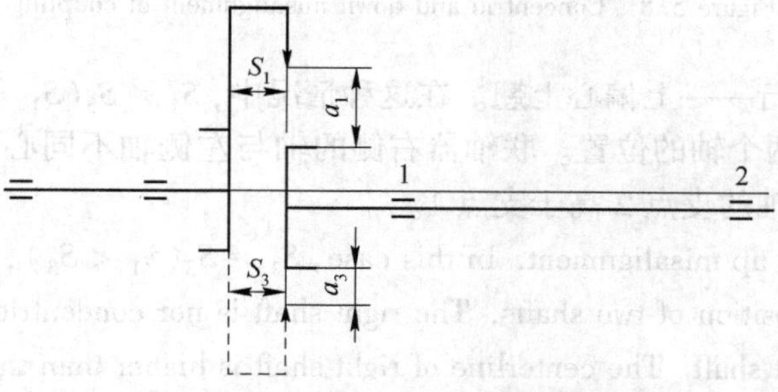

图 5.6　联轴器平行不同心——下偏心的状态

Figure 5.6　Parallel and down decentration of coupling

(4) 同心不平行——上翘。在这种情况中，$S_1 \neq S_3(S_1 < S_3)$，$a_1 = a_3$。图 5.7 给出了两个轴的位置。联轴器右侧的轴与左侧轴同心但不平行，右侧轴的支点 2 高于支点 1。

Concentric and up misalignment. In this case, $S_1 \neq S_3(S_1 < S_3)$, $a_1 = a_3$. Figure 5.7 gives the position of two shafts. The right shaft is concentric with the left shaft but not parallel with the left shaft. The height of support point 2 is higher than that of support point 1.

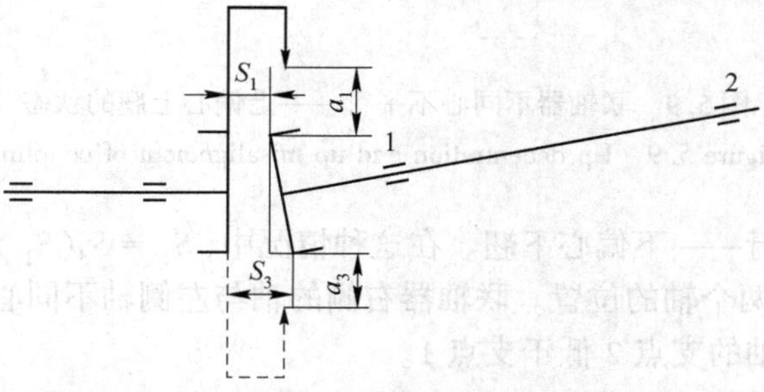

图 5.7　联轴器同心不平行——上翘的状态

Figure 5.7　Concentric and up misalignment of coupling

（5）同心不平行——下翘。在这种情况中，$S_1 \neq S_3(S_1 > S_3)$，$a_1 = a_3$。图 5.8 给出了两个轴的位置。联轴器右侧的轴与左侧轴同心但不平行，右侧轴的支点 2 低于支点 1。

Concentric and down misalignment. In this case, $S_1 \neq S_3(S_1 > S_3)$, $a_1 = a_3$. Figure 5.8 gives the position of two shafts. The right shaft is concentric with the left shaft but not parallel with the left shaft. The height of support point 2 is lower than that of support point 1.

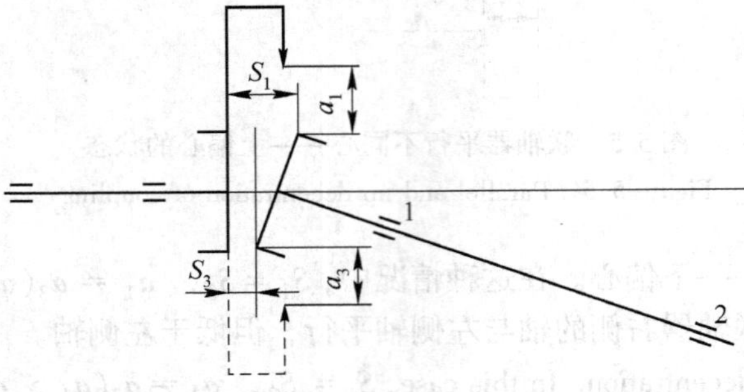

图 5.8 联轴器平同心不平行——下翘的状态

Figure 5.8 Concentric and down misalignment of coupling

（6）不同心不平行——上偏心上翘。在这种情况中，$S_1 \neq S_3(S_1 < S_3)$，$a_1 \neq a_3(a_1 < a_3)$。图 5.9 给出了两个轴的位置。联轴器右侧的轴与左侧轴不同心也不平行，右侧轴心高于左侧轴心，右侧轴的支点 2 高于支点 1。

Up concentric and up misalignment. In this case, $S_1 \neq S_3(S_1 < S_3)$, $a_1 \neq a_3(a_1 < a_3)$. Figure 5.9 gives the position of two shafts. The right shaft is nor concentric with the left shaft and not parallel with the left shaft. The centerline of right shaft is higher than that of the left shaft. The height of support point 2 is higher than that of support point 1.

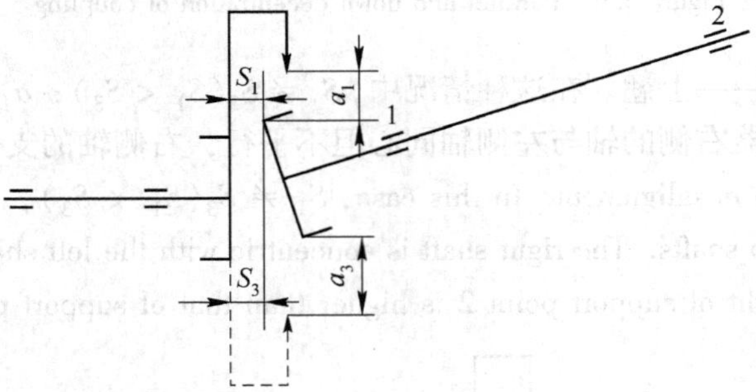

图 5.9 联轴器不同心不平行——上偏心上翘的状态

Figure 5.9 Up decentration and up misalignment of coupling

（7）不同心不平行——下偏心下翘。在这种情况中，$S_1 \neq S_3(S_1 > S_3)$，$a_1 \neq a_3(a_1 > a_3)$。图 5.10 给出了两个轴的位置。联轴器右侧的轴与左侧轴不同心也不平行，右侧轴心低于左侧轴心，右侧轴的支点 2 低于支点 1。

Down Concentric and down misalignment. In this case, $S_1 \neq S_3(S_1 > S_3)$, $a_1 \neq a_3(a_1 > a_3)$. Figure 5.10 gives the position of two shafts. The right shaft is nor concentric with the left shaft and not parallel with the left shaft. The centerline of right shaft is lower than that of the left shaft. The height of support point 2 is lower than that of support point 1.

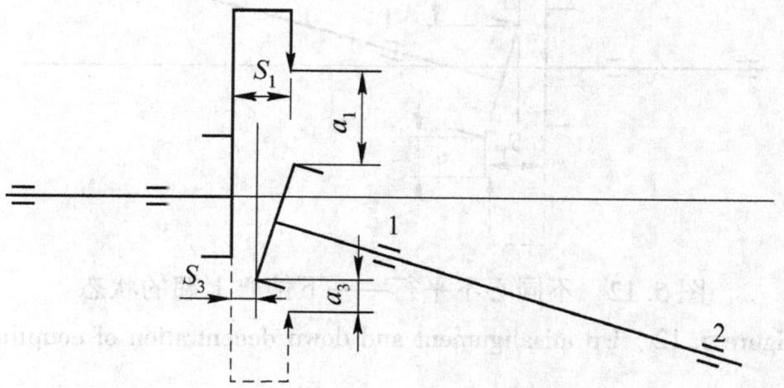

图 5.10　联轴器不同心不平行——下偏心下翘的状态
Figure 5.10　Down decentration and down misalignment of coupling

（8）不同心不平行——上偏心下翘。在这种情况中，$S_1 \neq S_3(S_1 > S_3)$，$a_1 \neq a_3(a_1 < a_3)$。图 5.11 给出了两个轴的位置。联轴器右侧的轴与左侧轴不同心也不平行，右侧轴心高于左侧轴心，右侧轴的支点 2 低于支点 1。

Down Misalignment and up decentration. In this case, $S_1 \neq S_3(S_1 > S_3)$, $a_1 \neq a_3(a_1 < a_3)$. Figure 5.11 gives the position of two shafts. The right shaft is nor concentric with the left shaft and not parallel with the left shaft. The centerline of right shaft is higher than that of the left shaft. The height of support point 2 is lower than that of support point 1.

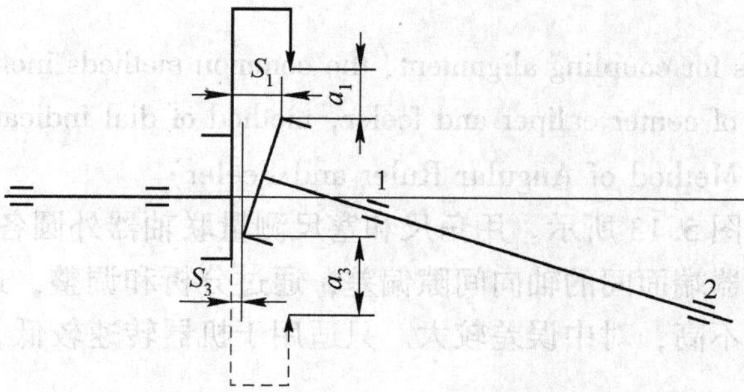

图 5.11　联轴器不同心不平行——上偏心下翘的状态
Figure 5.11　Down Misalignment and up decentration of coupling

（9）不同心不平行——下偏心上翘。在这种情况中，$S_1 \neq S_3(S_1 < S_3)$，$a_1 \neq a_3(a_1 > a_3)$。图 5.12 给出了两个轴的位置。联轴器右侧的轴与左侧轴不同心也不平行，右侧轴心低于左侧轴心，右侧轴的支点 2 高于支点 1。

Up misalignment and down decentration. In this case, $S_1 \neq S_3(S_1 < S_3)$, $a_1 \neq a_3(a_1 > a_3)$. Figure 5.12 gives the position of two shafts. The right shaft is nor concentric with the left shaft and not parallel with the left shaft. The centerline of right shaft is lower than that of the left shaft. The height of support point 2 is higher than that of support point 1.

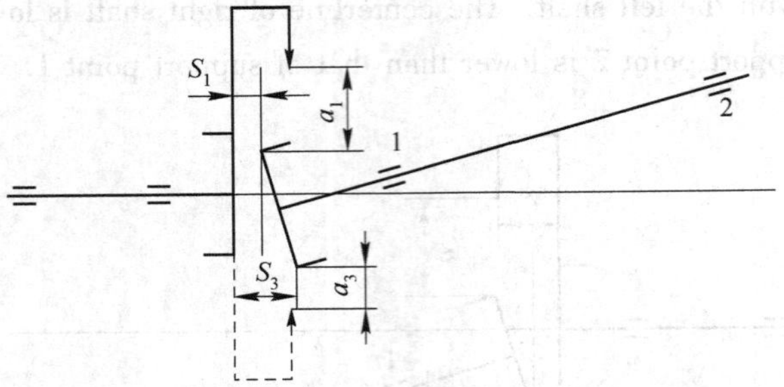

图 5.12　不同心不平行——下偏心上翘的状态

Figure 5.12　Up misalignment and down decentration of coupling

如果测量了点 1 和点 3 的轴向间隙 S 和径向间隙 a，就可以确定两个半联轴器的相对垂直位置。同理，如果测量了点 2 和点 4 的轴向间隙 S 和径向间隙 a，也可以确定两个半联轴器的相对水平位置。

If the axial clearance S and radial clearance a of point 1 and point 3 have been measured, the relative vertical position of two half-couplings can be determined. In the same way, if the axial clearance S and radial clearance a of point 2 and point 4 have been measured, the relative horizontal position of two half-couplings can be determined too.

5.1.2.3　联轴器装配间隙的测量方法（Methods of Measuring Coupling Assembly Clearance）

联轴器装配间隙的测量方法很多，主要包括：塞尺角尺法、中心卡塞尺法和百分表法。

There are many ways for coupling alignment, the common methods include method of angular ruler and feeler, method of center caliper and feeler, method of dial indicator.

A　塞尺角尺法（Method of Angular Ruler and Feeler）

角尺塞尺测量法如图 5.13 所示。用角尺和塞尺测量联轴器外圆各方位上的径向偏差，用塞尺测量两个半联轴器端面间的轴向间隙偏差，通过分析和调整，达到两轴对中。这种方法操作简单，但精度不高，对中误差较大。只适用于机器转速较低，对中要求不高的联轴器的安装测量。

Method of angular ruler and feeler is shown in Figure 5.13. The radial deviation on each point of the outer circle of the coupling is measured by the angular ruler and the feeler. The axial clearance deviation of each point on the face of the two half-couplings is measured by the feeler. This method is simple to operate, but the accuracy is not high, and the error of alignment is large. It is only suitable for the installation and measurement of coupling with low rotating speed and low requirement for alignment.

B　中心卡塞尺法（Method of Center Caliper and Feeler）

中心卡塞尺法优于角尺法塞尺法。中心卡的结构形式多种多样，应根据联轴器的结构选择合适的中心卡。中心卡没有统一规格，应该考虑测量和装卡的要求由钳工自行制作。

Method of center caliper and feeler is better than the method of angular ruler and feeler.

There are various structural forms of center caliper. According to the structure of the coupling, the appropriate center caliper should be selected. There is no uniform specification for the central card, so the central card shall be manmade according to the requirements of measuring and installing the center caliper.

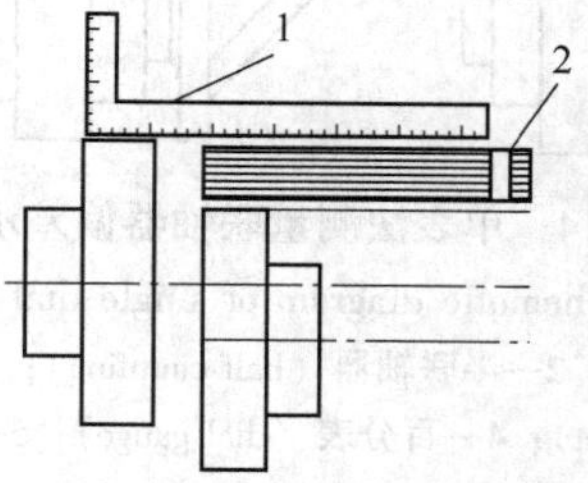

图 5.13 角尺和塞尺的测量方法

Figure 5.13 Measuring method of angular ruler and feeler

1—角尺（angular ruler）；2—塞尺（feeler）；

C 百分表法（Method of Dial Indicator）

百分表法是最好的方法。把专用的夹具（中心卡）或磁力表座装在作基准（常是装在主机转轴上）的半联轴器上，用百分表测量联轴器的径向间隙和轴向间隙的偏差值。此方法使联轴器找正的测量精度大大提高。

Method of dial indicator is the best method of all. The center caliper and magnetic dial indicator seat installed on the reference half-coupling (often installed on the shaft of main machine). The radial clearance and the axial clearance of the coupling can be read on the dial meters. This method greatly improves the measuring accuracy of coupling alignment.

常用的百分表测量方法有四种：单表法、双表法、三表法和五表法。

There are four common measuring methods of dial indicators: single dial indicator method, double dial indicators method, three dial indicators method, and five dial indicators method.

a 单表法（Single Dial Indicator Method）

单表法是近年来国外应用日益广泛的一种联轴器找正方法。这种方法只测定联轴器轮毂外圆的径向读数，不测量端面的轴向读数，测量操作时仅用一个百分表，故称单表法。其安装测量示意图如图 5.14 所示。

Single dial indicator method is widely used in the world in recent years. This method only measures the radial reading of the outer circle of the hub of the coupling, and does not measure the axial reading of the end face. Schematic diagram of single dial indicator method is shown in Figure 5.14.

此种方法用一块百分表就能判断两轴的相对位置并可计算出轴向和径向的偏差值。也可以根据百分表上的读数用图解法求得调整量。用此方法测量时，需要特制一个找正用表架，其尺寸、结构由两个半联轴器间的轴向距离及轮毂尺寸大小而定。表架自身质量要小，并有足够的刚度。表架及百分表均要求固紧，不允许有松动现象。

The relative positions of two shafts can be determined with one dial indicator, and the axial and radial deviation values of can be calculated. The adjustment values can also be obtained by

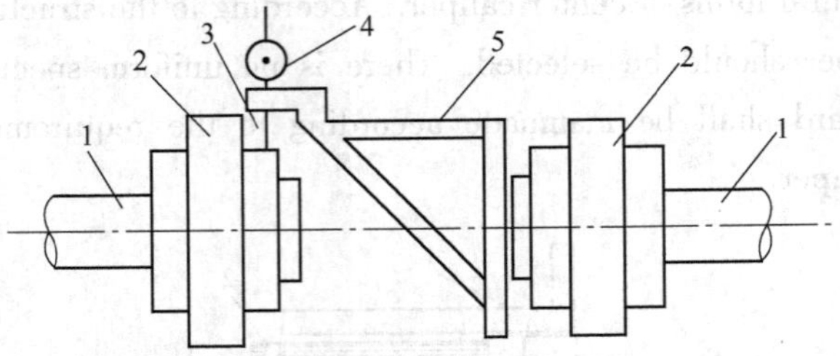

图 5.14 单表法测量联轴器偏差示意图

Figure 5.14 Schematic diagram of single dial indicator method

1—轴（shaft）；2—半联轴器（half coupling）；3—百分表触头（dial gauge contact）；4—百分表（dial gauge）；5—支架（bracket）

graphical method from the reading on the dial indicator. When measuring with this method, a special dial indicator seat is needed, whose size and structure are determined by the axial distance between the two half-couplings and the size of the hub. The mass of dial indicator rack should be small, and have enough stiffness. The dial indicator rack and dial indicator are required to be tightened without any loose.

b 双表法（Double Dial Indicators Method）

双表法是广泛采用的联轴器装配质量检测和调整方法。

Double dial indicators method is widely used to determine the quality check and adjustment of coupling assembly.

双表法就是用两块百分表测量同一方向上的轴向和径向间隙。这种方法也称为径面法。

Double dial indicators are used to measure cylindrical and face deviations on the same direction with double dial indicators. This method is also called rim-face method.

双表法又称单点测量法，即测量某一方向的径向读数，同时测量同一方向的轴向读数。

It is also known as one-point measurement method, that is, the radial reading in a certain direction is measured, at the same time, the axial reading in the same direction is measured.

双表法的测量步骤是用方尺对联轴器进行初步测量和调整。然后，在从动轴（左轴）一侧的半联轴器上安装中心卡尺和百分表。表盘指示头指向传动轴（右轴）上半联轴器的外圆和端面，如图 5.15 所示。

The measure procedure of double dial indicators method is that a square ruler is used to preliminary measure and adjust the coupling. Then, a center caliper and dial indicators are installed on the half-coupling on the side of the driven shaft (left shaft) as the reference. The head of the dial indicator points to the outer circle and end face of the half-coupling on the drive shaft (right shaft), as shown in Figure 5.15.

测量时，先测 0°方位的径向读数 a_1 及轴向读数 S_1。为了分析计算方便，常把 a_1 和 S_1 调整为零，然后两个半联轴器同时转动，每转 90°读一次表中数值，并把读数值填到记录图中。圆外记录径向读数 a_1、a_2、a_3 和 a_4，圆内记录轴向读数 S_1、S_2、S_3 和 S_4，当百

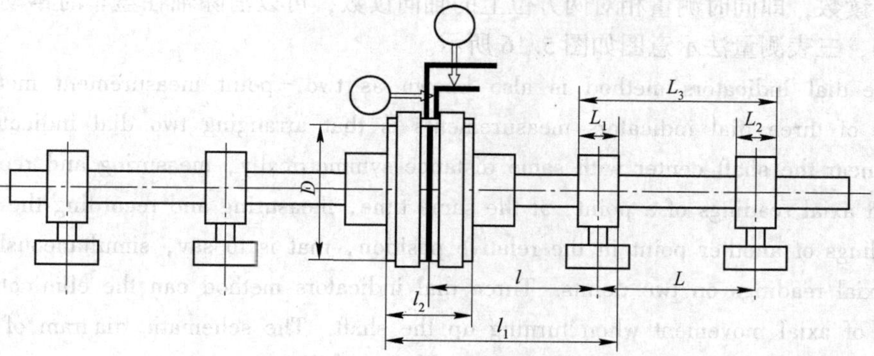

图 5.15　双表法测量联轴器偏差示意图

Figure 5.15　Double dial indicators measuring method

分表转回到零位时，必须与原零位读数一致，否则需找出原因并排除之。

When measuring, radial reading a_1 and axial reading S_1 in 0° directions are first measured. Before measuring, it is necessary to distinguish the relationship between the rotation direction of dial indicator hand and the movement direction of dial indicator head. After recording a_1 and S_1, the two half-couplings rotate together. The values in the dial indicator are read from point 1 to point 2, point 3, point 4. The readings a_1, a_2, a_3, a_4, S_1, S_2, S_3, and S_4 are filled in the tables. When the coupling turns back to point 1, the readings of point 1 must be the same as the original readings. Otherwise, the reasons should be found out and corrected.

常见的原因是轴窜动或地脚螺栓松动，测量的读数必须符合下列条件才正确，即：

A common reason is shaft movement or anchor bolts loosen. Readings must meet the following criteria, i. e. :

$$a_1 + a_3 = a_1 + a_4;\quad S_1 + S_3 = S_2 + S_4$$

通过对测量数值的分析计算，确定两轴在空间的相对位置，然后按计算结果进行调整。

The relative position of the two axes in space is determined through the analysis and calculation of the measurement value. Then the coupling can be adjusted according to the calculation result.

这种方法应用比较广泛，可满足一般机器的安装精度要求。主要缺点是对有轴向窜动的联轴器，在盘车时其端面的轴向度数会产生误差。因此，这种测量方法适用于由滚动轴承支撑的转轴，轴向窜动比较小的中小型机器。

This method is widely used and can meet the installation accuracy requirements of general machines. The main disadvantage is that the axial degree of its end surface will produce errors when turning for coupling with axial movement. Therefore, this measurement method is suitable for small and medium-sized machines with a rotating shaft supported by rolling bearings and relatively small axial movement.

c　三表法（Three Dial Indicators Method）

三表法又称两点测量法。三表测量法与两表测量法不同之处是在与轴中心等距离处对称布置两块百分表，在测量一个方位上径向读数和轴向读数的同时，在相对的一个方位上测其轴向读数，即同时测量相对两方位上的轴向读数，可以消除轴在盘车时窜动对轴向读数的影响，三表测量法示意图如图 5.16 所示。

Three dial indicators method is also known as two - point measurement method. The procedure of three dial indicators measurements is that arranging two dial indicators at the positions near the shaft center with same distance symmetrically, measuring and recording the radial and axial readings of a point, at the same time, measuring and recording the radial and axial readings of another point in the relative position, that is to say, simultaneously measure relative axial readings on two points. Three dial indicators method can the eliminate dynamic influence of axial movement when turning up the shaft. The schematic diagram of three dial indicator measurement is shown in Figure 5.16.

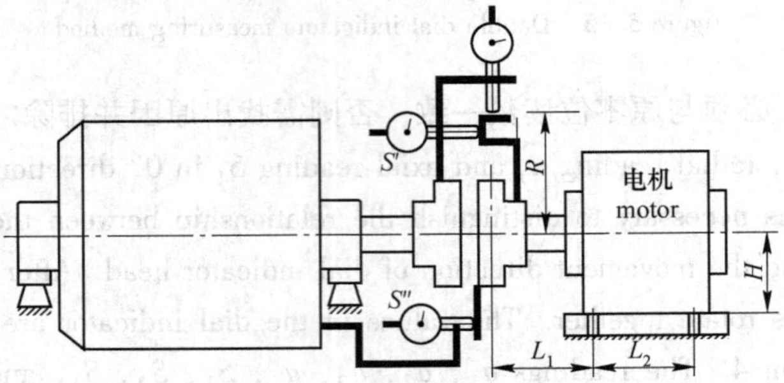

图 5.16 三表法测量联轴器偏差示意图

Figure 5.16 The schematic diagram of three dial indicators method

根据测量结果，计算轴向读数在 0°～180°和 180°～0°测量方向上的平均值，如式（5.5）所示。

According to the measurement results, the average value of the axial reading in the measurement directions of 0°～180° and 180°～0° is calculated. They are shown as Formula (5.5).

$$S_1 = (S_1' + S_1'')/2,\ S_3 = (S_3' + S_3'')/2 \tag{5.5}$$

计算 90°～270°和 270°～90°测量方向轴向读数的平均值，如式（5.6）所示。

The average value of the axial readings in the measurement directions of 90°～270° and 270°～90° is calculated too. They are shown as Formula (5.6).

$$S_2 = (S_2' + S_2'')/2,\ S_4 = (S_4' + S_4'')/2 \tag{5.6}$$

用 S_1、S_2、S_3 和 S_4 的平均值作为每个点的轴向读数与 a_1，a_2，a_3和 a_4 的径向读数记录在同一个记录图中。根据图中的数据对联轴器的偏差进行分析、计算和调整。

The average values of S_1, S_2, S_3 and S_4 are used as the axial readings for each point are recorded in the same recording diagram with the radial readings of a_1, a_2, a_3 and a_4. The offset of the coupling is analyzed, calculated and adjusted according to the data in the diagram.

该测量方法精度高，适用于汽轮机、离心压缩机等需要精确校直或高速运转的机械。相比之下，三表法的操作和计算比双表法稍微复杂一些。

This measurement method has high precision and is suitable for machines that need accurate alignment or high - speed operation, such as steam turbines and centrifugal compressors. In contrast, the three dial indicators method is a little more complicated than the double dial indicators method in operation and calculation.

d 五表法（Five Dial Indicators Method）

五表法（又称四点测量法）是指测量并记录四个点的轴向读数，同时测量并记录一个点的径向读数。取同一位置四个轴向读数的平均值作为轴向读数与径向读数的结合，分析计算联轴器的偏差。五表法的特点与三表法相同。

Five dial indicators method (also known as the four points measurement method) refers to the axial readings of four points are measured and recorded and the radial readings of a point is measured and recorded at the same time. The average values of four axial readings on the same position are taken as the axial readings combined with the radial readings to analysis and calculate the deviation of coupling. Five dial indicators method has the same features with three dial indicators.

D 联轴器偏差的计算和调整（Deviation Calculation and Adjustment）

以双表法为例分析联轴器的偏差。以不同心不平行——下偏心下翘（图 5.10）为例，介绍联轴器偏差的计算和调整方法。

The double dial indicators method is taken as an example to analyze the deviation of coupling. The calculation and adjustment method of coupling deviation is introduced by taking a position of down decentration and down misalignment (Figure 5.10) as an example.

根据测量数据可以确定两根轴的相对位置，如图 5.17 所示。

The relative position of two shafts can be determined according to the readings (as shown in Figure 5.17).

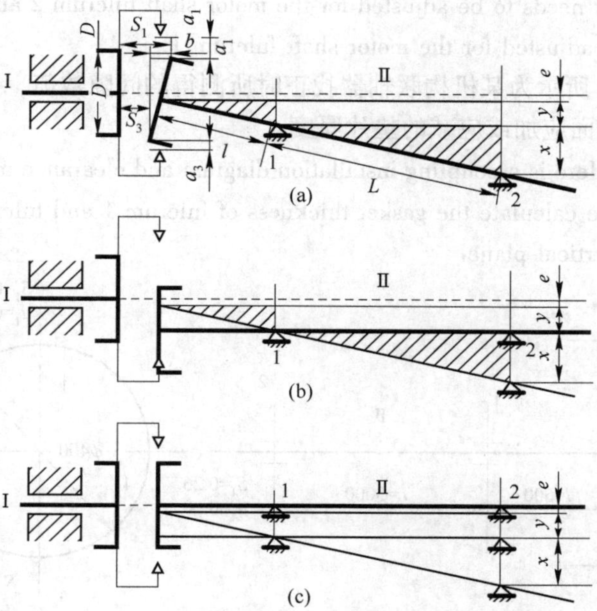

图 5.17 联轴器找正计算和垫片调整方法

Figure 5.17 Coupling alignment calculation and gasket adjustment method

如图 5.17（a）所示，令 $b = |S_3 - S_1|$。

From Figure 5.17（a）, let $b = |S_3 - S_1|$.

根据图 5.17（a）中两个三角形相似，可得到 x 和 y，如式（5.7）和式（5.8）所示。

According to the similarity of the two triangles in Figure 5.17（a）, get x and y, shown as Formula（5.7）and Formula（5.8）.

$$\frac{x}{L} = \frac{b}{D}$$

$$x = \frac{bL}{D} \tag{5.7}$$

$$\frac{y}{l} = \frac{b}{D}$$

$$y = \frac{bl}{D} \tag{5.8}$$

而 e 的值可由径向间隙差求出，结果是绝对值，如式（5.9）所示。

The value of e can be calculated from the radial gap difference, and the result is absolute value, shown as Formula（5.9）.

$$e = \frac{|a_1 - a_3|}{2} \tag{5.9}$$

为了使两轴既平行又同心，需要计算电机轴支点 2 需要调整的垫片厚度 $x + y + e$ 和电机轴支点 1 需要调整的垫片厚度 $y + e$。

In order to make the two axes parallel and concentric, it is necessary to calculate the gasket thickness $x + y + e$ that needs to be adjusted for the motor shaft fulcrum 2 and the gasket thickness $y + e$ that needs to be adjusted for the motor shaft fulcrum 1.

例 5.2　图 5.18 所示为某机构联轴器找正时所测得的间隙数值，试求轴Ⅱ垂直平面上支点 1 和支点 2 的下面应加或应减的垫片厚度。

Example 5.2　Here is a coupling installation diagram and clearance measured data as shown in Figure 5.18. Please calculate the gasket thickness of fulcrum 1 and fulcrum 2 of shaft Ⅱ shall be plus or minus in vertical plane.

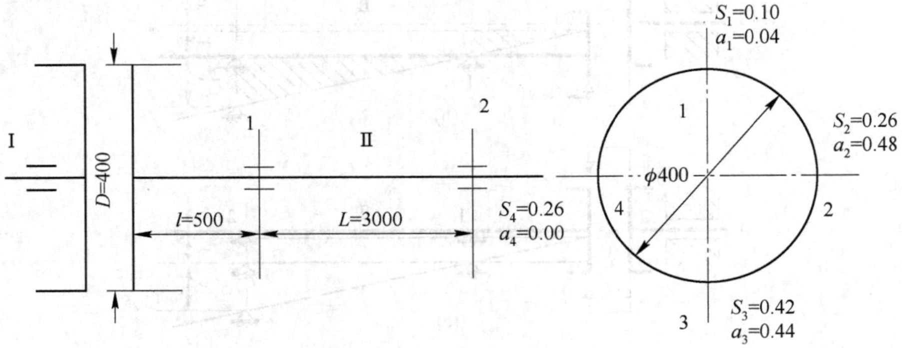

图 5.18　某机构联轴器找正时所测得的间隙数值和结构参数

Figure 5.18　Clearance measured data and structure parameters of a coupling

解（Solution）：

从测量数据可以看出，联轴器在0°与180°两个位置上 $S_3 > S_1$， $a_3 > a_1$。这表示两半联轴节在垂直平面内既不平行又不同心。根据这些数据便可做出联轴器偏移的示意图，即找正计算图（如图5.19所示）。

From the measured data, it can be seen that the coupling is in positions $S_3 > S_1$, $a_3 > a_1$ at 0° and 180°. This means that the two shafts are neither parallel nor centered in the vertical plane. According to these data, a schematic diagram of two shafts position, namely, the alignment calculation diagram, can be made (as shown in Figure 5.19).

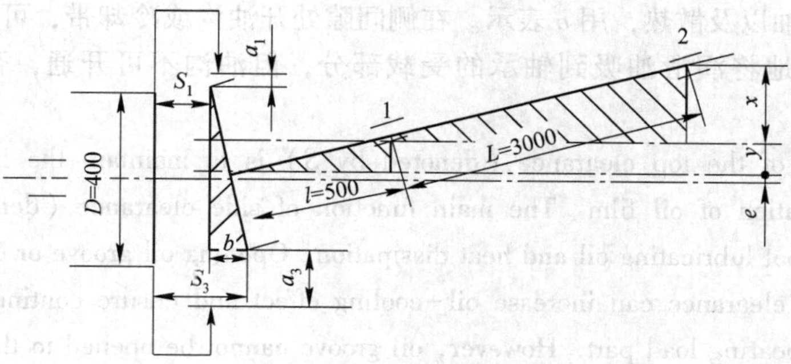

图5.19 联轴器找正计算图

Figure 5.19 Picture of coupling alignment calculation

$$b = |S_3 - S_1| = |0.42 - 0.1|\text{mm} = 0.32\text{mm}$$

$$x = \frac{b}{D}L = \frac{0.32}{400} \times 3000\text{mm} = 2.4\text{mm}$$

$$y = \frac{l}{L}x = \frac{500}{3000} \times 2.4\text{mm} = 0.4\text{mm}$$

$$e = \frac{|a_3 - a_1|}{2} = \frac{|0.44 - 0.04|}{2}\text{mm} = 0.2\text{mm}$$

为了使两半联轴节在垂直平面内既平行又同心，需要在支点2减3mm（$y + e + x = 0.6 + 2.4 = 3$mm）厚的垫片，在支点1减0.6mm（$y + e = 0.4 + 0.2 = 0.6$mm）厚的垫片。

In order to make the two axes parallel and concentric, it is necessary to subtract 3mm ($y + e + x = 0.6 + 2.4 = 3$mm) from the 2 fulcrum and subtract 0.6mm ($y + e = 0.4 + 0.2 = 0.6$mm) from the 1 fulcrum.

5.1.3 滑动轴承的安装（Assembly of Sliding Bearing）

滑动轴承的种类很多，常见的主要有剖分式滑动轴承和整体式滑动轴承两种。

There are many kinds of sliding bearings, including split sliding bearings and integral sliding bearings.

滑动轴承装配的间隙应特别注意。装配完成后，如果间隙在 $S_{初}$ 和 S_{max} 之间，滑动轴承就可以正常工作。

Special attention should be paid to clearance in sliding bearing assembly. If the clearance is between $S_{initial}$ and S_{max} after assembly, the sliding bearing can work normally.

5.1.3.1 滑动轴承的间隙（Clearances of Sliding Bearing）

轴颈与轴瓦的配合间隙有两种，一种是径向间隙，一种是轴向间隙。径向间隙包括顶间隙和侧间隙。

There are two kinds of matching clearance between journal and bushing of sliding bearing, one is radial clearance, and the other is axial clearance. Radial clearance includes top clearance and side clearance.

顶间隙的作用是保持液体摩擦，以利形成油膜，用 S 表示；侧间隙的主要作用是为了积聚和冷却润滑油以及散热，用 b 表示。在侧间隙处开油沟或冷却带，可增加油的冷却效果，并保证连续地将润滑油吸到轴承的受载部分，但油沟不可开通，否则运转时将会漏油。

The function of the top clearance (denoted by S) is to maintain the liquid friction and facilitate the formation of oil film. The main function of side clearance (denoted by b) is to accumulate and cool lubricating oil and heat dissipation. Opening oil groove or cooling zone at the position with side clearance can increase oil-cooling effect and ensure continuous absorption of lubricating oil to bearing load part. However, oil groove cannot be opened to the end of bearing. Otherwise, oil will leak during operation.

轴向间隙的作用是使轴在温度变化时有自由伸长的余地。

The function of the axial clearance is to allow the axis to extend freely when the temperature changes.

在实际工作中，顶间隙可以通过计算决定，也可以根据经验决定。侧间隙两侧应相等，单侧间隙应为顶间隙的一半或三分之二。在固定端，轴向间隙不得大于 0.2mm；在自由端，轴向间隙不应小于轴受热膨胀时的间隙。

In practical work, the top clearance can be determined by calculation or by experience. Two sides of the lateral clearance shall be equal, and the unilateral clearance shall be half or two-thirds of the top clearance. The axial clearance at the fixed end shall not be greater than 0.2mm, and the axial clearance at the free end shall not be smaller than that when the shaft is heated and expanded.

5.1.3.2 滑动轴承装配间隙的测量（Measuring of Sliding Bearing Assembly Clearance）

滑动轴承装配间隙的测量方法有两种：压铅法和塞尺测量法。

There are two methods for measuring the assembly clearance of sliding bearing: lead pressing method and feeler measuring method.

压铅法是用 0.6~1mm 软铅丝或软铅条按如图 5.20 所示位置先放铅丝，然后盖上轴承盖，均匀拧紧螺丝，并用塞尺检查轴瓦接合面间的间隙是否均匀相等。

In lead pressing method, 0.6~1mm soft lead wire or soft lead strip is put in the position as shown in Figure 5.20, then cover the bearing cover, tighten the screws evenly, and check whether the clearance between the bearing bushing connection surfaces is uniform and equal with the feeler.

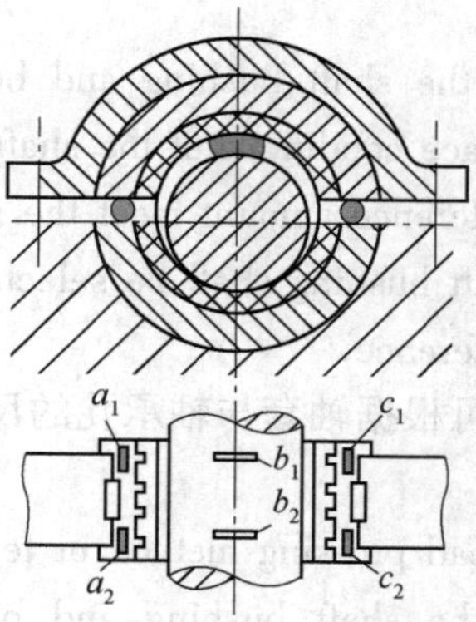

图 5.20　压铅法测量滑动轴承顶间隙和侧间隙

Figure 5.20　Measure the assembly clearances of sliding bearing with lead pressing method

再打开轴承盖，用外径千分尺测量压扁部位的铅丝厚度，侧间隙 A_1 和 A_2 可用式（5.10）和式（5.11）计算。

The bearing cover is opened again and the lead wire thickness is measured at the flattening position with the outside micrometer. Side clearances A_1 and A_2 can be calculated by Formula（5.10）and Formula（5.11）.

$$A_1 = \frac{a_1 + c_1}{2} \tag{5.10}$$

$$A_2 = \frac{a_2 + c_2}{2} \tag{5.11}$$

顶间隙的平均值可用公式（5.12）计算。

Average of the top clearance can be calculated by Formula（5.12）.

$$S_{\text{Average}} = \frac{(b_1 - A_1) + (b_2 - A_2)}{2} = \frac{b_1 + b_2}{2} - \frac{1}{4}(a_1 + a_2 + c_1 + c_2) \tag{5.12}$$

对整体式轴承的间隙测量均用塞尺，要求精确时可用百分表。

The clearance measurement of integral bearing is measured with feeler. Dial indicator can be used when precision is required.

滑动轴承的轴向间隙的要求是，固定端间隙值为 0.1~0.2mm，自由端的间隙值应大于轴的热膨胀伸长量。

The requirements of axial clearance of sliding bearing are 0.1~0.2mm for the fixed end, greater than the thermal expansion and elongation of shaft for the free end.

5.1.3.3　*滑动轴承装配过程*（Sliding Bearing Assembly Process）

整体式滑动轴承的装配过程包括轴套与轴承孔的清洗、检查、轴套的安装等。

The assembly process of integral sliding bearing includes the cleaning, checking and installation of shaft bushing and bearing block hole.

（1）清洗检查。用煤油或清洗剂将轴套与轴承孔清洗干净后，检查轴套与轴承孔的表面情况以及配合过盈量是否符合要求，然后再根据尺寸以及过盈量的大小选择轴套的装配方法。

Cleaning and checking. After the shaft bushing and bearing block hole are cleaned with kerosene or cleaning agent, the surface conditions of the shaft bearing bushing and bearing block hole and whether the matching interference amount meet the requirements shall be checked, and then the assembly method of the shaft bushing shall be selected according to the size of the shaft bushing and the amount of the interference.

（2）轴套装配。轴套的安装可根据轴套与轴承孔的尺寸以及过盈量的大小选择压入法或温差法。

Shaft bushing assembly. The lead pressing method or temperature difference method can be selected according to the size of the shaft bushing and bearing block hole and the size of interference.

（3）轴瓦装配。轴瓦和轴承座配合采用小过盈配合（0. 01～0. 05mm），或滑动配合；轴瓦固定：防止转动有定位销；防止轴向窜动，轴瓦有翻边或止口；为使轴颈和轴瓦有理想的配合面，要进行研瓦、开瓦口和油槽。

Bearing bushing assembly. Small interference fit（0. 01～0. 05mm）or sliding fit is adopted for the matching of bearing bushing and bearing block. Bearing bushing is fixed. Positioning pin is set up to prevent the bearing bushing rotating with the shaft. The bearing bushing is flanged or checked to prevent the bearing bushing from axial moving. In order to make the shaft journal and bushing have the ideal mating surface, the inner surface and outer surface of the bearing bushing should be ground. The edge of bearing bushing should be erased. The oil groove should be shoveled.

研瓦首先研轴瓦表面，使其有微小的凹面，以储油和润滑，一般接触角为70°～90°。接触斑点指标：一般瓦，1～2点/cm^2；高精度瓦，2～3点/cm^2。

The surface of bearing bushing should be ground first in order that small concave formed on the surface of bearing bushing to store oil and lubricate. The general contact angle is 70°～90°. The contact spot is 1～2 points/cm^2 for general bearing bushing, 2～3 points/cm^2 for high precision bushing.

5.1.4 滚动轴承的装配（Assembly of Rolling Bearing）

滚动轴承是一种精密元件，其套圈和滚动体有较高的精度和光洁度，因此装配时应认真仔细，不要损坏轴承的零件。实践证明，轴承装配不正确，是轴承过早损坏的原因之一。轴承损坏的另一个原因是轴承发热，其结果导致滚动体表面剥落、刮伤和出现麻坑或是套圈出现裂纹，磨成深坑而报废。

Rolling bearing is a kind of component with high precision. The precision and smoothness of rings and rollers are very high. Therefore, the assembly of rolling bearing should be careful to prevent the bearing parts from damage. Practice has proved that improper bearing assembly is one of the causes of premature bearing damage. Another cause of bearing damage is the temperature of bearing increase, which would lead to rolling surface peeling, scratches and pockmarks or ring cracks, grinding into a deep pit and scrap.

5.1.4.1 装配前的准备（Preparation for Rolling Bearing Assembly）

滚动轴承装配的准备工作包括装配工具和测量工具的准备、零部件的清洗和检查。轴承必须安装在干燥和清洁的环境条件下。安装前应仔细检查轴与轴承座匹配面、轴肩端面、坡口及连接面的加工质量。

Preparation for rolling bearing assembly includes preparation of assembly tools and measuring tools, parts or components cleaning and checking. Bearings must be installed under dry and clean environmental conditions. Before installation, the machining quality of matching surface of shaft and bearing block, the end face of shaft shoulder, groove and connection surface should be carefully checked.

所有配合表面必须仔细清洁和去毛刺。铸件粗糙的表面必须清除型砂。安装前应用汽油或煤油清洗轴承。干燥后才能使用。必须保证良好的润滑。

All mating surfaces must be carefully cleaned and deburred. The rough surface of the casting must be cleaned of molding sand. The bearing should be cleaned with gasoline or kerosene before installation. It can be used only after drying. Good lubrication must be ensured.

油润滑也可用于滚动轴承。选用无杂质、抗氧化、抗锈、耐极压等性能优异的润滑脂。注油量应为轴承和轴承座体积的30%~60%，不能太多。

Oil lubrication can also be used for rolling bearings. Grease with excellent performances such as no impurities, oxidation resistance, rust resistance, extreme pressure, etc. should be chosen. The amount of grease to be filled should be 30%~60% of the volume of bearing and bearing block, not too much.

具有密封结构的双列圆锥滚子轴承已注满油脂，用户可直接使用，不能再次清洗。

Double row taper roller bearing with sealing structure have been filled with grease, which can be used directly by users and cannot be cleaned again.

5.1.4.2 滚动轴承装配的方法（Method Selection of Rolling Bearing Assembly）

一般滚动轴承内圈和轴采用基孔制配合，外圈和轴承座采用基轴制配合，通过改变轴承座的公差和轴的公差进行装配。

Generally, the inner ring of rolling bearing and shaft are matched by the base hole system. While the outer ring and bearing block are matched by the base shaft system, and the assembly is carried out by changing the tolerance of bearing block and the tolerance of shaft.

滚动轴承装配的方法通常有冷装和热装两种。

The rolling bearing assembly method usually has the room temperature assembly method and hot charging assembly method.

A 冷装（Room Temperature Assembly Method）

滚动轴承的冷装是指室温装配，包括锤击法和压入法。

Cold assembly of rolling bearings refers to room temperature assembly. Room temperature assembly method includes hammering method and press-in method.

锤击法就是用铜棒或软金属套筒打内、外圈，适用小型轴承装配。压入法就是用压力压入。

Hammering method is suitable for small bearing fighting inner and outer ring with copper or soft metal sleeve to finish assembling. Pressure should be used in press-in method to assemble rolling bearing.

轴承的内圈与轴配合紧密，而外圈与轴承座的孔配合松散。首先要把轴承压在轴上，然后把轴和轴承装入轴承座的孔中。

The inner ring of the bearing fits tightly with the shaft, while the outer ring fits loosely with the bore of the bearing housing. The bearing needs to be pressed onto the shaft first, and then the shaft and the bearing are loaded into the bore of the bearing housing.

受压时，应将软金属制成的套筒放置在轴承的端面上。轴套的内径应大于轴颈的直径，外径应小于轴承内圈法兰的直径，以免损坏保持架。

When pressing, a sleeve made of soft metal should be placed on the end face of the bearing. The inner diameter of the sleeve should be larger than the diameter of the shaft journal, and the outer diameter should be smaller than the diameter of the flange of the inner ring of the bearing, so as to avoid damaging the cage.

此外，装配时要注意对中，防止轴承歪斜。否则，不仅装配困难，还会产生压痕，造成轴和轴承的过早损坏。

In addition, when assembling, attention should be paid to the alignment, to prevent the bearing askew. Otherwise, it is not only difficult to assemble, but also can produce indentation, causing premature damage of shaft and bearing.

对于外圈与轴瓦孔配合紧、内圈与轴配合松的轴承，采用外径略小于轴瓦孔直径的套，先将轴承压入轴瓦孔内，再安装轴。

For bearings with tight fit between outer ring and bearing block hole and loosen fit between inner ring and shaft, sleeve with outer diameter slightly smaller than the diameter of bearing block hole is adopted to press the bearing into bearing block hole first, and then install the shaft.

当轴承的内圈与轴配合紧密，外圈与座孔配合也紧密时，可采用特殊的外壳将轴承同时压入轴颈和座孔。

When the inner ring of the bearing fits the shaft and the outer ring fits the seat hole tightly too, special casing can be used to press the bearing into the journal and the seat hole at the same time.

安装轴承时，必须在环端面的圆周上施加相等的压力，将环压入轴承内。不要用工具直接撞击轴承端面，以免损坏轴承。在过盈量小的情况下，可在室温下用套筒压紧轴承环端面，并用锤头撞击套筒，使套筒均匀地穿过套筒。

When installing the bearing, equal pressure must be applied on the circumference of the end face of the ring to press the ring into it. No tool shall be used to directly hit the end face of the bearing to avoid damage to the bearing. In the case of small interference amount, the end face of bearing ring can be pressed with sleeve at room temperature, and the sleeve can be hit with hammer head to press the sleeve evenly through the sleeve.

批量安装时，可采用液压机。压紧时，应将外圈端面和壳体肩面，内圈端面和轴肩面压紧。不允许有间隙。

When the mass installation, hydraulic machine can be used. When pressing, the end face of the outer ring and the shoulder face of the shell, and the end face of the inner ring and the shoulder face of the shaft should be pressed tightly. No gap is allowed.

B 热装（Heat Bearing Assembly Method）

当过盈量较大时，可采用油浴或电感应加热的方式安装轴承。轴承热装方式有油加热和感应加热两种。

When interference is large, oil bath or electricity induction heating method can be used to install bearings. Bearing hot charging method has oil heating and induction heating method.

油加热是将轴承放入油箱，在100℃左右均匀加热。轴承受热后由离底面50～70mm高度的格栅吊起。该方法适用于大中型轴承装配。

Oil heating method means putting the bearing into the oil tank and evenly heat it at about 100℃. When heated, the bearings are hoisted by the grid with a height of 50～70mm from the bottom surface. This method is suitable for large and medium-sized bearing assembly.

感应加热适用于外圈、挡圈、滚动体、内圈分离的轴承。内圈被加热，然后组装在轴上。加热温度为100℃。恒温时间由电磁继电器和恒温器控制。

Induction heating method is applicable to bearings with outer ring, retainer, rolling body and inner ring separated. The inner ring is heated and then assembled on the shaft. Heating temperature is 100℃. The constant temperature and time are controlled by electromagnetic relay and thermostat.

5.1.4.3 滚动轴承间隙调整（Clearance Adjustment of Rolling Bearing Assembly）

滚动轴承的游隙包括径向游隙和轴向游隙。径向游隙是指内圈和外圈在直径方向上产生的最大游隙，轴向游隙是指内圈和外圈在轴向上产生的最大游隙。

The clearance of rolling bearing includes radial clearance and axial clearance. Radial clearance refers to the maximum clearance generated between inner and outer rings in the direction of diameter, while axial clearance refers to the maximum clearance generated between inner and outer rings in the direction of axis.

滚动轴承间隙的作用是弥补制造和装配误差，保证轴和轴承受热膨胀时滚动的正常运行，延长轴承的使用寿命。

The role of rolling bearing clearance is to make up for manufacturing and assembly errors, to ensure the normal operation of rolling when the shaft and bearing are expanded by heat, and to extend the service life of the bearing.

A 滚动轴承间隙的确定（Clearance Determination of Rolling Bearing Assembly）

滚动轴承的间隙分为两种：不可调和可调。当滚动轴承的装配间隙不能调整时，考虑轴受热膨胀和轴承与端盖之间的轴向间隙 a，a 可由式（5.13）计算。

The clearance of rolling bearing is divided into two kinds: non-adjustable and adjustable. When the assembly clearance is not adjustable for rolling bearing, consider the expansion of shaft under heat and the axial clearance a between bearing and end cover, which can be calculated by Formula (5.13).

$$a = Lk_{\mathrm{d}}\Delta t + 0.15\mathrm{mm} \tag{5.13}$$

式中，a 为轴向间隙，mm；L 为轴承间的距离，mm；k_d 为线膨胀系数，℃$^{-1}$；Δt 为工作温度与环境温度最大差值，℃。

Where, a is axial clearance; L is distance between bearings, mm; k_d is linear expansion coefficient, ℃$^{-1}$; Δt is maximum difference between operating temperature and ambient temperature, ℃.

装配间隙可调整滚动轴承时，例如单列圆锥滚柱轴承，轴向间隙可用式（5.14）和式（5.15）计算。

When assembling a rolling bearing with adjustable clearance, single row tapered roller bearings is taken as an example. Axial clearance and radial clearance can be calculated by the following Formula (5.14) and Formula (5.15).

$$a = \frac{C}{\sin\beta} \tag{5.14}$$

$$\lambda = \arctan\beta = \frac{C}{\sin\beta}\tan\beta = \frac{C}{\cos\beta} \tag{5.15}$$

式中，β 为圆锥角；λ 为径向间隙；C 为内外圈垂直距离。

Where, β is conical angle; λ is radial clearance; C is vertical distance between inner and outer rings.

根据 GB 4604—93，将滚动轴承径向游隙分为 2 组、0 组、3 组、4 组、5 组，游隙值由小到大，其中 0 组为标准游隙。基本径向间隙组适用于一般操作条件，常温和过盈配合。

According to GB 4604—93, the radial clearance of rolling bearing is divided into five groups, group 2, group 0, group 3, group 4 and group 5, and the clearance value is successively from small to large, among which group 0 is the standard clearance. The basic radial clearance group is suitable for general operating conditions, normal temperature and interference fit.

在高温、高速、低噪声、低摩擦等特殊条件下工作的轴承建议使用较大的径向游隙。精密主轴和机床主轴轴承应采用小径向游隙。

Large radial clearance is recommended for bearings working under special conditions such as high temperature, high speed, low noise and low friction. Small radial clearance should be used for precision spindle and machine tool spindle bearings.

对于滚子轴承，可以保持少量的工作间隙。此外，分离轴承没有间隙。最后，轴承安装后的工作间隙小于安装前的原间隙，这是因为轴承必须承受一定的载荷旋转，以及由于轴承协调和载荷引起的弹性变形。

A small amount of working clearance can be maintained for roller bearings. In addition, there is no clearance for separated bearings. Finally, the working clearance after bearing installation is smaller than the original clearance before installation, because the bearing must bear certain load rotation, as well as the elastic deformation caused by bearing coordination and load.

B 滚动轴承间隙的调整方法（Clearance Adjustment of Rolling Bearing Assembly）

滚动轴承间隙的调整方法主要有垫片调整法、螺钉调整法、止推环调整法和内外套调整法。

The adjustment methods of the clearance of rolling bearing mainly include gasket adjustment method, screw adjustment method, thrust ring adjustment method and inner and outer sleeve adjustment method.

a　垫片调整法（Gasket Adjustment Method）

最常用的方法是调整轴承压盖处的垫片，如图 5.21 所示。首先从轴承盖上拆下所有原垫片，然后慢慢拧紧轴承盖上的螺栓，使轴慢慢转动。当轴不能转动时，停止拧紧螺栓。此时，表示轴承中没有间隙。

The most commonly used method is to adjust the gasket at the bearing gland, as shown in Figure 5.21. First remove all the original gaskets from the bearing gland, and then slowly tighten the bolts on the bearing gland, making the shaft rotate slowly. When the shaft cannot rotate, stop tightening the bolts. At this point, it indicates that there is no clearance in the bearing.

用一个塞尺来测量轴承压盖和盒子端面之间的间隙 K。将测量的间隙 K 加到所需的轴向间隙 C 上，$K+C$ 为需要填充的垫片厚度。一套垫片应由各种厚度的垫片组成，垫片应光滑、清洁，内、外边缘无毛刺。除塞尺法外，间隙测量法还可采用压铅法和千分尺法。

A feeler is used to measure the clearance K between the bearing gland and the end face of the box. Add the measured clearance K to the required axial clearance C, and $K+C$ is the gasket thickness to be filled. A set of gaskets shall be composed of gaskets of various thicknesses and the gaskets shall be smooth and clean without burrs on the inner and outer edges. In addition to the feeler method, the gap measurement method can also be used by the lead pressure method and the micrometer method.

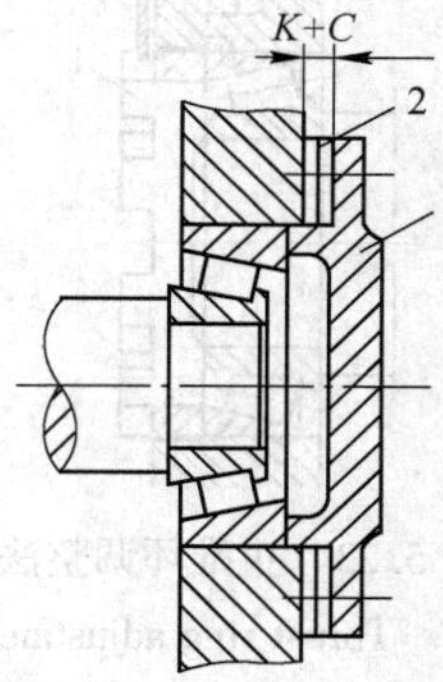

图 5.21　垫片调整法

Figure 5.21　Gasket adjustment method

1—端盖（gland）；2—垫片（gasket）

b　螺钉调整法（Screw Adjustment Method）

先松开调节螺丝上的锁紧螺母，再拧紧调节螺母，如图 5.22 所示。将止推阀压在轴承的外圈上，直到轴不能转动为止。最后根据轴向间隙的大小进行调整。

First loosen the locking nut on the adjusting screw, and then tighten the adjustment, as shown in Figure 5.22. Press the thrust disc against the outer ring of the bearing until the shaft fails to turn. Finally, it will be adjusted according to the value of axial clearance.

螺杆反转一定角度，达到规定的轴向间隙后，拧紧锁紧螺母，防止调节螺杆松脱。

Screw reverses a certain angle, reaches the specified axial clearance, and then tighten the locking nut to prevent the adjustment screw from loosening.

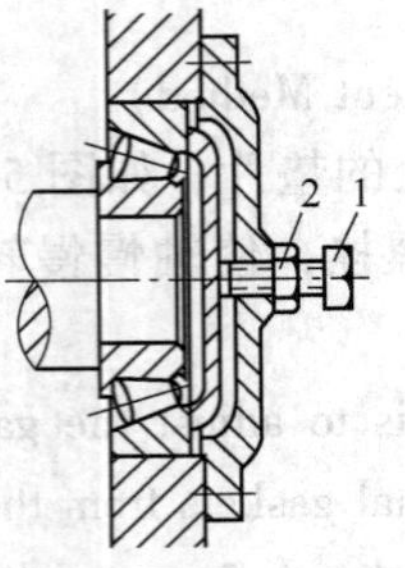

图 5.22 螺钉调整法

Figure 5.22 Screw adjustment method

1—调整螺钉（adjusting screw）；2—锁紧螺母（locknut）

c 止推环调整法（Thrust Ring Adjustment Method）

如图 5.23 所示，先用外螺纹拧紧止推环 1，直到轴无法转动，然后根据轴向间隙值将止推环反向一定角度，最后与止动板 2 固定。

As shown in Figure 5.23, first tighten the thrust ring 1 with external thread until the shaft fails to rotate, then reverse the thrust ring by a certain angle according to the value of axial clearance, and finally fix it with the stop plate 2.

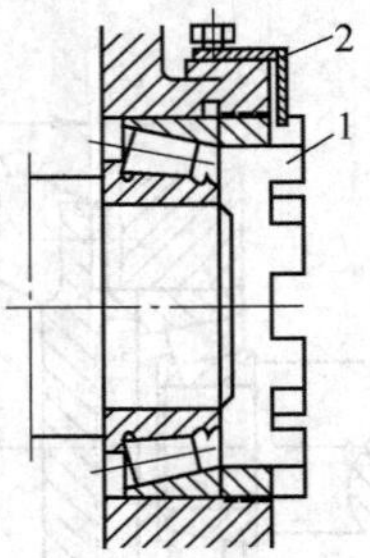

图 5.23 止推环调整法

Figure 5.23 Thrust ring adjustment method

1—止推环（thrust ring）；2—止动板（stop plate）

d 内外套调整法（Inner and Outer Sleeve Adjustment Method）

当两个圆锥滚子轴承安装在同一轴上时，经常使用内套和外套来调节轴向间隙，如图 5.24 所示。这种调整方法是在轴承不安装在轴上时进行的，内套和外套的长度根据轴承的轴向间隙来确定。

When two tapered roller bearings are installed on the same shaft, the inner and outer sleeves are often used to adjust the axial clearance, as shown in Figure 5.24. This adjustment method is carried out when the bearing is not installed on the shaft, and the length of the inner and outer sleeves is determined according to the axial clearance of the bearing.

5.1.5 齿轮的安装（Assembly of Gears）

齿轮、蜗杆装置的装配质量对传动工作的耐久性和可靠性起重要作用。齿轮装配质量

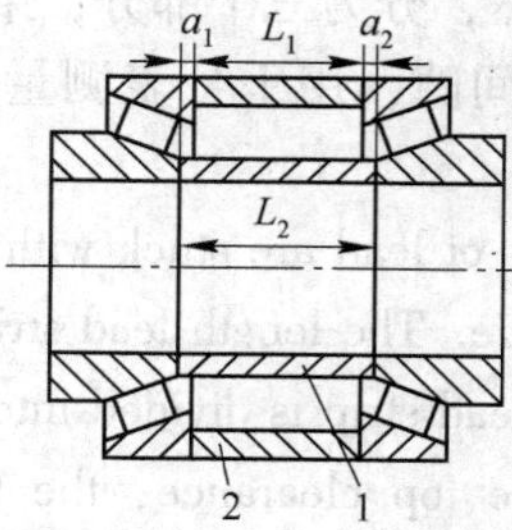

图 5.24　内外套调整法

Figure 5.24　Inner and outer sleeve adjustment method

1—内套（inner sleeve）；2—外套（outer sleeve）

表现在是否具有一定的侧间隙，齿轮表面接触情况和齿轮位置的正确性。

The assembly quality of gears and worm device plays an important role in the durability and reliability of transmission work. Gears assembly quality is shown in whether there is a certain side clearance (backlash), gear surface contact and correct gear position.

5.1.5.1　齿侧间隙的定义和作用（Definition and Function of Backlash）

齿侧间隙指一对相互啮合齿轮的非工作表面沿法线方向的距离，如图 5.25 所示，用 C_n 表示。

Backlash C_n refers to the distance between the non-working surface of a pair of mutually engaged gears along the normal direction (as shown in Figure 5.25).

齿侧间隙的作用是补偿装配和制造误差；补偿由于温度变化产生变形和受力引起的弹性变形；贮有一定的润滑油以改善齿面的摩擦条件。

The function of backlash is to compensate assembly and manufacturing errors, compensate the elastic deformation caused by temperature change and stress. Some lubricating oil is stored to improve the friction condition of tooth surface.

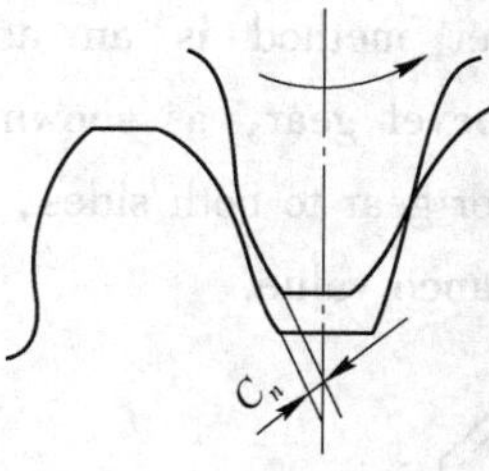

图 5.25　齿侧间隙

Figure 5.25　Backlash

5.1.5.2　测量齿侧间隙的方法（Methods of Measuring Backlash）

齿侧间隙的测量可以采用压铅法、塞尺法和千分表法。

The measurement of backlash can be made by means of lead pressing method, feeler method and micrometer method.

(1) 压铅法。用干油将 1~3 条铅丝粘在齿轮上，直径由间隙大小来定，长度为压下三个齿的长度。对一个齿来说，很显然，分为三个部分，第一部分为工作侧间隙，第二部分为齿顶间隙，第三部分为非工作侧间隙，用千分表测量厚度以判别齿啮合情况，其厚度 $a+b$ 为齿侧间隙。如图 5.26 所示。

Lead pressing method. 1～3 strips of lead are stuck with dry oil on the gear. The diameter of lead strip can be decided by the gap size. The length lead strip is long enough to press down three teeth. For one tooth, apparently, the lead strip is divided into three parts: the first part is working side clearance, the second part is the top clearance, the third part is the non－working side clearance. The thickness of the lead strip is measured with a dial indicator. The sum of thickness of a and b is backlash. As shown in Figure 5.26.

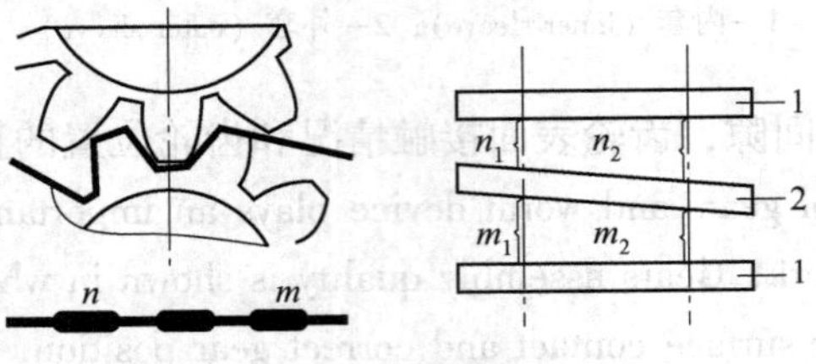

图 5.26 压铅法测量齿侧间隙

Figure 5.26 Measuring gear backlash with the method of pressing-in lead

1—大齿轮的齿（teeth of big gear）；2—小齿轮的齿（teeth of small gear）

（2）塞尺法。塞尺法可用来测量齿轮的齿隙。一个轮齿不动，另一个轮齿压紧，用塞尺塞入，测出齿侧间隙。

Feeler method. Feeler can be used to measure the backlash of gear. One tooth does not move. The other tooth is pressed on the surface of another gear. Feeler is inserted into the side clearance. The thickness of feeler is the backlash of gear.

（3）千分表法。千分表法是测量圆锥齿轮侧间隙的精确测量方法，如图 5.27 所示。先将一个齿轮固定，另一个齿轮向两侧摆动，接触在齿廓表面的千分表可读出齿侧间隙值。

Micrometer method. The micrometer method is an accurate measurement method for measuring the clearance on the side of bevel gear, as shown in Figure 5.27. The method is finished by fixing one gear, swing the other gear to both sides, and contact the micrometer on the tooth profile surface to read the side clearance value.

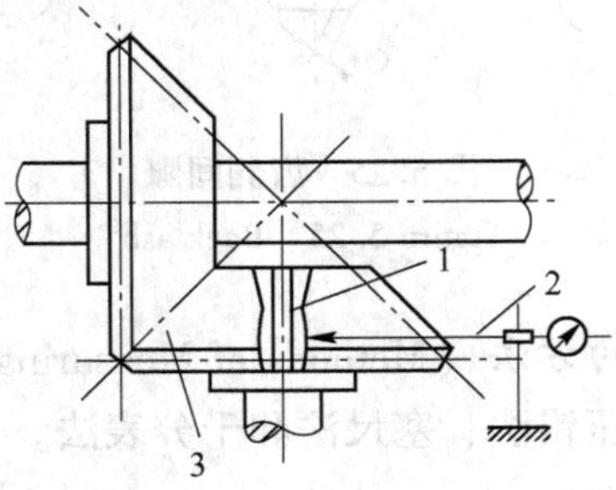

图 5.27 用千分表测量圆锥齿轮的齿侧间隙

Figure 5.27 Measuring backlash of bevel gear with micrometer

1—齿面（tooth surface）；2—百分表（dial indicator）；3—斜齿轮（bevel gear）

5.1.5.3 接触斑点法检查齿面接触情况（Methods of Measuring the Tooth Surface Contact Points Distribution）

根据齿轮的精度等级，通过涂色法检查齿面接触斑点分布，判断两齿轮啮合情况（如图 5.28 所示）。

According to the precision grade of gear, the tooth surface contact points distribution is checked by painting method to judge the meshing condition of two gears (as shown in Figure 5.28).

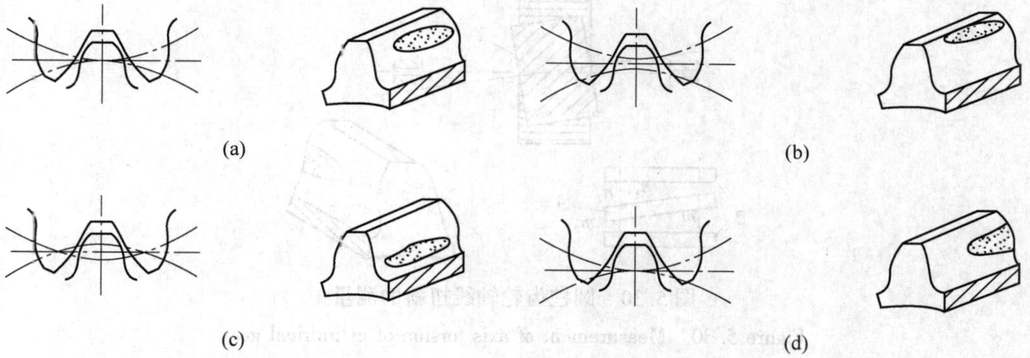

图 5.28 齿轮接触斑点的分布情况

Figure 5.28 Distribution of gear contact spots

(a) 啮合正确（Mesh correctly）；(b) 中心距过小（Smaller center distance）；
(c) 中心距过大（Larger center distance）；(d) 轴线扭斜（Axis skewed）

5.1.5.4 齿轮中心距偏差的测量方法（Methods of Measuring Gear Center Distance Deviation）

齿轮中心距离偏差的测量方法有两种。可使用内径千分尺和方水平检查中心距偏差，如图 5.29 所示，也可通过涂色法测量接触点的位置来判断。

There are two kinds of method for measuring gear center distance deviation. Inside micrometer or microcallipers and square level can be used (as shown in Figure 5.29) and judged by the position of contact spots measured by coloring or painting with marking compound.

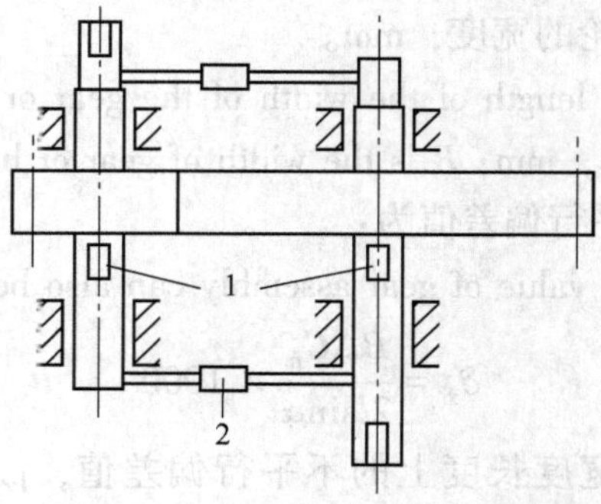

图 5.29 齿轮中心距和平行度测量

Figure 5.29 Gear center distance and parallelism measurements

1—方水平（square level）；2—内径千分尺（inside micrometer）

5.1.5.5 齿轮轴线不平行度的测量（Methods of Measuring Nonparallelism）

不平行度的测量与中心距的测量方法相同。扭斜的测量可用涂色法及压铅法，用压铅法还可以计算出扭斜及不平行度的大小。圆柱齿轮轴线扭斜的测量如图 5.30 所示。

The measurement of nonparallelism is the same as that of center distance. The measurement of torsion can be made by painting method and lead pressing method. The measurement of axis torsion of cylindrical gear is shown in Figure 5.30.

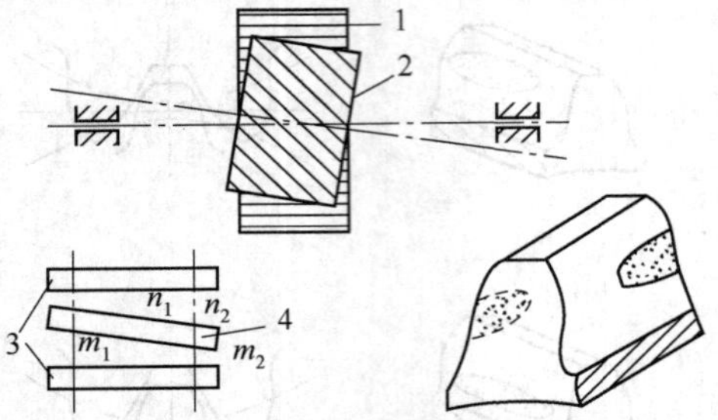

图 5.30　圆柱齿轮轴线扭斜的测量

Figure 5.30　Measurement of axis torsion of cylindrical gear

1—大齿轮（bigger gear）；2—小齿轮（smaller gear）；3—大齿轮的齿（teeth of bigger gear）；4—小齿轮的齿（teeth of smaller gear）

图 5.30 中：n_1、n_2 为工作侧铅丝厚度；m_1、m_2 为非工作侧铅丝厚度。

In Figure 5.30, n_1 and n_2 are the thickness of lead strip on the working side; m_1 and m_2 are the thickness of lead strip on the non-working side.

$$C_n = n_1 + m_1 = m_2 + n_2 \tag{5.16}$$

由于齿轮装配扭斜 $n_1 < n_2$，$m_1 > m_2$，故齿轮装配的扭斜偏差值为：

Because of $n_1 < n_2$, $m_1 > m_2$, gears are skewed. The torsion deviation value of gear assembly is:

$$\delta_y = \frac{B(n_2 - n_1)}{l} \times 1000 \tag{5.17}$$

式中，δ_y 为齿轮或半个人字齿轮宽度的长度上的扭斜偏差值，μm；l 为两铅丝间的距离，mm；B 为齿轮或半个人字齿轮的宽度，mm。

Where, δ_y is deviation from the length of the width of the gear or half-personal gear, μm; l is distance between two lead wires, mm; B is the width of gear or half personal gear, mm.

可以求出齿轮装配的不平行偏差值为：

The non-parallel deviation value of gear assembly can also be calculated as:

$$\delta_x = \frac{B\Delta C_n}{2l\sin\alpha} \times 1000 \tag{5.18}$$

式中，δ_x 为齿轮或半个齿轮宽度长度上的不平行偏差值，μm；α 为齿轮的啮合角，$\alpha = 20°$；ΔC_n 为沿齿宽不同位置侧间隙的差值，mm。

Where, δ_x is the unparallel deviation on the width and length of gears or half gears, μm; α is mesh angle of gears, $\alpha = 20°$; ΔC_n is difference value of lateral clearance along different tooth width locations, mm.

$$\Delta C_n = n_2 - n_1 \text{ 或 } \Delta C_n = m_1 - m_2 \tag{5.19}$$

5.2 机械设备的安装（Installation of Mechanical Equipment）

机械设备的安装是指按照一定的技术条件，将机器设备或其单独部件正确地安放或牢固地固定在设计位置上，使它们在空间上获得设计的坐标位置，称为机械设备的安装。

The installation of mechanical equipment refers to that mechanical equipment or its accessories are properly placed or firmly fixed in the design position according to certain technical conditions so that they can obtain the coordinate position of design in space, which is called the installation of mechanical equipment.

机械设备的安装是工厂基本建设的重要环节，是机械设备从制造到投入使用的必要过程。机械设备安装的好坏，直接影响机械设备的使用性能和生产的顺利进行，关系到能否正常投产和投产后能否达到高产的重要问题，因此，做好机械设备的安装具有重要的意义。

The installation of mechanical equipment is an important link in the capital construction of a factory. And it is a necessary process from manufacturing to putting into use. The quality of mechanical equipment installation has a direct impact on the performance of mechanical equipment and the smooth production. There is an important relationship between the correct installation of mechanical equipment and the normal operation and the high output after the operation. Therefore, the installation of mechanical equipment is of great significance.

机械设备的安装工艺过程包括基础的验收、安装前的物质和技术准备、设备的吊装、设备安装位置的检测和校正、基础的二次灌浆及试运转等。

The installation process of mechanical equipment includes acceptance of foundation, material and technical preparation before installation, lifting of equipment, detection and correction of equipment installation position, secondary grouting of foundation and test run, etc.

5.2.1 机械安装前的准备工作（Preparation before Installation of Machines）

机械设备的安装之前，有许多准备工作要做。工程质量的好坏、施工速度的快慢都和施工的准备工作有关。

There is much preparatory work, such as construction organization, technical preparation, and supply preparation, to be done before the installation of the mechanical equipment. The quality of the project, the speed of construction are related to the preparation of the construction.

5.2.1.1 组织技术和供应准备（Organization Technical and Supply Preparation）

在进行一项大型设备的安装之前，应该根据当时的情况，结合具体条件成立适当的组织机构。应当分工明确、紧密协作，以使安装工作有步骤地进行。

Before the installation of a large equipment, an appropriate organizational structure should be established according to the current situation and specific conditions. The division of labor should be clear and cooperation should be close, so that the installation work can be carried out step by step.

施工组织是施工管理的指导文件，它是根据工厂基建总体施工组织设计、机械设备图纸等技术资料及领导部门下达的文件为依据制定的。

Construction organization is the guiding document of construction management, which is formulated on the basis of the overall construction organization design of factory infrastructure, mechanical equipment drawings and other technical data and documents issued by management departments.

施工组织的内容包括工程概况、总体规划、工艺流程图、总体施工现场平面图。

The contents of construction organization includes project overview, overall planning, process flow diagram, and overall construction site plan.

工程概况包括工程简明技术性能、工作量、开工和竣工日期、安装时重大技术措施和施工条件等。

The general situation of the project includes the technical performance, workload, commencement date and completion date, major technical measures during installation and construction conditions, etc.

施工统筹图包括施工顺序、设备到货情况、基础支付情况、场地条件、运输条件、施工机械条件。

The overall plan process flow chart includes construction sequence, equipment arrival, foundation delivery, site conditions, transportation conditions, and construction mechanical equipment conditions.

施工总平面图包括场地使用、机具布置、运输通路等综合平面图，注意地下管网、隧道和地面建筑等综合平衡。

The overall construction site plan includes the comprehensive plan of site use, arrangement of mechanical equipment and tools, transportation access, etc. Pay attention to the comprehensive balance of underground pipe network, tunnel and ground building.

A　技术准备（Technical Preparation）

技术准备是机械设备安装前的一项重要准备工作，其主要内容如下：

Technical preparation is an important preparation work before the installation of mechanical equipment. The main contents of technical preparation are as follows:

（1）研究机械设备的图样、说明书、安装工程的施工图、国家部委颁发的机械设备安装规范和质量标准。在施工之前，必须对施工图样进行会审，对工艺布置进行讨论审查，注意发现和解决问题。

The engineers must study the drawings and specifications of mechanical equipment, construction drawings of installation project, and installation specifications and quality standards of mechanical equipment issued by Nation Ministries and Commissions. Before the construction, the construction drawings must be checked, the process layout must be discussed and checked. Problems should be found and solved.

（2）熟悉设备的结构特点和工作原理，掌握机械设备的主要技术数据、技术参数、使用性能和安装特点。

The engineers must be familiar with the structural characteristics and working principles of equipment, master the main technical data, technical parameters, performance and installation characteristics of mechanical equipment.

（3）对安装工人进行必要的技术培训。

Technical training must be arranged for the install workers.

（4）编制安装工程施工作业计划。安装工程施工作业计划应包括安装工程技术要求、安装工程的施工程序、安装工程的施工方法、安装工程所需机具和材料及安装工程的试车步骤、方法和注意事项。

The construction plans of installation project must be determined. The construction plans of installation project shall include the technical requirements of the installation project, the construction procedures of the installation project, the construction methods of the installation project, the machines, tools and materials required for the installation project, test run steps, methods and precautions of the installation project.

B 供应准备（Supply Preparation）

供应准备是安装中一个重要方面。供应准备主要包括机具准备和材料准备。

Supply preparation is an important aspect of installation. Supply preparation includes material preparation and equipment and tools preparation.

（1）机具准备。机具准备是根据设备的安装要求准备各种规格和精度的安装检测机具和起重运输机具。应当认真地进行检查，以免在安装过程中才发现不能使用或发生安全事故。

Equipment and tools preparation. According to the installation requirements of the equipment, various specifications and accuracy of the installation testing equipment and lifting transport equipment should be prepared. And check should be done carefully, so as to avoid equipment and tools cannot be used in the installation process or avoid safety accident.

常用的安装检测机具包括：水平仪、经纬仪、水准仪、准直仪、拉线架、平板、弯管机、电焊机、气焊及气割工具、扳手、万能角度尺、卡尺、塞尺、千分尺、千分表和其他检验测试设备等。

Commonly used installation check tools include: level, theodolite, collimator, cable holder, flat plate, pipe bender, electric welder, gas welding and gas cutting tool, wrench, universal angle ruler, caliper, feeler, micrometer, micrometer, and other check and testing equipment.

起重运输机具包括：双梁、单梁桥式起重机、汽车吊、坦克吊、卷扬机、起重杆、起重滑轮、葫芦、绞盘、千斤顶等起重设备；汽车、拖车、拖拉机等运输设备；钢丝绳、麻绳等索具。

Lifting transport equipment includes: double beam, single beam bridge crane, car crane, tank crane, hoist, lifting rod, lifting pulley, hoist, winch, jack and other lifting equipment, vehicles, trailers, tractors and other transport equipment, rope, hemp rope, and other rigging.

（2）材料准备。安装中所用的材料要事先准备好。对于材料的计划与使用，应当是既要保证安装质量与进度，又要注意降低成本，不能有浪费现象。安装中所需材料主要包括：各种型钢、管材、螺栓、螺母、垫圈、铜皮、铝丝等金属材料；石棉、橡胶、塑料、沥青、煤油、机油、润滑油、棉纱等非金属材料。

Material preparation. The materials used in the installation should be prepared in advance. The planning and use of materials should not only ensure the quality and progress of installation, but also pay attention to reduce the cost, and there should be no waste. The materials required for installation mainly include: Various types of steel sections, pipes, bolts, nuts, washers, copper skin, aluminum wire, and other metal materials; Asbestos, rubber, plastics, asphalt, kerosene, oil, lubricating oil, cotton yarn, and other non-metallic materials.

5.2.1.2 机械的开箱检查与清洗（Equipment Unpack Checking and Cleaning）

机械设备安装前，要和供货方一起进行设备的开箱检查。检查后应做好记录，并且要双方人员签字。设备的检查工作主要包括以下几项。

Before the installation of mechanical equipment, the unpack checking of the equipment should be carried out together with the supplier. Records should be made after checking and signed by both parties. The checking of the equipment mainly includes the following items.

（1）设备表面及包装情况。

Checking the surface and packaging of equipment.

（2）设备装箱单、出厂检查单等技术文件。

Checking technical documents such as packing list of equipment, factory checking list, etc.

（3）根据装箱单清点全部零件及附件。

Checking all parts and accessories according to the packing list.

（4）各零件和部件有无损坏、变形或锈蚀等现象。

Whether there is damage, deformation or rust and other phenomena on the parts and components.

（5）机件各部分尺寸是否与图样要求相符合。

Whether the dimension of the machine parts is the same with those of requirements on drawings.

开箱检查后，为了清除机器、设备部件加工面上的防锈剂，残存在部件内部的铁屑、锈斑及运输保管过程中的灰尘、杂质，必须对机器和设备的部件进行清洗。

After the unpacking check, the parts of the machine and equipment must be cleaned in order to remove the anti-rust agent on the processing surface of the machine and equipment parts, iron scraps and rust spots remaining in the parts, and dust and impurities in the transportation and storage process.

清洗的步骤一般是：粗洗，细洗和精洗。粗洗，主要清除掉部件上的油污、旧油、漆迹和锈斑；细洗，也称油洗，是用清洗油将脏污冲洗干净；精洗，采用清洁的清洗油最后洗净，主要用于安装精度和加工精度都较高的部件。

The steps of cleaning are generally: rough washing, fine washing, and fine cleaning. Rough washing aims to remove the oil, old oil, paint and rust spots on the parts. Fine washing, also known as oil washing, aims to clean the dirt by using of cleaning oil. Fine cleaning aims to clean using cleaning oil finally. Fine cleaning is mainly used for the parts with higher installation accuracy and higher machining accuracy.

5.2.1.3 预装配和预调整（Pre-assembly and Pre-adjustment）

为了缩短工期减少安装时的组装、调整工作量，常常在安装前预先对设备的若干零部件进行预装和预调整，把若干零部件组装成大部件。用这些预先组装好的大部件进行安装，可以大大加快安装进度。预装配和预调整可以提前发现设备存在的问题，及时加以处理，以确保安装的质量。

In order to shorten the construction period and reduce the work load of assembly and adjustment during installation, some parts and components of the equipment are usually pre-installed and pre-adjusted before installation, and some parts and components are assembled into large parts. Pre-installation and pre-adjustment can detect the problems of equipment in advance and deal with them in time to ensure the quality of installation.

5.2.2 基础的设计与施工（Design and Construction of Foundation）

5.2.2.1 机器基础的作用（The Function of the Machine Foundation）

机器基础的作用不仅是把机器牢固地固定在要求的位置上，还是把机器本身的重量和工作时的作用力传递到土壤中云，并吸收振动。所以机器基础是设备中重要的组成部分，如果机器基础设计和施工不正确，不但会影响机器设备本身的精度、寿命和产品的质量，甚至会损害周围厂房和设备结构。

The function of the machine foundation not only holds the machine firmly in the required position, but also transfers the weight and working force of the machine into the soil and vibration can be absorbed. Therefore, the machine foundation is an important part of the equipment. If the design and construction of the machine foundation are not correct, it will not only affect the accuracy, life and product quality of the machine equipment itself, but also damage the surrounding plant and equipment structure.

5.2.2.2 机器基础的设计内容（Content of Design the Machine Foundation）

机器基础的设计包括根据机器的结构特点、动力作用的性质，选择基础的类型，在坚固和经济条件下，确定基础最合适的尺寸和强度等。

The design of the machine foundation includes selecting the type of foundation according to the structural characteristics of the machine and the nature of dynamic action, and determining the most suitable dimension and strength of the foundation under the limits of strength and economy.

5.2.2.3 机器基础的分类（Classification of Machine Foundation）

根据基础的结构，机器基础可以分为两类：一类是大块式（刚性）基础；另一类是构架式（非刚性）基础。大块式基础建成大块状、连续大块状或板状，其中开有机器、辅助设备和管道安装所必需的以及在使用过程中供管理用的坑、沟和孔。

According to the structure of the foundation, machine foundation can be divided into two categories: one is bulk (rigid) foundation, the other is architectural (non-rigid) foundations. Bulk foundations are built in large, continuous blocks or plate with pits, ditches, and holes necessary for the installation of machines, auxiliary equipment and pipelines and for management.

大块式基础应用最为广泛，可以安装所有类型的机器设备，尤其是有曲柄连杆的机器，还适用于安装绝大部分的破碎机、大部分电机（主要是小功率和中功率的电机）等。

锻锤则只能建造大块式基础，而构架式基础一般仅用来安装高频率的机器设备。

Bulk foundations is widely used for installing all kinds of machine equipment, especially the machine with crank connecting rod, also apply to install the majority of crushers, most motor (mainly small power and medium power motor) and so on. Large blocks of foundation can be built for forging hammers only. Frame foundations are usually used only for high frequency equipment.

5.2.2.4 机器基础的检查与验收（Check and Acceptance of the Machine Foundation）

基础浇灌凝固后应进行质量检查与验收。主要检查基础的尺寸位置偏差是否符合机器安装的要求。

Quality check and acceptance of foundation shall be carried out after solidification of foundation grouting. Whether the dimension and position deviation of the foundation meet the requirements of machine installation is mainly checked.

安装重型机器设备时，为了防止安装后基础下沉或倾斜而破坏机器的正常运转，应在安装前对基础进行预压。基础养生期满后，在基础上压重物（钢板或铸件等），其重量为设备自重的两倍与最大机件重量之和。用水准仪每天观测，直到测出基础不再下沉。

When installing heavy machines, the foundation shall be preloaded before installation in order to prevent the foundation sinking or tilting after installation from damaging the normal operation of the machine. After the expiration of the basic curing period, a heavy weight (steel plate or casting, etc.) is pressed on the foundation, whose weight is twice the weight of the equipment plus the maximum machine weight. A level gauge is used to daily observe until the foundation is no longer sinking.

机器设备正式安装前要认真清理基础表面，除去表面灰土、浮浆和油污。在基础的上部表面，除放置垫板的位置外，需要二次灌浆的地方都应铲麻面以保证基础与二次灌浆层能结合牢固。铲麻面要求每 $10000mm^2$ 面积有 2~3 个小坑。小坑深 10~20mm。

Before the formal installation of the machine and equipment, the foundation surface should be carefully cleaned to remove the surface dust, floating slurry and grease. On the upper surface of the foundation, in addition to the position where the backup plate is placed, all areas requiring secondary grouting should be shoveled to ensure that the foundation can be firmly combined with the secondary grouting layer. Shoveling surface requires 2~3 small pits per $10000mm^2$ area. The pit is 10~20mm deep.

5.2.2.5 基础螺丝的种类（Classification of Foundation Bolt）

基础螺丝又称地脚螺丝，作用是固定所安装的机器设备。根据地脚螺栓与基础的连接形式不同，地脚螺栓分为固定地脚螺栓、活动地脚螺栓和锚定地脚螺栓三种类型。

The foundation bolt is used to fix the machine equipment installed. According to the different connection form of the anchor bolt and the foundation, the foundation bolt is divided into three types: fixed foundation bolt, movable foundation bolt and anchor foundation bolt.

A 固定式基础螺栓的种类（Classification of Fixed Foundation Bolt）

固定式基础螺栓常用的形式有全埋式、半埋式和预留孔式三种，如图 5.31 所示。

There are three commonly used types of fixed foundation bolt. They are fully buried type, semi-buried type and reserved hole type, as shown in Figure 5.31.

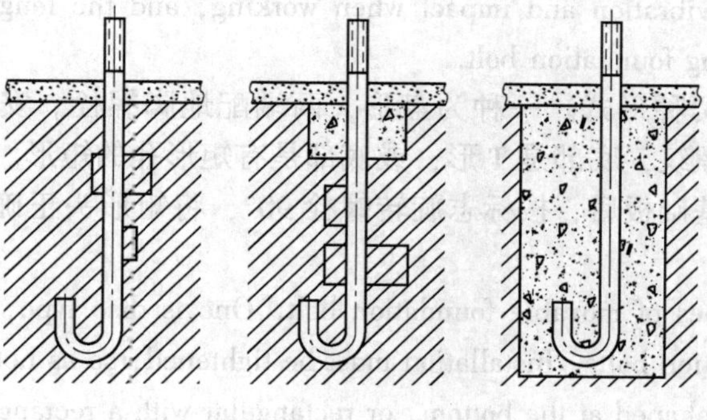

图 5.31 地脚螺丝的形式

Figure 5.31 Type of fixed foundation bolt

（a）全埋式（Full-buried）；（b）半埋式（Half buried）；

（c）预留孔式（The reserved hole）

（1）全埋固定式基础螺栓。全埋式是把地脚螺丝和金属固定架先连在一起（焊接或结扎），再把它们都浇灌在基础混凝土之中。全埋式的缺点是固定架留在基础中消耗大量钢材。若浇捣基础时地脚螺丝浇偏，事后不易校正。

Fully buried fixed foundation bolt. In the fully buried type, the anchor screws are first connected (welded or ligated) tc the metal mounting and then poured into the base concrete. The disadvantage of the fully buried type is that a lot of steel is needed to fixed frame to keep foundation bolts staying in the foundation. If there are deviations of foundation bolts position when foundation grouting, it is difficult to correct.

（2）半埋固定式基础螺栓。半埋式是在基础上部留一定深度的调整孔，可以弥补地脚螺丝浇偏不易校正的缺点。

Semi-buried fixed foundation bolt. The semi-buried type is to leave a certain depth of adjustment hole in the upper part of the foundation, which can make up for the disadvantage that the grouting deviation of the anchor screw is not easy to correct.

（3）预留孔固定式基础螺栓。预留孔式是在浇灌基础时把地脚螺丝孔位置全部留出（放置木壳板即可），待机器安装找正后再用水泥砂浆补灌螺丝孔。预留孔式虽施工简单，但牢度较差，不宜用于矿山、冶金等重型机器设备的安装。

Reserved hole fixed foundation bolt. The reserved hole type is to leave all the position of the anchor screw hole when grouting the foundation (just place the wooden shell plate), and then use cement mortar to fill the screw hole after the machine is installed and rectified. Although the reserved hole type is simple in construction, it has poor fastness and is not suitable for the installation of heavy machines such as mining and metallurgy.

B 活动地脚螺栓的分类（Classification of Movable Foundation Bolt）

活动地脚螺栓（如图 5.32 所示）一般用于安装工作时振动、冲击较大的重型设备，长度一般为 1~4m，属于长地脚螺栓。

The movable foundation bolt (as shown in Figure 5.32) is generally used to fix heavy equipment with violent vibration and impact when working, and the length is generally 1 ~ 4m, which belongs to the long foundation bolt.

活动地脚螺栓有两种形式。一种为盘形，两端配地脚螺栓。安装必须紧固，以免松动。另一种是顶部有螺纹，底部是T形，或底部是有矩形孔的矩形。安装时，在螺栓顶面标明安装方向。插入基础板后，按标志旋转螺栓90°，将矩形头正确放入基础板槽中。其材料一般为灰铸铁。

There are two shapes of movable foundation bolt. One is disc type, and it is matched with both ends of the foundation bolts. Installation must be tightened, so as not to loosen. The other is threaded at the top, T-shaped at the bottom, or rectangular with a rectangular hole at the bottom. During installation, mark the direction on the top surface of the bolt. After inserting the foundation plate, rotate the bolt by 90° according to the mark to put the rectangular head into the foundation plate slot correctly. Its material is generally gray cast iron.

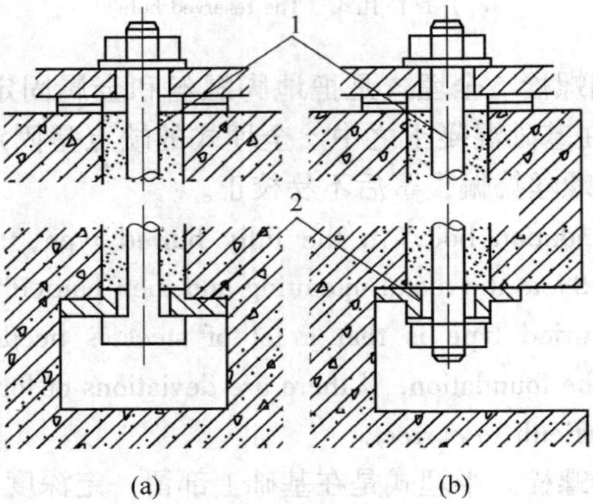

图 5.32 活动地脚螺栓类型

Figure 5.32 Type of movable foundation bolt

(a) 锤式（Hammer type）；(b) 螺柱式（Stud type）

1—活动地脚螺栓（movable foundation bolt）；2—基础板（foundation plate）

活动地脚螺栓拆卸方便，不与地基一起灌浆。在基础上预留活动地脚螺栓孔，孔下埋设基础板。移动设备或更换地脚螺栓时，可移动地脚螺栓易于拆卸。

Movable foundation bolts are easy to be removed, and are not grouted together with the foundation. Holes for movable foundation bolts are reserved in the foundation, and foundation plates are embedded under the holes. Movable foundation bolts can be easily removed when moving equipment or replacing anchor bolts.

锚式地脚螺栓（膨胀螺栓）是一种新型地脚螺栓。这种地脚螺栓结构复杂，生产成本高。基础施工中，无预埋地脚螺栓，无预留孔。安装时，先在基础上钻孔，再安装地脚螺栓。

Anchor foundation bolt (expansion bolt) is a new type of foundation bolt. This kind of foundation bolt has complex structure and high production cost. In foundation construction, there is no embedded foundation bolt and no reserved hole. When installing, drill holes on the foundation and install anchor bolt again.

5.2.2.6 安装地脚螺栓的要求（Requirements for Mounting Anchor Bolts）

安装地脚螺栓的要求如下：

Requirements for mounting anchor bolts are as follows:

（1）地脚螺栓垂直余量为0.01mm。

Vertical allowance of foundation bolts is 0.01mm.

（2）地脚螺栓与地脚螺栓孔壁的距离应大于15mm。

The distance from the foundation bolt to the hole wall of the foundation bolt should be greater than 15mm.

（3）地脚螺栓上的弯钩下端不得接触孔底。

The bottom end of the bent hook on foundation bolt shall not touch the bottom of the hole.

（4）地脚螺栓到达孔壁时的脚螺栓应大于15mm时应润滑地脚螺栓。

The foundation bolt should be lubricated when the foundation bolt to the hole wall of the foundation bolt should be greater than 15mm.

（5）地脚螺栓安装时应润滑螺纹部分，清理杆上的油脂和污垢。

The threaded part of the foundation bolt shall be lubricated during installation, and the grease and dirt on the rod should be cleaned.

（6）待地脚螺栓混凝土固化达到设计强度的75%后，拧紧地脚螺栓螺母。拧紧螺母时，均匀用力，按对称顺序拧紧数次。螺母、垫片、设备底座接触良好。拧紧螺母后，地脚螺栓必须高于螺母1.5~5倍螺距。

Tighten the nuts of the foundation bolts after the concrete solidification of the anchor bolts reaches 75% of the design strength. When tightening nuts, use uniform force in symmetrical order, and tighten for several times. The contact of nuts, gaskets and equipment base shall be good. After the nut is tightened, the foundation bolt must be higher than nut with 1.5~5 times pitch.

（7）安装时设备应均匀落地，地脚螺栓螺纹部分不得损坏。

The equipment should be evenly falling during installation, and the threaded part of the foundation bolt should not be damaged.

5.2.3 安装过程（Installation Procedure）

机械设备的安装，重点要注意设置安装基准、设置垫板、设备吊装、找正找平找标高、二次灌浆、试运行等几个问题。机器设备在基础上安装的情况如图5.33所示。

In the installation of mechanical equipment, it is important to pay attention to setting installation references, setting backup plate, lifting equipment, leveling and elevation finding, secondary grouting, test run and other problems. The installation of the machine on the base are shown in Figure 5.33.

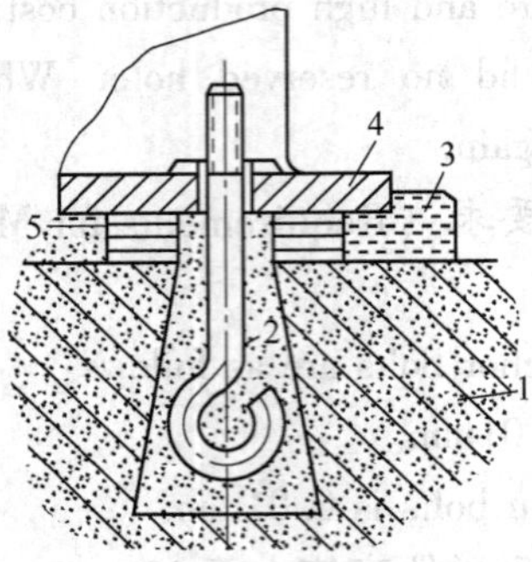

图 5.33　机器在基础上的安装

Figure 5.33　The installation of the machine on the base

1—混凝土基础（concrete foundation）；2—基础螺丝（foundation bolt）；
3—二次灌浆（secondary grouting）；4—机座（base of machine）；5—垫板（backup plate）

5.2.3.1　测量安装偏差的依据（References for Measuring the Deviation of Installation）

机器安装时，根据纵横中心线来调整其前后左右的位置，根据标高按基准点来调整上下的位置。这样就可利用中心线和基准点来确定机器在空间的坐标。

When the machine is installed, its front, back, left and right positions should be adjusted according to the vertical and horizontal center line, and the upper and lower positions should be adjusted according to the elevation and the reference point. In this way, the center line and reference point can be used to determine the coordinates of the machine in space.

决定中心线位置的标记称为中心标板，标高的标记称为基准点。

The mark that determines the position of the center line is called the center plate, and the mark of elevation is called the reference point.

(1) 中心标板。中心标板是一段长度为 150~200mm 的钢轨或工字钢、槽钢、角钢等。用水泥浆把它非常牢固埋设在相应设备的基础表面，然后用经纬仪测出机器设备安装中心线并向中心标板上投出中心标点。通常以冲孔的点表示中心标点并用红漆或白漆划圈于点外以示明显标志。中心标板如图 5.34 所示。

Center plate. Center plate is a length of 150~200mm rails or beams, channels, angles and so on. Water and mud buried it firmly in the base surface of the corresponding device. Then measured using theodolite equipment installed center-line to cast a center punctuation mark center board. Usually indicates the center point of punching punctuation circle with a red or white paint on the outside to show clear signs point. Figure 5.34 is the schematic diagram of center plate.

(2) 基准点。基准点在机器设备的基础表面靠近边缘处埋设铆钉，并根据每个工厂的永久性基准点（某一海拔高度），测出这个铆钉的标高（用红漆标出）。该铆钉及其标高就作为机器设备安装时找标高的依据，这种作用的铆钉称为基准点，如图 5.35 所示。基准点是机器设备安装时找标高的依据，可以用水准仪测量。

Reference point. Reference point on the basis of the surface near the edge of the machine equipment buried rivet and according to a permanent reference point for each plant (a certain height above sea level), measured the elevation of the rivet (marked with red paint). The rivet and its elevation to find the elevation of the machine is taken as installation reference. Figure 5.35 is the schematic diagram of reference point. This effect is called the reference point of the rivet. The reference point is the basis for finding the elevation when the machine and equipment are installed.

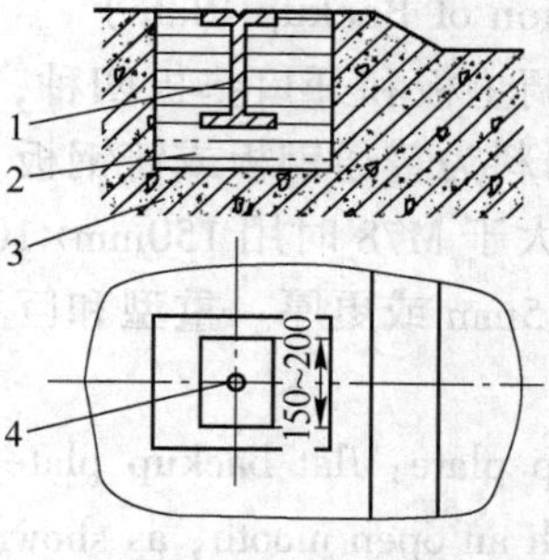

图 5.34　中心标板

Figure 5.34　The schematic diagram of center plate

1—I 型钢（I -beam）；2—水泥砂浆（cement martar）；3—基础（foundation）；4—中心标板（center plate）

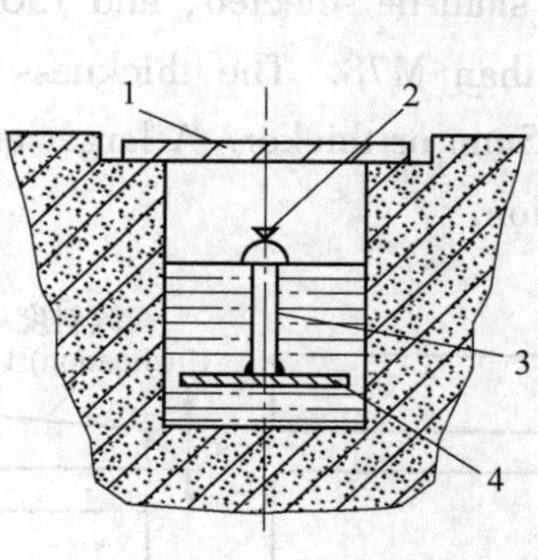

图 5.35　基准点

Figure 5.35　The schematic diagram of reference point

1—保护盖板（protection cover plate）；2—参考点（reference point）；3—铆钉（rivet）；4—基板（baseboard）

5.2.3.2　设置垫板（Setting up the Backup Plate）

垫板放置在机器底座与基础表面之间。

The backup plate is placed between the machine base and the surface of foundation.

A　垫板的作用（Function of Backup Plate）

垫板有以下四个作用：

The function of the backup plate includes the following four items:

（1）利用调整垫板的高度可调节机器设备的标高和水平。

The elevation and level of machine can be adjusted by adjusting the height of backup plates.

（2）通过垫板把机器重量和工作载荷均匀地传给基础。

The work load and the weight of the machine can be passed to foundation evenly through backup plate.

（3）使机器底面与基础之间保持一定距离，以便二次灌浆能充满机器底部空间。

To keep a certain distance between the machine base and foundation, so that the secondary grouting can fill the space between the bottom of the machine and the surface of foundation.

（4）在特殊情况下可通过垫板校正底座的变形。

The deformation of machine base can be corrected through backup plate under special circumstances.

B　垫板的分类（Classification of Backup Plate）

垫板分平垫板、斜垫板、可调垫板和开口垫板四种，如图 5.36 所示。斜垫板的斜度为（1∶15）～（1∶50）。垫板材料为普通钢板或铸钢板。地脚螺丝直径小于 M78 时，选用长×宽为 100mm×75mm 垫板，大于 M78 时用 150mm×100mm 垫板。垫板厚度有 0.5mm、1.0mm、2.0mm、3.0mm，直到 15mm 或更厚。重型和巨型设备安装，可采用更大面积的垫板。

There are four kinds of backup plate: flat backup plate, oblique backup plate, adjustable backup plate, and backup plate with an open mooth, as shown in Figure 5.36. The inclination of the oblique backup plate is 1∶15～1∶50. The backup plate is made of ordinary steel plate or cast steel plate. When the diameter of the foundation screw is smaller than M78, the 100mm×75mm length × width backup plate shall be selected, and 150mm×100mm backup plate shall be used when the diameter is larger than M78. The thickness of backup plate can be 0.5mm, 1.0mm, 2.0mm, 3.0mm, up to 15mm or thicker. A larger area of backup plate can be used for heavy and huge equipment installation.

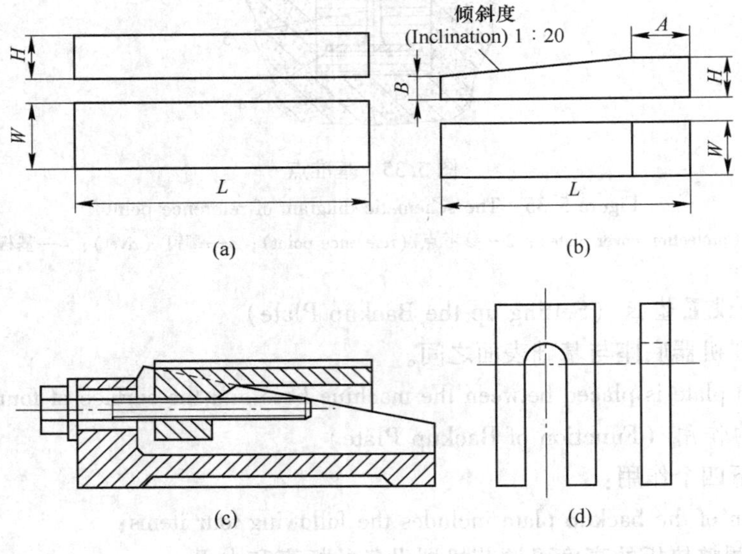

图 5.36　垫板的种类

Figure 5.36　The type of backup plate

（a）平垫板（Flat backup plate）；（b）斜垫板（Oblique backup plate）；
（c）可调垫板（Adjustable backup plate）；（d）开口垫板（Backup plate with an open mouth）

垫板通常在地脚螺栓的两侧对称布置，也可以根据地脚螺栓的间距和设备基础的形状与重量酌情布置，如图 5.37 所示。

Backup plates are usually arranged symmetrically on both sides of the foundation bolts, or according to the spacing of the foundation bolts and the shape and weight of the equipment foundation, as shown in Figure 5.37.

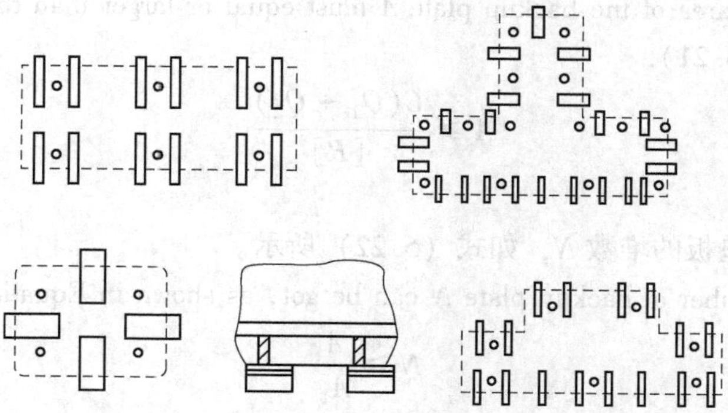

图 5.37　垫板的布置

Figure 5.37　The layout of backup plate

C　垫板的计算（Calculation of Backup Plate）

垫板对混凝土基础的单位压力 p 应等于或小于混凝土基础表面的抗压强度［R］，如式（5.20）所示。

The unit pressure p of backup plate to concrete foundation should be equal or smaller than the compress strength on the surface of the concrete foundation［R］, as shown in Equation（5.20）.

$$p = \frac{C(Q_1 + Q_2)}{A} \leqslant [R] \tag{5.20}$$

式中，C 为安全系数，$C=1.5\sim3.0$，轻型机器安装取较小值；Q_1 为机器设备总重量，N；Q_2 为基础螺丝紧固力，N，$Q_2=[\sigma]F$；$[\sigma]$ 为基础螺丝材料许用拉应力，MPa；F 为基础螺丝总有效截面积，mm^2，$F=n\pi d^2/4$；n 为螺丝的个数；d 为螺丝的直径，mm；N 为垫板的堆数；L 为垫板的长度，mm；B 为垫板的宽度，mm；$[R]$ 为混凝土抗压强度（取混凝土设计标号），MPa。

Where, C is safety factor, $C=1.5\sim3.0$, 1.5 is for light machine, 3.0 is for heavy machine; Q_1 is weight of machine, N; Q_2 is fastening force of base screw, N, $Q_2=[\sigma]F$; $[\sigma]$ is the ultimate tensile stress of foundation bolt material, MPa; F is total effective cross-sectional area of foundation screws, mm^2, $F=n\pi d^2/4$; n is the number of screws; d is the diameter of screws, mm; N is the pile number of backup plate; L is the length of backup plate, mm; B is the width of backup plate, mm; $[R]$ is the ultimate foundation compressive strength (taken from the design label of concrete), MPa.

混凝土设计标号就是浇捣好了的混凝土经养生 28 天后可达到的抗压强度（MPa）。通常有 10MPa、15MPa、20MPa、25MPa、30MPa、40MPa、50MPa 七种抗压强度可选择。重要设备的基础抗压强度应大于 30MPa。

The design label of concrete is the compressive strength (MPa) of grouted concrete after 28 days of curing. Usually there are seven kinds of compressive strength to choose from: 10MPa,

15MPa，20MPa，25MPa，30MPa，40MPa，50MPa. The basic compressive strength of important equipment should be more than 30MPa.

因此，垫板的总面积 A 必须等于或大于所需的面积，如式（5.21）所示。

Therefore，total area of the backup plate A must equal or larger than the area of needed，as shown in Equation（5.21）.

$$A \geqslant \frac{C(Q_1 + Q_2)}{[R]} \tag{5.21}$$

因此，可求得垫板的堆数 N，如式（5.22）所示。

So，the pile number of backup plate N can be got，as shown in Equation（5.22）.

$$N \geqslant \frac{A}{A_1} \tag{5.22}$$

式中，A_1 为一块垫板的面积，$A_1 = L \times B$；L 为一块垫板的长度，mm；B 为一块垫板的宽度，mm。

Where，A_1 is the area of a backup plate，$A_1 = L \times B$；L is the length of backup plate，mm；B is the width of backup plate，mm.

N 应该是一个比计算结果大的整数。

N should be an integer larger than calculation results.

D 垫板的位置和放置要求（Position and Placement Requirements of Backup Plate）

垫板的位置和放置要求如下：

Position and placement requirements of backup plate：

（1）每堆垫板的组合可以是厚度不同的平垫板组合、一对斜垫板组合、一对斜垫板加平垫板组合。

Different thickness of the backup plates can be combined into a combination of backup plate，such as a combination of flat backup plates，a combination of oblique backup plates or a combination of flat backup plates and oblique backup plates.

（2）要求垫板长边与地脚螺栓边缘的距离应为地脚螺栓直径的两倍。要求垫板短边露出机器底座边缘的长度应该为25~30mm，以便于调整。

The required distance between the long edge of backup plate and the edge of the foundation bolt should be twice the diameter of foundation bolt and the required distance between the short edge of backup plate and the edge of the machine base should be 25 ~ 30mm in order to be adjusted easily.

（3）每堆垫板块数应尽可能少，厚的在下层，以保证刚度和可靠性。

The numbers of backup plate in each pile should be as small as possible. The thicker backup plate should be in the lower layer to ensure stiffness and reliability.

（4）垫板高度应在50~120mm，以便于二次灌浆。

The height of the backup plate should be 50~120mm to facilitate secondary grouting.

（5）垫板应磨去飞边和毛刺，可保证平整与良好接触，以免机器投产后垫板松动。

The trimming and burr of backup plate should be ground off to ensure smooth and good contact between backup plates. Thus, the loose of the backup plate can be avoided after the machine put into production.

（6）二次灌浆前必须把每准垫板组点焊在一起。

Before the second grouting, each pile of backup plate groups must be point welded together.

垫板与基础表面的研磨先用磨石（或砂轮片）研磨基础表面，再用垫板与研磨表面磨合并使接触面积达 70%以上（用涂色法检查）。基础研磨面水平性要求为 0.1～0.5mm/m（安装轧机为 0.1mm/m）。

The surface of foundation should be ground with the grindstone (or grinding wheel) first. And then the surface of backup plate is ground to make the contact area of the surface of backup plate and the surface of foundation reach more than 70% (checked with the painting method). The horizontal requirement of foundation grinding surface is 0.1～0.5mm/m (0.1mm/m for installation of rolling mill).

例 5.3 齿轮座地脚板如图 5.38 所示，机器设备总重量 Q_1=600kN，用地脚螺丝直径为 60mm，数量是 8 个，地脚螺丝的许用拉伸应力 $[\sigma]$ = 100MPa，基础抗压强度 $[R]$ = 25MPa，安全系数为 3，要选一块面积为 90mm×240mm 垫板多少堆合适？画出垫板放置的示意图，并在图上标出垫板长边与地脚螺栓边缘的距离以及垫板短边露出机器底座边缘的长度。

Example 5.3 The distribution of gear box foundation bolts is shown in Figure 5.38. The weight of gear box Q_1 is 600kN. The number and the diameter of foundation bolt is 8 and 60mm. The ultimate tensile stress of foundation bolt material $[\sigma]$ is 100 MPa. The ultimate foundation compressive strength $[R]$ is 25MPa. The safety factor is 3. The dimension of one backup plate is 90mm×240mm. How many piles of the backup plate will be needed? Please draw all of the backup plates and the required distance between the long edge of backup plate and the edge of the foundation bolt and the required distance between the short edge of backup plate and the edge of the machine base on the Figure 5.38.

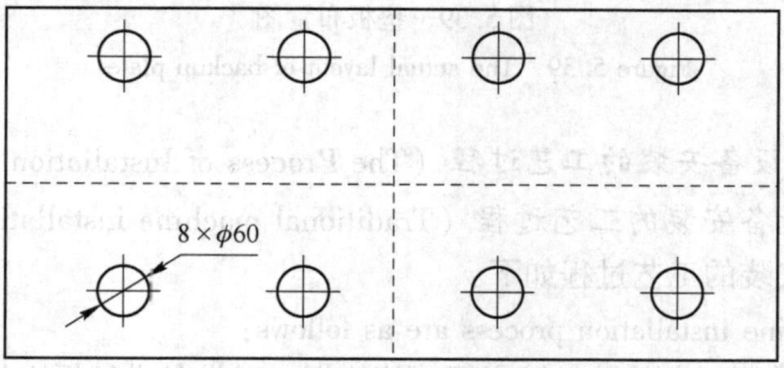

图 5.38 设备底板

Figure 5.38 The equipment base

解（Solution）：

（1）计算地脚螺丝紧固力 Q_2（Calculation of bolt fastening force Q_2）：

$$Q_2 = \frac{\pi d^2}{4}[\sigma] \cdot n = \frac{\pi}{4} \times 60^2 \times 100 \times 8 = 2261946.7N \approx 2261.9kN$$

（2）计算垫板的总面积 A（Total area of backup plates）：

$$A = \frac{Q_1 + Q_2}{[R]} C = \frac{(600 + 2261.9) \times 10^3 \times 3}{25} \text{mm}^2 = 343428\text{mm}^2$$

（3）计算一块垫板的面积（Area of one backup plate A_1）：

$$A_1 = 240 \times 90 = 21600\text{mm}^2$$

（4）计算垫板的堆数（Piles of the backup plate）：

$$N \geqslant \frac{A}{A_1} = \frac{A}{21600} = \frac{343428}{21600} = 15.899$$

取 $N=16$。

$N=16$ is the result.

垫板放置的示意图，以及垫板长边与地脚螺栓边缘的距离和垫板短边露出机器底座边缘的长度均标注如图 5.39 所示。

The distribution of the backup plate and the required distance between the long edge of backup plate and the edge of the foundation bolt and the required distance between the short edge of backup plate and the edge of the machine are on the Figure 5.39.

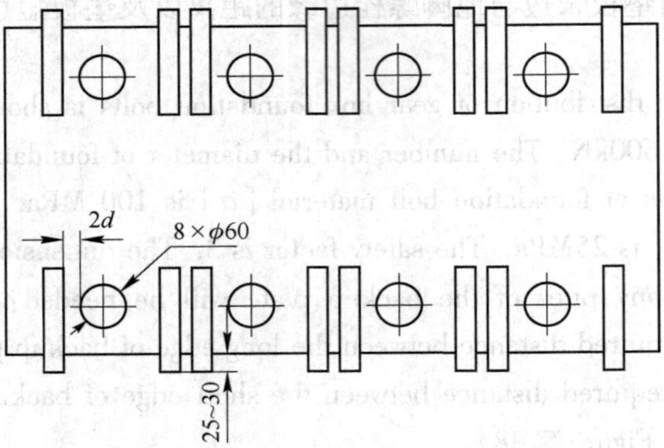

图 5.39 垫板布置图

Figure 5.39 The actual layout of backup plates

5.2.3.3 机器设备安装的工艺过程（The Process of Installation of Machine）

A 传统机器设备安装的工艺过程（Traditional machine installation process）

传统机器设备安装的工艺过程如下：

Traditional machine installation process are as follows:

（1）验收基础质量，在基础上铲麻面、放垫板、对设备进行拆洗检查或预装。

Check the foundation quality, shovel the pockmarked surface, place the backup plates, disassemble and wash the equipment and pre-assemble.

（2）安装并调整设备使之具有正确的位置（找平找正找标高）、紧固基础螺丝并复查位置的正确性。

Install and adjust the equipment so that it has the correct position (alignment, level and elevation), tighten the foundation screw and check the correct position again.

（3）二次灌浆。

Secondary grouting.

（4）试运转。

Test run.

B 机器设备安装的新工艺（New Technology of Machines Installation）

机器设备安装的新工艺包括：环氧树脂砂浆黏结基础螺丝、三点安装法、座浆法安装、垫板安装和无垫板安装技术。

New technology of machines installation includes epoxy resin mortar bond foundation bolt, three-paint installation method, slurry installation method, backup plate installation method and installation without backup plate technology.

（1）环氧树脂砂浆黏结基础螺丝。环氧树脂砂浆黏结基础螺丝不需要固定架，可节省大量钢材和劳动量。

Epoxy resin mortar bond foundation bolt. Epoxy resin mortar bond foundation bolt does not require a fixed frame and can save a lot of steel and labor.

（2）三点安装法。不同于垫板安装法，可利用三点决定平面的原理安装。在机器底座下适当位置先放置三对斜垫板或三个斜铁器，可使“三找”很快达到要求。此时，在需要放置垫板的其他位置打入相应高度的平垫板组，收紧基础螺丝再复查、正、平、高，确认无误后即可二次灌浆。

Three-point installation method. Three-point installation method is different from the mounting method of backup plate. Three points can determine a plane of installation. Three backup plates are placed in proper position under the machine base, the requirements can be met very quickly through adjusting the alignment, level and elevation. At this time, the flat backup plate groups with the corresponcing height are inserted in other places where the backup plate needs to be placed. The foundation bolts are tightened and then the alignment, level and elevation are checked again. The second grouting can be finished if there is no mistake.

（3）座浆法安装。直接用高强度微膨胀座浆混凝土埋置垫板。在座浆垫板上叠加垫板来调整机器的标高和水平工效很高。

Slurry installation method. The slurry installation method refers to that backup plates are embedded directly with high-strength micro-expansion slurry concretes. The backup plates are added on the slurry backup plates to adjust the elevation and level of a machine. The efficiency of this method is very high.

（4）垫板安装。利用斜铁找正机器（包括中心、水平、标高），然后用座浆混凝土二次灌浆，混凝土养护后抽去斜铁，其空洞再次补灌。采用三点安装原理和座浆混凝土二次灌浆，则可使安装工程达到多快好省。

Backup plate installation method. Backup plate installation method relies on second grouting after adjusting the alignment, the level and the elevation of machine. The oblique backup plates are taken away after curing. Three-point installation principle and secondary grouting are adopted to make the installation project quicker and more economical.

（5）无垫板安装技术。无垫板安装技术即设备安装不用垫板的技术。过去安装设备必须用垫板，而垫板埋于二次灌浆层里，不能回收，耗量大。以1700热连轧机为例，一台轧机的底座就用了6.4t经机械加工的垫板。粗略估计，在正常建设年份，全国一年用于垫板的钢材追近10000t。

Installation without backup plate technology. Installation without backup plate technology is the technology that backup plate is not necessary for installation of machine. In the past, it was necessary to use the backup plate for installation of equipment, but the backup plate buried in the secondary grouting layer could not be recycled, and the consumption was too large. For example, for 1700 hot continuously rolling mill, 6.4t of machined backup plates are used for the base of one rolling mill. It is roughly estimated that in the normal construction year, nearly 10000t of steel are used for backup plate nationwide.

无垫板安装技术的关键是采用了新开发的早强高标号微膨胀且能自流灌浆的浇筑料，将此浇筑料填充到二次灌浆层后，由于浇筑料的微膨胀，二次灌浆层与设备底座下平面贴实，从而起到承载作用，因此垫板的承载作用便可被代替。

Installation without backup plate technology is the key to adopt the new development of latest strength grade micro expansion and casting of gravity grouting material with grouting by itself. The secondary grouting layer and the surface under the equipment base are pressed to withstand the weight and the force of machine after the second grouting layer due to the micro expansion of the pouring material. Therefore, the function of backup plates can be replaced.

垫板的另一找平、找标高作用，则可以用微调千斤顶或斜铁器来代替，将它们放在原来该放垫板之处，用以调整机器的空间位置。调整完毕，紧固地脚螺栓，在它们的周围搭设木模板再进行二次灌浆，3天后脱去木模板，取出微调千斤顶或斜铁器，以便回收利用，将它们遗留的空穴以普通混凝土填充，再将二次灌浆层周边用水泥砂浆抹平。

Another leveling and elevation adjusting function of the backup plate can be replaced by fine-tuning jack or inclined iron, which can be placed where the backup plate was originally placed to adjust the spatial position of the machine. After the adjustment, the anchor bolts are tightened. The wooden formworks are set up around them and the second grouting is finished. After 3 days, the wooden formworks are removed. The fine-tuning jack or inclined iron are taken out for recycling. The holes left are filled with ordinary concrete. And the surrounding area of the second grouting layer is smoothed with cement mortar.

5.2.3.4 机械设备安装位置的检测与调整（Installation Position Checking and Adjusting of Mechanical Equipment）

（1）机械设备安装位置的检测与调整的目的：调整机器设备的中心、水平和标高的实际偏差达到允许偏差之内。

The purpose of checking and adjusting the installation position of machine is adjusting the actual deviation of the alignment, level and elevation of the machine within the allowable deviation.

（2）机械设备安装位置的检测与调整的内容。机械设备安装中反复检测调整过程称为找正、找平、找标高。

Content of checking and adjusting the installation position of machine. The content of check repeatedly and adjustment includes alignment check and adjusting, leveling check and adjusting and elevation check and adjusting.

找正是使机器设备的中心线对正安装中心线的过程。常用于找正的方法是激光法如图5.40所示。拨动设备的工具可采用撬棍、大锤和楔铁，也可以使用千斤顶等。

Alignment testing and adjusting is a process of aligning the center line of the machine and equipment with the center line. Common method to check alignment is laser method. as shown in Figure 5.40. Crowbars, sledgehammers, wedge irons, or jacks and be used as common tools to move the device in order to adjust alignment.

找平就是把设备调整到要求的水平度（或垂直度）的过程。常用于找平的工具是水平仪。

Checking and adjusting leveling is the process of adjusting the equipment to the required level (or perpendicularity). Common tool to check leveling is using the level.

找标高就是把设备的高度位置调整到设计高度的过程。常用于找标高的方法是改变垫板高度。

Checking and adjusting elevation finding are the process of adjusting the equipment height to the design height. Common methods to adjust elevation is changing the height of the backup plate.

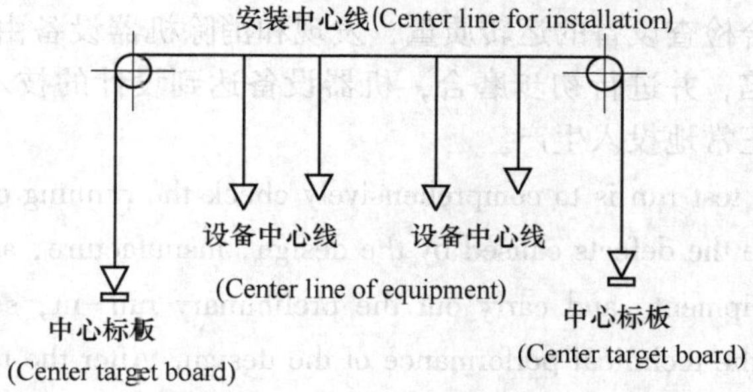

图5.40　机械设备安装找正示意图

Figure 5.40　Installation alignment diagram of mechanical equipment

5.2.3.5　二次灌浆（Secondary Grouting）

由于垫板的作用，基础表面与机器底座下部会形成空洞，这些空洞必须在机器投产前用混凝土填满，这一作业称为二次灌浆。

Because of the backup plate, there will be space between the surface of the foundation and the machine base. The space must be filled with concrete before the machine is put into production. This job is called secondary grouting.

灌浆时的注意事项如下：

Tips for secondary grouting are as follows:

(1) 要清除二次灌浆处的混凝土表面上的油污、杂物和浮灰。

The grease, debris and floating ash on the concrete surface must be removed before the second grouting.

（2）用清水冲洗表面。

The surface must be washed with water.

（3）小心放置模板，以免碰动已找正的设备。

The templates used for secondary grouting should be carefully placed to avoid destroying the alignment of machine.

（4）灌浆工作应连续完成。

Grouting work should be completed continuously.

（5）灌浆后要浇水养护。

The surfaces after the second grouting should be watered for curing.

（6）拆模板时要防止已调整好设备的位置变动，拆除模板后要将二次灌浆层周边用水泥砂浆抹平。

The location of the equipment has been adjusted should not be changed when removing the templates. The cement mortar of the secondary grouting layer should be smoothed.

5.2.3.6 机械设备安装后的试运转（Test Run after Installation of Mechanical Equipment）

试运转是机械设备安装中最后的，也是最重要的阶段。

The test run is the last and most important stage in the installation of mechanicd equipment.

试运转目的是综合检查设备的运转质量，发现和消除机器设备由于设计制造、装配和安装等原因造成的缺陷，并进行初步磨合，机器设备达到设计的技术性能。经过试运转，机械设备就可按要求正常地投入生产。

The purpose of the test run is to comprehensively check the running quality of the equipment to find out and eliminate the defects caused by the design, manufacture, assembly and installation of the machine and equipment, and carry out the preliminary run-in, so that the machine and equipment can achieve the technical performance of the design. After the test run, the mechanical equipment can be put into production normally as required.

由于机械设备种类和型号繁多，试运转涉及的问题面较广，所以安装人员在试运转之前一定要认真熟悉有关技术资料，掌握设备的结构性能和安全操作规程，以便做好试运转工作。

Due to the various types of mechanical equipment, the test run involves a wide range of problems, so the installation personnel must be carefully familiar with the relevant technical data before the test run, grasp the structural performance and safe operation rules of the equipment, in order to do a good job in the test run.

A 试运转前的检查内容（Content of Check before Test Run）

机械设备在试运转前一般还需要进行下列的检查：

Mechanical equipment in the test before the general need to carry out the following check:

（1）机械设备周围应全部清扫干净。

Floors and spaces around mechanical equipment should be cleaned.

（2）机械设备上不得放有任何工具、材料及其他妨碍机械运转的东西。

Any tools, materials and other things that hinder the operation of the mechanical equipment should not be put on the machine.

(3) 机械设备各部分的装配零件必须完整无缺，各种仪表都要经过试验，所有螺钉、销钉之类的紧固件都要拧紧并固定好。

All parts of the mechanical equipment assembly parts must be intact. All kinds of instruments must be tested. All screws, pins and other fasteners must be tightened and fixed.

(4) 所有减速器、齿轮箱、滑动面及每个应当润滑的润滑点，都要按照产品说明书上的规定，保证质量地加上润滑油。

All reducer, gear box, sliding surface and each lubrication point should be lubricated according to meet the quality requirements of the product manual.

(5) 检查水冷、液压、气动系统的管路、阀门等，该开的是否已经打开，该关的是否已经关闭。

Check the water cooling, hydraulic, pneumatic system pipelines, valves, etc. Those pipelines and valves should be opened has been opened and should be closed has been closed.

(6) 在设备运转前，应先开动液压泵将润滑油循环一次，以检查整个润滑系统是否畅通，各润滑点的润滑情况是否良好。

Before the operation of the equipment, hydraulic pump oil cycle should be first started to check whether the whole lubrication system is smooth and whether the lubrication situation of each lubrication point is good.

(7) 检查各种安全设施（如安全罩、栏杆、围绳等）是否都已安设妥当。

Check all kinds of safety facilities (such as safety cover, balustrade, rope, etc.) have been installed properly.

(8) 只有确认设备完好无疑，才允许进行试运转，并且在设备起动前还要做好紧急停车的准备，确保试运转时的安全。

The test run can be permitted only if the equipment is intact and emergency stop has been prepared before the start of the equipment to make to ensure the safety of the test run.

B 试运转的步骤（Test Run Steps）

试运转的步骤一般是：

The test run steps are generally followings:

(1) 先无负荷，后有负荷。

No load first, load later.

(2) 先低速，后高速。

Low speed first, high speed later.

(3) 先单机，后联动。

Single machine first, all machines later.

(4) 每台单机要从部件开始，由部件到组件，由组件到单台设备。

Each single machine should start from components, from components to modules, from modules to a single equipment.

(5) 对于数台设备联成一套的联动机组，要将每台设备分别调试好后，才能进行整个机组的联动试运转。

The whole unit can be carried out only if each machine has tested separately for a set of machine unit.

（6）前一步骤未合格，不得进行下一步骤的试运转。

The next step of test run shall not be carried out if the previous step is not qualified.

（7）设备试运转前，电动机应单独试验，以判断电力拖动部分是否良好，并确定其正确的回转方向。

Before the test run of the equipment, the motor should be tested separately to judge whether the electric driving part is good or not, and to determine the correct direction of rotation.

（8）其他如电磁制动器、电磁阀限位开关等各种电气设备，都必须提前做好试验调整工作。

Other electrical equipment, such as electromagnetic brake, electromagnetic valve limit switch, must be checked and adjusted in advance.

（9）试运转时，能手动的部件先手动然后再机动。

During the test run, the parts that can be manually operated are manually operated and then driven by motor.

（10）对于大型设备，可利用盘车器或吊车转动两圈以上，没有卡阻等异常现象，方可通电运转。

For large equipment, the wheel or crane can be used to rotate more than two turns. It can be electrified operation if there is no jam and other abnormal phenomenon.

思考题（Questions）

5.1 过盈配合的装配方法有哪些？

How many methods can be used for assembly of interference fit?

5.2 滑动轴承装配间隙的测量有哪几种方法？

How many methods can be used for measuring assembly clearance of sliding bearing?

5.3 滑动轴承油沟的作用是什么？

What is the function of oil groove?

5.4 滑动轴承顶间隙的作用是什么？

What is the function of the top clearance of sliding bearing?

5.5 滑动轴承侧间隙的作用是什么？

What is the function of the side clearance of sliding bearing?

5.6 滑动轴承装配间隙的测量有哪几种方法？

How many methods can be used for measuring the assembly clearance of sliding bearing?

5.7 滚动轴承的装配方法有哪几种？

How many methods can be used for assembling rolling bearings?

5.8 齿轮的齿侧间隙的测量方法有哪些？

How many methods can be used for measuring gear side clearance?

5.9　用双表检查联轴器记录如图 5.41 所示（单位：mm），进行 1、3 点找正计算，说明垂直方向支点 1 和支点 3 怎么调整？

Double gauges are used to check the coupling record as shown in Figure 5.41 (unit：mm), and perform alignment calculations at points 1 anad 3. Explain how to adjust point 1 and point 3 in the vertical direction?

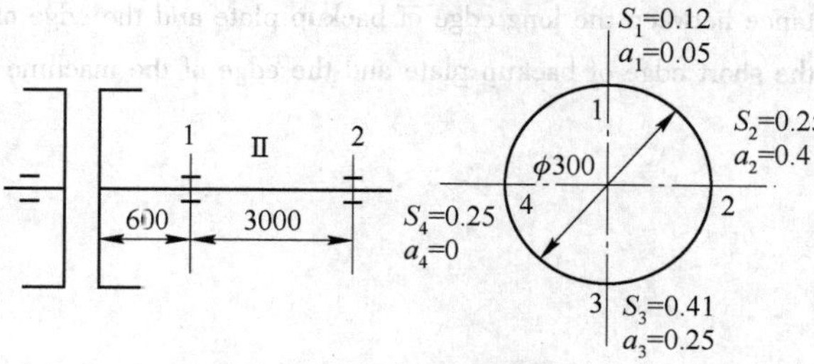

图 5.41　题 5.9 的图

Figure 5.41　Picture of question 5.9

5.10　如何设置中心标板？

How to set center target board?

5.11　如何设置基准点？

How to set reference point?

5.12　写出传统机器设备安装的工艺过程？

Describe the traditional installation process of the machine.

5.13　埋设基础螺丝常用的形式有哪些？

What are the common types of foundation bolts?

5.14　机器设备安装的新工艺有哪些？

What are the new processes for the installation of mechanical equipment?

5.15　机械设备安装时，垫板的作用是什么？

How many functions of the backup plate are there?

5.16　机械设备安装后的试运转的目的是什么？

What is the purpose of the test run after installation of the mechanical equipment?

5.17　设备安装反复检查调整的内容有哪些？

How many contents of repeated check and adjustment while installing the mechanical equipment?

5.18　齿轮座地脚板如图，机器设备总重量 $Q_1 = 500$kN，用地脚螺丝直径为 50mm，许用拉伸应力 $[\sigma] = 120$MPa，基础抗压强度 $[R] = 24$MPa，安全系数为 3，要选一块面积为 90mm×220mm 垫板多少堆合适？画出垫板放置的示意图，并在图 5.42 上标出垫板长边与地脚螺栓边缘的距离以及垫板短边露出机器底座边缘的长度。

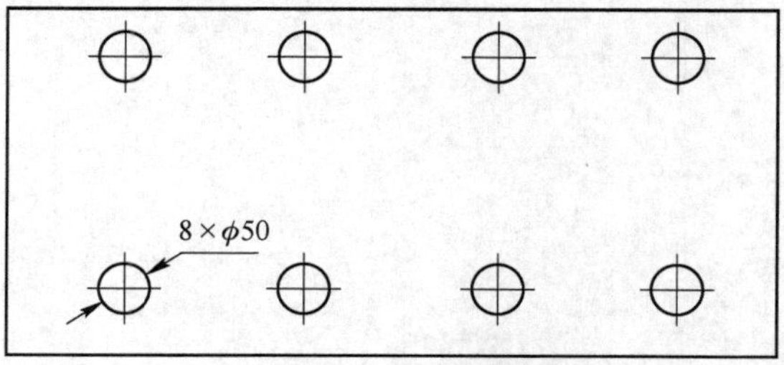

图 5.42　题 5.18 的图

Figure 5.42　Picture of Question 5.18

The total weight of the machine equipment Q_1 is 500kN. The number and the diameter of the foundation bolt is 8 and 50mm, as shown in the Figure 5. 11. The ultimate tensile stress of foundation bolt material [σ] is 120 MPa. The ultimate foundation compressive strength [R] is 24 MPa. The safety factor is 3. The dimension of a block of backup plate is 90mm×220mm. How many piles of the backup plate will be needed? Please draw all of the backup plates and the required distance between the long edge of backup plate and the edge of the foundation bolt and the required distance between the short edge of backup plate and the edge of the machine base on the Figure 5. 42.

6 机械设备的润滑与密封

Chapter 6 Lubrication and Sealing of Mechanical Equipment

润滑和密封对于机械设备健康稳定运转均具有重要作用，密封也会影响润滑，一旦密封装置被破坏，设备的润滑状态也会随之受到破坏。机械设备的润滑与密封的设计与安装非常重要。

Lubrication and sealing play an important role in the healthy and stable operation of mechanical equipment. Sealing also affects lubrication. Once the sealing device is damaged, the lubrication state of the equipment will also be damaged. The design and installation of lubrication and sealing of mechanical equipment is very important.

6.1 机械设备的润滑（Lubrication of Mechanical Equipment）

润滑是指在机件作相对运行的接触面间加入润滑介质，使其间形成一层润滑膜，以减小摩擦和磨损，延长机械设备的使用寿命。此外，某些种类的润滑剂还能起到冲洗、冷却摩擦表面、阻尼振动、防锈和密封等作用。

Lubrication refers to the action that lubrication medium is added between the contact surfaces of the parts in relative operation, so as to form a layer of lubricant film, in order to reduce friction and wear and extend the service life of mechanical equipment. In addition, certain types of lubricants can be used to wash and cool friction surfaces, damp vibrations, prevent rust and seal.

6.1.1 润滑的分类（Classification of Lubrication）

根据相对运动的两表面间润滑的多少，可以将润滑分为无润滑、边界润滑、流体润滑、半流体润滑和半干润滑。

According to the amount of lubricant between the two surfaces in relative movement, lubrication can be divided into: no lubrication, semi-dry lubrication, boundary lubrication, semi-fluid lubrication and lubrication fluid.

（1）无润滑。在具有相对运动的两表面间完全没有任何润滑介质存在，处于干摩擦状态，称为无润滑。干摩擦系数 μ 可以高达 0.5 以上。

No lubrication. No lubrication means there is no lubricant between two surfaces in relative movement. Two surfaces in relative movement are absolutely in dry friction condition. Friction coefficient μ of no lubrication is more than 0.5.

这种状态除机械的制动外，一般应是尽量避免的。润滑系统的故障、润滑剂失效或流失也会出现这种状态，从而造成机械设备的损坏。

This state in addition to mechanical braking, generally should be as far as possible to avoid. Failure of the lubrication system, failure or loss of lubricant can also occur in this state, resulting in mechanical equipment damage.

（2）边界润滑。边界润滑是指介于有润滑和无润滑之间的一种临界状态的润滑形式。边界润滑的摩擦系数μ在0.03和0.1之间。

Boundary lubrication. Boundary lubrication refers to a form of critical state of lubrication between lubricated and non-lubricated. Friction coefficient μ of boundary lubrication is between 0.03 and 0.1.

（3）流体润滑。流体状态的润滑剂在做相对运动的两摩擦表面之间形成一层足够厚的润滑膜，把两摩擦表面完全隔开，称为流体润滑。这是一种理想的润滑状态。摩擦系数一般在0.001和0.01之间或更低。

Fluid lubrication. Fluid lubrication means a layer sufficiently thick lubrication film is formed between the two friction surfaces in relative movement. The two friction surfaces are completely separated. This is an ideal state of lubrication. Friction coefficient μ of fluid lubrication is generally between 0.001 and 0.01, or lower.

（4）半流体润滑。半流体润滑是指在流体润滑状态下，若流体膜遭受破坏的比例不大，则属于流体润滑与边界润滑之间的一种润滑状态。

Semi-fluid lubrication. Semi-fluid lubrication means the ratio of the fluid film suffered little damage in the fluid lubrication. It belongs to a state between fluid lubrication and boundary lubrication and is called semi-fluid lubrication.

（5）半干润滑。半干润滑是指在边界润滑状态下，若边界膜遭到破坏的程度不太严重，就出现边界润滑与干摩擦之间的一种润滑状态。

Semi-dry lubrication. Semi-dry lubrication means the border membrane suffered little damage in boundary lubrication. It belongs to a state between boundary lubrication and dry friction and is called semi-dry lubrication.

6.1.2 润滑原理（Lubrication Theory）

不同的润滑方式有不同的润滑原理。主要的润滑原理包括流体润滑原理、边界润滑原理和固体润滑原理。

Different lubrication principles are based on different types of lubrication. Lubrication theory mainly includes theory of liquid lubrication, theory of boundary lubrication and theory of solid lubrication.

6.1.2.1 流体润滑原理（The Theory of Liquid Lubrication）

流体润滑原理包括流体动压润滑原理、流体静压润滑原理和流体静动压润滑原理。

The theory of liquid lubrication includes hydrodynamic pressure lubrication theory, hydro-static pressure lubrication and hydro-static and hydrodynamic pressure lubrication theory.

A 流体动压润滑原理（Liquid Hydrodynamic Pressure Lubrication Theory）

在具有收敛型油楔的滑动摩擦副中，由于收敛的楔形间隙对流体的升压作用，流体膜具有足够的压力把两摩擦表面分隔开来，形成流体润滑。由于流体膜的压力来自摩擦副两表面的相对运动，所以称为流体动压润滑。

In a convergent oil wedge sliding friction pair, two friction surface are separated by enough pressure of the fluid film because the convergence wedge gap can increase the pressure of the fluids. In this condition, fluid lubrication is formed. Because the pressure of the fluid film comes from the relative movement of the two surfaces, it is called hydrodynamic pressure lubrication.

实现流体动压润滑必须具备下列条件：

The following conditions are requirement for achieving hydrodynamic lubrication:

（1）做相对运动的对偶摩擦表面，必须沿运动方向形成收敛的楔形间隙。

A convergent wedge gap must be formed along the direction of motion for the mating friction surfaces being in relative motion.

（2）对偶表面必须具有一定的相对速度。

The mating surfaces must have a certain relative velocity.

（3）润滑流体必须具有适当的黏度，并且供油充足。

The lubrication fluid must have appropriate viscosity and sufficient oil supply.

（4）外载荷必须小于油膜所能承受的极限值。

The external load must be less than the limit value that the oil film can bear.

（5）对偶表面的加工精度应较高，表面粗糙度应小，这样可以在较小的油膜厚度下实现流体动压润滑。

The machining accuracy of the mating surfaces should be high, and the surface roughness should be small, so as to achieve hydrodynamic lubrication under a smaller oil film thickness.

径向滑动轴承摩擦副建立流体动压润滑的过程如图 6.1 所示。图 6.1（a）为轴承静止状态时轴与轴承的接触状况，在轴的下部正中，轴与轴承接触，轴的两侧形成楔形间隙。

The process of establishing hydrodynamic lubrication for friction pairs of radial sliding bearings is shown in Figure 6.1. Figure 6.1 (a) shows the contact between the shaft and the bearing when the bearing is stop. The shaft contacts the bearing in the middle of the lower part of the shaft. And wedge clearance is formed on both sides of the shaft.

起动开始时，轴滚向一侧（如图 6.1（b）所示），具有一定黏度的润滑油黏附在轴颈表面，随着轴的转动油被带入楔形间隙。油在楔形间隙中只能沿轴向溢出，但轴颈有一定长度，而油的黏度使其沿轴向溢出受阻而流动不畅，这样，油就聚集在楔形间隙的尖端互相挤压而使油压升高。

At the beginning of starting, the shaft is rolled to one side (as shown in Figure 6.1 (b)). The lubricating oil with a certain viscosity sticks to the surface of the shaft journal, and the oil is brought into the wedge clearance as the shaft rotates. In the wedge gap, the oil can only overflow along the axial direction, but the shaft journal has a certain length, and the viscosity of the oil makes it flow blocked along the axial flow. In this way, the oil gathers in the tip of the wedge gap and squeezes each other, making the oil pressure rise.

随着轴的转速升高，楔形间隙中油压也升高，形成一个压力油隙逐渐把轴抬起，如6.1（c）所示。但此时轴尚处于不稳定状态，轴心位置随着轴被抬起的过程而逐渐向轴承中心的另一侧移动。

As the shaft speed increases, so does the oil pressure in the wedge, a pressure oil gap is formed that gradually lifts the shaft, as shown in Figure 6.1 (c). However, at this time, the shaft is still in an unstable state. As the shaft is lifted, the shaft position gradually moves to the other side of the bearing center.

当达到一定转速后，轴就趋于稳定状态，如图 6.1（d）所示。此时油楔作用于轴上的压力总和与轴的负载（包括轴的自重）相平衡，轴与轴承表面完全被一层油膜隔开，实现了流体润滑，这就是流体动压润滑的油楔效应。

When the rotation speed reaches a certain value, the shaft tends to be in a stable state, as shown in Figure 6.1 (d). At this time, the total pressure acted by the oil wedge on the shaft is balanced with the load of the shaft (including the weight of the shaft). The shaft and the bearing surface are completely separated by a layer of oil film to achieve fluid lubrication, which is the oil wedge effect of hydrodynamic lubrication.

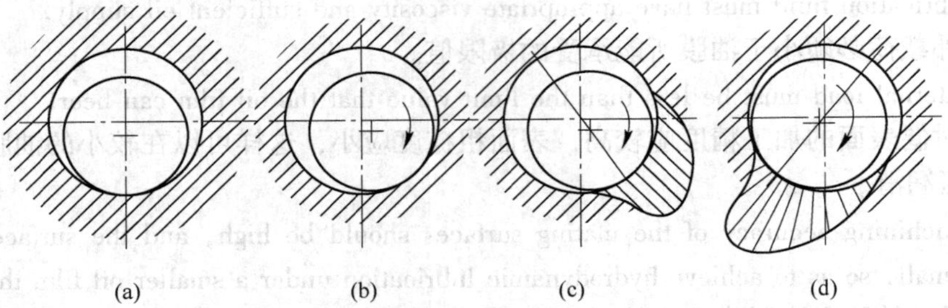

图 6.1 径向滑动轴承摩擦副建立流体动压润滑的过程

Figure 6.1 Establishment process of friction-pressure lubricating oil film for radial sliding bearing

(a) 静止阶段（Static state）；(b) 启动阶段（Start up）；

(c) 加速阶段（Speed up）；(d) 稳定运行阶段（Steadily running state）

接触表面弹性变形形成赫兹区，赫兹压力将两表面隔开，从而实现流体动压润滑，称为弹性流体动压润滑。

Elastic deformation of contact surface forms Hertz zone, and Hertz pressure separates the two surfaces to achieve hydrodynamic lubrication. This is called elastohydrodynamic pressure lubrication.

摩擦副表面磨损的主要形式表现为疲劳剥落，因而寿命很长。一般齿轮和滚动轴承摩擦副中也存在着部分的弹性流体动压润滑现象。

The main form of surface wear of friction pair is fatigue spalling. Some elastohydrodynamic pressure lubrication phenomena also exist in the friction pairs of gears and rolling bearings.

判断摩擦副是否处于弹性流体动压润滑状态，主要看油膜厚度是否足够可靠地将摩擦副对偶表面隔开，通常用膜厚比 λ 来判断。

The problem of whether the friction pair is in the elastohydrodynamic pressure lubrication state depends mainly on whether the oil film thickness is reliable enough to separate the dual surfaces of the friction pair. Usually, the film thickness ratio λ is used to judge it.

膜厚比 λ 就是最小油膜厚度与摩擦副对偶表面的平均均方根粗糙度的比值。可靠的弹性流体动压润滑的条件是 $\lambda>3$。

The film thickness ratio λ is the ratio of the minimum oil film thickness to the average root mean square roughness of the mating surface of the friction pair. Reliable elastohydrodynamic pressure lubrication conditions are $\lambda>3$.

当 $\lambda<1$ 时，摩擦副处于边界润滑状态；当时，摩擦副中只有部分的弹性流体动压润滑存在，属于半流体润滑。

When $\lambda<1$, the friction pair is in the boundary lubrication state. At that time, only part of the elastohydrodynamic pressure lubrication exists in the friction pair. This belongs to semi-fluid lubrication.

B　流体静压润滑（Hydro-static Pressure Lubrication Theory）

从外部将高压（几到几十兆帕）流体强制送入摩擦副的油腔中去，利用流体的压强使两摩擦表面在未开始做相对运动前就被隔开，从而形成流体润滑状态，称为流体静压润滑。

The high pressure (a few to tens of MPa) fluid is forced into the oil groove of friction pair from the outside. Two friction surfaces are separated by fluid pressure before they relatively move. Thereby a fluid lubrication is formed. This kind of lubrication is called hydro-static pressure lubrication.

流体静压润滑的优点如下：

Advantages of hydro-static pressure lubrication are as followings:

（1）应用范围广、承载能力高。

Wide application range and high bearing capacity.

（2）摩擦系数低并且稳定。

Low and stable friction coefficient.

（3）几乎没有磨损，所以寿命极长。

Almost no wear, life of parts is very long.

（4）由于表面不直接接触，所以对轴承材料要求不高，只需比轴颈稍软即可。

Because the surface does not contact directly, so the requirements of bearing material are not high, just slightly softer than that of the shaft journal.

流体静压润滑的缺点是需要一套昂贵的供油系统，油泵长期工作要耗费大量能源。操作人员担心高压管路的工作可靠性和寿命，因为一旦断油将立即导致重大事故。

Disadvantages of hydro-static pressure lubrication are hydro-static lubrication requires an expensive oil supply system and requires a lot of energy to operate the pump for a long time. Operators worry about the reliability and service life of high-pressure pipelines. Once the oil is cut off, catastrophe failure will occur.

C 流体静动压润滑（Hydro-static and Hydrodynamic Pressure Lubrication Theory）

流体动压轴承如果在起动制动时采用流体静压润滑，而在达到额定转速后，高压油泵停止供油，轴承靠流体动压润滑，这样，就克服了流体静压润滑的缺点而又避免了静压系统长期工作的大量能耗，因此，流体静动压润滑近年来已在各类轧机轧辊轴承上得到普遍应用。

If hydro-static pressure lubrication is used when starting and braking, and hydrodynamic pressure lubrication is used after reaching the rated speed while high pressure oil pump can be cut off. This way is called hydro-static and hydrodynamic pressure lubrication can avoid the disadvantages of hydro-static lubrication and a large number of energy consumption after working for a long time. Therefore, in recent years, hydro-static and hydrodynamic pressure lubrication has been widely applied in all kinds of mill roller bearings.

6.1.2.2 边界润滑原理（The Theory of Boundary Lubrication）

边界润滑是一种极为普遍的润滑状态，几乎各种摩擦副在做相对运动时都存在着边界润滑状态。

Boundary lubrication is a very common lubrication state. Almost all kinds of friction pairs have boundary lubrication state when they are in relative motion.

在边界润滑状态下，摩擦系数的大小并不主要决定于润滑油的黏度，而主要是与边界膜的特性有关。边界润滑理论包括物理吸附膜、化学吸附膜和化学反应膜。

In the boundary lubrication state, the friction coefficient is not mainly determined by the viscosity of the lubricating oil, but mainly related to the characteristics of the boundary film. The theory of boundary lubrication includes physisorption film, chemisorbed film and chemical reaction film.

6.1.2.3 固体润滑原理（The Theory of Solid Lubrication）

在摩擦表面加入一层剪切强度很低的薄膜，直到润滑作用减弱。如果薄膜被固体物质填充，则该薄膜为固体润滑膜。

A thin film of very low shear strength is added to the friction surface until the lubrication effect is reduced. If the film is filled by a solid substance, the film is a solid lubricating film.

固体润滑有两个必要条件。首先，固体润滑分子必须具有低剪切强度和易滑移的特性。其次，与固体润滑剂的摩擦表面有较强的亲和力，在摩擦过程中，摩擦总是让始终保持表面的固体润滑剂的薄膜及固体润滑膜的层不腐蚀金属表面。金属表面通常是机械附着和化学附着的形式。

There are two necessary conditions for solid lubrication. First of all, the solid lubrication molecules should have low shear strength and easy to slip. Second, solid lubricant to be and the friction surface has the strong affinity, in the process of friction, always make friction always stay on the surface of a film of solid lubricant, and the layer of solid lubrication film not corrosion of metal surface, usually in the metal surface is mechanical attachment, but also has the form of chemical adhesion.

6.1.3 润滑材料（Lubrication Material）

凡是能够在做相对运动的、相互作用的对偶表面间起到减小摩擦降低磨损作用的物质，均可以称作润滑材料。

Any material that can reduce the friction and wear between the opposite moving and interacting dual surfaces can be called lubrication material.

6.1.3.1 润滑材料分类（Classification of Lubrication Material）

润滑材料大致分为四大类：液体润滑材料、塑性体及半流体润滑材料、固体润滑材料和气体润滑材料。

Lubrication material is broadly divided into four categories：liquid lubrication material，solid lubrication material，gas lubrication material and plastic and semi-fluid lubrication material.

（1）液体润滑材料主要是矿物油和各种植物油、乳化液和水等，近年来性能优异的合成润滑油发展很快，如氟油、脂肪酸及合成烃等。

Liquid lubrication material mainly includes mineral oils，various vegetable oils，emulsions and water，etc. In recent years，synthetic lubricants with excellent performance have developed rapidly，such as fluorine oils，fatty acids and synthetic hydrocarbons.

矿物油和润滑脂目前应用最广、使用量最大，原因是来源稳定且价格低廉。

Mineral oils and greases are the most widely used and used oils because of their stable sources and low cost.

动植物油脂主要用作润滑脂的添加剂和某些有特殊要求的润滑部位。

Animal and vegetable oils are mainly used as additives for lubrication grease and some special lubrication parts.

乳化液主要用作机械加工和冷轧金属带材的工艺冷却润滑液。

Emulsion is mainly used as cooling lubricant for machining and cold rolling of metal strip.

水只用于某些塑料轴瓦（如胶木）的冷却润滑。

Water is used only for cooling and lubrication of certain plastic bearings，such as fabroil.

（2）塑性体及半流体润滑材料主要是由矿物油及合成润滑油稠化而成的各种润滑脂和动物脂，以及近年来试制的半流体润滑脂等。

The plastic and semi-fluid lubrication material mainly includes various greases and animal greases thickened by mineral oils and synthetic lubricants，as well as semi-fluid greases produced in recent years.

（3）固体润滑材料是一类很有发展前途的新型润滑材料，可以单独使用，也可以作润滑油和润滑脂的添加剂，利用其些具有润滑性能的固体粉末、薄膜或组合材料隔离摩擦副的对偶表面，起到和油脂一样的润滑作用。石墨、二硫化钼、二硫化钨、氮化硼及聚四氟乙烯、塑料基或金属基自润滑复合材料均属于固体润滑材料。

Solid lubrication material is a promising new type of lubrication material. They can be used alone and can also be used as an additive of lubricating oil and grease. Some lubricating properties of solid powder，film or combined materials are used to isolate the dual surface of the friction pair in order to play the same lubrication role as the oil. Graphite，molybdenum disulfide，tungsten disulfide, boron nitride and polytetrafluoroethylene, plastic-based or metal-based self-lubricating composites are all solid lubricating materials.

固体润滑材料分为无机的、有机的和自润滑复合材料三大类。

Solid lubrication materials are divided into three categories: inorganic, organic and self-lubrication composite materials.

（4）气体润滑材料目前主要用于航空、航天及某些精密仪表的气体静压润滑轴承。如气体轴承中使用的空气、氮气和二氧化碳等气体。图 6. 2 是合成机油样品。

Gas lubricated materials are mainly used in aerostatic lubricated bearings of aviation, aerospace and some precision instruments. Gas lubrication material mainly includes air, nitrogen, carbon dioxide and other gases used in gas bearings. Figure 6. 2 is a sample of full synthetic motor oil.

图 6. 2　合成机油样品

Figure 6. 2　A sample of full synthetic motor oil

图 6. 3 是汽车发动机散装润滑油样品。

Figure 6. 3 is a sample of bulk lubricant oil for vehicle engines bulk.

图 6. 3　汽车发动机散装润滑油样品

Figure 6. 3　A sample of bulk lubricant oil for vehicle engines

图 6. 4 是锂基润滑脂样品。

Figure 6. 4 is a sample of lithium base lubricant grease.

图 6.4　锂基润滑脂样品
Figure 6. 4　A sample of lithium base lubricant grease

图 6. 5 是工业用通用钙脂样品。
Figure 6. 5 is a sample of excellent general calcium grease for industry.

图 6. 5　工业用通用钙脂样品
Figure 6. 5　A sample of hot sale excellent general calcium grease for industry from China

图 6. 6 是高温润滑脂样品。
Figure 6. 6 is a sample of high temperature grease.

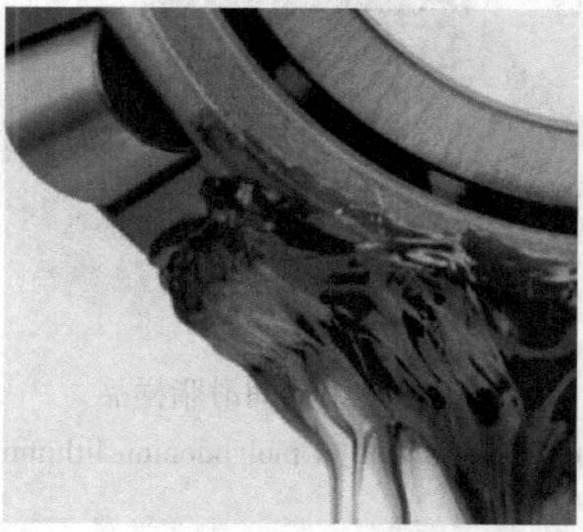

图 6. 6　高温润滑脂样品
Figure 6. 6　A sample of high temperature grease

图 6.7 是多功能铜脂防粘润滑剂样品。
Figure 6.7 is a sample of multipurpose copper grease anti seize lubricant.

图 6.7 多功能铜脂防粘润滑剂样品
Figure 6.7 A sample of multipurpose copper grease anti seize lubricant

图 6.8 是二硫化钼样品。
Figure 6.8 is a sample of molybdenum disulfide.

图 6.8 二硫化钼样品
Figure 6.8 A sample of molybdenum disulfide

图 6.9 是钼锂润滑脂样品。
Figure 6.9 is a sample of molybdenum lithium grease.

图 6.9 钼锂润滑脂样品
Figure 6.9 A sample of molybdenum lithium grease

6.1.3.2 润滑材料的基本要求（Requirements of Engineering Lubrication Material）

润滑材料应满足工程上对润滑的一些基本要求：

Lubrication material shall meet some basic requirements of engineering lubrication:

（1）较低的摩擦系数，从而减少动力消耗，降低磨损速度，提高设备使用寿命。

Lower friction coefficient, so as to reduce power consumption, reduce wear speed and improve the service life of equipment.

（2）具有良好和吸附及楔入能力，以便能渗入摩擦副微小的间隙内，并能牢固地黏附在摩擦表面上，不易被相对运动形成的剪切力刮掉。

It has good adsorption and wedging ability, so that it can penetrate into the tiny gap of friction pair and adhere to the friction surface firmly, which is not easy to be scraped off by the shear force formed by relative movement.

（3）有一定的内聚力（黏度），以便能抵抗较大的压力而不至于从摩擦副中被挤出，以保持足够的润滑膜厚度。

It has a certain cohesion (viscosity), so that it can resist larger pressure without being extruded from the friction pair, so as to maintain sufficient lubrication film thickness.

（4）具有较高的纯度及抗氧化安定性，没有研磨和腐蚀性，不致因迅速与水或空气接触生成酸性化合物或胶质沥青而变质。

With high purity and anti-oxidation stability, no grinding and corrosion, not easy to form acidic compounds or colloidal asphalt and deterioration when quickly contact with water or air.

通常，设备说明书上有制造厂规定的润滑保养规程，其中对润滑材料有要求，但当设备使用条件改变时，原来的材料不一定适用；另外新型材料不断涌现，要求操作者掌握各种润滑材料的性能、选用和使用方法。

Normally, requirements for lubrication materials and the manufacturer's lubrication maintenance rules should be given on the equipment manual. But when the equipment operating conditions change, the original materials may not be applicable. In addition, new materials keep emerging, operators are required to master the performance, selection and use of various lubrication materials.

6.1.3.3 润滑油（Lubricating Oil）

矿物润滑油是目前最重要的一种润滑材料，占润滑材料总耗量的90%以上，是利用石油提炼中蒸馏出的高沸点物质再经精炼制成的。

Mineral lubricating oil is one of the most important lubrication materials at present, accounting for more than 90% of the total consumption of lubrication materials.

石油经过初馏和常压热馏提取汽油、煤油和柴油，剩下的称为常压渣油。再经减压蒸馏，按沸点不同而依次切取一线、二线、三线、四线馏分油，经精制而获得黏度较低的润滑油。这种油含沥青质少，油性分子含量低因而油性也差。对减压渣油蒸馏残留下来的减压渣油进行精制，获得高黏度润滑油，这种油胶质沥青含量较多，油性比前一种润滑油好。将各线馏分油与减压渣油按不同比例调配而制成各种不同牌号、黏度的润滑油。

Atmospheric residue is the remaining after petroleum is extracted from gasoline, kerosene and diesel by primary distillation and atmospheric hot distillation. After atmospheric residue is vacuum distillated, the first-line distillate, second-line distillate, third-line distillate and four-line distillate are taken successively according to different boiling points. This oil is low in asphaltene and therefore poor in oiliness. Vacuum residuum is a high viscosity lubricating oil refined by the vacuum residuum distillation. Various line distillates and vacuum residuum are mixed as different ratio to make various grades of lubricating oil

除矿物油外，还有以软蜡、石蜡等为原料用人工方法生产的合成润滑油。植物油和蓖麻籽油用于制取某些特种用途的高级润滑油。

Besides mineral oil, there are some synthetic lubricants produced by artificial methods using soft wax and paraffin as raw materials. Vegetable oil and castor oil are used to make high grade lubricants for some special purposes.

A 润滑油的物理化学性能及主要质量指标（Physical and Chemical Properties of the Lubricating Oil and the Main Quality Indicators）

润滑油的物理化学性能及主要质量指标有：外观、黏度、闪点、凝点、抗乳化性、抗氧化安定性、热氧化安定性和抗磨性。

Physical and chemical properties of the lubricating oil and the main quality indicators includes appearance, viscosity, flash point, solidifying point/condensation point, anti-emulsification, anti-oxidation stability, thermal oxidation stability, and abrasion resistance.

（1）润滑油的外观。润滑油的外观主要指油品的颜色，外观往往可以反映其精制程度和稳定性。对于基础油来说，一般精制程度越高，其烃的氧化物和硫化物脱除得越干净，颜色也就越浅。但是，即使精制的条件相同，不同油源和基属的原油所生产的基础油，其颜色和透明度也可能是不相同的。对于新的成品润滑油，由于添加剂的使用，颜色作为判断基础油精制程度高低的指标已失去了它原来的意义。

Appearance of lubricating oil. The appearance of lubricating oil mainly refers to the color of oil. The appearance can often reflect the degree of refinement and stability of oil. For base oils, the more refined they are, the cleaner and lighter the hydrocarbon oxides and sulfides are removed. However, even under the same conditions of refinement, the color and transparency of base oils produced from different sources and base genera of crude oil may vary. For the new finished lubricating oil, because of additives, the color as an index to judge the degree of refinement of base oil has lost its original significance.

（2）黏度。黏度反映了润滑油的稀稠程度。黏度越高，流动性越差，不易渗入间隙较小的摩擦副中，但也不易被从摩擦面间挤出来，因而油膜承载能力高。高黏度油的摩擦阻力大，油温容易升高，设备的功率损耗也高。黏度低的润滑油正好相反。

Viscosity. Viscosity reflects the thick and thin degree of lubricating oil. The higher the viscosity, the worse the fluidity. It is not easy to penetrate into the friction pair with small gap, but it is not easy to be extruded from the friction surface, so the oil film bearing capacity is high. High viscosity oil friction resistance, oil temperature is easy to rise, the power loss of equipment is high. For low viscosity lubricants, the situations are the opposite.

黏度是润滑油的一项很重要的指标。在选择润滑油时，通常以黏度为主要依据。黏度可以用绝对黏度和相对黏度表示。绝对黏度又有动力黏度和运动黏度两种表示方法。动力黏度实质上反映流体内摩擦阻力的大小。运动黏度是在同一温度下流体的动力黏度与密度的比值。相对黏度也称条件黏度，各国采用的测定相对黏度的黏度计不同，因而条件黏度有恩氏、赛氏、雷氏黏度等几种。我国采用恩氏黏度，在规定的温度下让体积为200mL的液体从恩氏黏度计流出所需的时间与同体积蒸馏水在20℃时从恩氏黏度计流出所需时间的比值，即为恩氏黏度。

Viscosity is an important index of lubricating oil. When choosing lubricating oil, viscosity is usually the main basis. Viscosity can be expressed by absolute viscosity and relative viscosity. Absolute Viscosity is divided into dynamic viscosity and kinematic viscosity. Dynamic viscosity essentially reflects the magnitude of frictional resistance in the flow body. Kinematic viscosity is a ratio of dynamic viscosity and density at the same temperature of the fluid. Relative viscosity is also known as the condition viscosity. Condition viscosity Includes Engler viscosity, Saybolt viscosity, Reye viscosity and so on. Relative viscosity is different because relative viscosity viscometer adopted by each country is different. Engler viscosity refer to a ratio of time that 200mL fluid flow out from Engler viscosity at the specified temperature and 200mL distilled water flow out from Engler viscosity at 20℃.

润滑油的黏度随温度升高而降低，润滑油的这种特性被称为粘温特性。

Lubricating oil viscosity decreases with temperature increasing. This feature is called the lubricant viscosity-temperature characteristics.

（3）闪点。在规定条件下加热润滑油，当油蒸汽与空气的混合气体与火焰接触时发生闪火现象的最低温度，称为闪点。闪点是润滑油的一项安全指标，要求润滑油的工作温度低于闪点20~30℃。

Flash point. Lubricating oil is heated under specified conditions. The lowest temperature at which a flash fire phenomenon happens as soon as mixed gas of steam and air contact with the flame. This lowest temperature is called flash point. Flash point is a safety index of lubricating oil. The operating temperature of lubricating oil is required to be 20~30℃ below the flash point.

（4）凝点。润滑油在规定条件下冷却到失去流动性时的最高温度，称为凝点（也有称为倾点的）。

Solidifying point/Condensation point. Under the specified conditions, lubricating oil is cooled to lose liquidity as the temperature decrease. The highest temperature is called condensation point.

（5）抗乳化性。润滑油与水接触并搅拌后，能迅速分离的能力，称为抗乳化性。对工作在潮湿环境和可能有水进入摩擦副的润滑用油，应考虑此项指标。

Anti-emulsification ability. The ability of lubricating oil being stirred and quickly separated from water while contacting with water is called anti-emulsification. This indicator should be considered for lubricating oils that work in humid environments and may have water entering the friction pair.

（6）抗氧化安定性。润滑油在使用和贮存过程中，抵抗氧化变质的能力，称为抗氧化安定性。

Oxidation resistance ability. The oxidation resistance ability of lubricating oil in the process of use and storage is called anti-oxidation stability.

（7）热氧化安定性。润滑油膜在较高的工作温度下易与空气中的氧化合，生成胶质膜，使油迅速变质。热氧化安定性反映了润滑油在高温下抑制胶质膜生成的能力。

Thermal oxidation resistance ability. Lubricating oil film is easy to oxidize with air at higher working temperature and generate gelatinous membranes. In this case, the oil rapidly deteriorated. The thermal oxidation resistance ability reflects the ability of lubricating oil to inhibit the formation of gum films at high temperatures.

（8）抗磨性。反映边界润滑状态下油膜的承载能力。它包括油性和极性两个方面。油性对非极压状态的中温中等负载的摩擦副边界润滑膜的形成和润滑性能影响极大；而在极压状态下，油的极压性能对边界膜的润滑性能起关键作用。油的抗磨性要在专门的试验机如梯姆肯或四球试验机上测定。

Abrasion resistance ability. Carrying capacity of the oil film in the boundary lubrication state is called abrasion resistance ability. It includes oiliness and polarity. Oiliness has a great influence on the formation and lubrication properties of boundary lubrication film of friction pair with medium temperature and medium load at non-extreme pressure. In the extreme pressure state, the extreme pressure performance of oil plays a key role in the lubrication performance of boundary film. The abrasion resistance of oil shall be determined by a special testing machine such as a Timken or four ball tester.

B　常用润滑油的性能和用途（The Performance and Usage of Common Lubricating Oil）

采用 ISO 标准，按 40℃时的运动黏度划分油的标号，标号的数值就是润滑油在 40℃时的运动黏度的中心值。新标号前加“N”，如 N5、N7、N15 等。

ISO standard is adopted to classify the oil label according to its kinematic viscosity at 40℃. The value of the label is the central value of the kinematic viscosity of the lubricating oil at 40℃. “N” is added before the new label, such as N5, N7, N15, etc.

常用润滑油包括普通机械油、机床工业通用润滑油、工业齿轮油、多级 QB 汽油和多级普通汽车齿轮油（GL-3）、30 号 CC-1 柴油机润滑油。

Common lubricating oil include ordinary machine oil, general lubricating oil for machine tool industry, industrial gear oil, multipole QB gasoline oil and multistage ordinary vehicle gear oil (GL-3), and No. 30 CC-1 diesel engine oil.

（1）普通机械油。普通机械油是没有添加油性或极压添加剂的润滑油，仅适用于载荷、转速及温度均不高的一般无特殊要求的轴承、纺织机锭子、齿轮及其他工作条件类似的机械的润滑。

Ordinary machine oil. Ordinary machine oil is a kind of lubricating oil without oil or extreme pressure additive, suitable only for bearing, spindle of textile machine, gear and other similar machines with low load, rotation speed and temperature.

（2）机床工业通用润滑油。通用型机床工业用润滑油里加入了一定量的油性及抗氧防锈等添加剂，所以抗磨、抗氧化、防锈及抗乳化性能均优于普通机械油，是取代普通机械油的新油品。

General lubricating oil for machine tool industry. General lubricating oil for machine tool industry is a kind of lubricating oil in which a certain amount of oil and anti-oxygen and anti-rust additives have been added. So, the anti-wear, anti-oxidation, anti-rust and anti-emulsification properties of general lubricating oil for machine tool industry are better than ordinary mechanical oil. General lubricating oil for machine tool industry is a new oil to replace the ordinary mechanical oil.

（3）工业齿轮油。工业齿轮油分普通工业齿轮油和极压工业齿轮油。普通工业齿轮油主要是在普通机械油中加入了防锈抗氧添加剂制成，今后可能被通用机床用润滑油取代。极压工业齿轮油分铅型和硫磷型，铅型正被淘汰，硫磷型重点推广。

Industrial gear oil. Industrial gear oil is grouped in general industrial gear oil and extreme pressure industrial gear oil. Ordinary industrial gear oil is mainly made by adding anti-rust and anti-oxygen additives in the ordinary machine oil. In the future, ordinary industrial gear oil may be replaced by general machine lubricants. Extreme pressure industrial gear oil is divided into lead type and sulfur-phosphorus type. Lead type is being eliminated, sulfur-phosphorus type is promoted.

（4）多级 QB 汽油和多级普通汽车齿轮油（GL-3）。多级 QB 汽油机油及多级普通车辆齿轮油（GL-3）是用于汽车发动机和车辆齿轮的润滑油。它们的性能优于旧牌号的汽油机油和齿轮油。多极 QB 汽油机油对于我国大多数地区均可冬夏通用，只是在严寒地区的冬季暂时用 5W/20 汽油机油，夏春秋三季用 10W/30 汽油机油。关于车辆齿轮油，国内目前仍大量使用渣油型齿轮油，这种油黏度高、流动性不好、摩擦阻力大，抗氧化安定性差。新试制的车辆齿轮油性能有较大的改善，试用效果良好。

Multipole QB gasoline oil and multistage ordinary vehicle gear oil (GL-3). Multipole QB gasoline oil and multistage ordinary vehicle gear oil (GL-3) is a kind of developed lubricating oil for automobile engine and vehicle gear. Their performance is superior to old grade gasoline oil and gear oil. Multi-pole QB gasoline oil can be used in most areas of China in winter and summer, but temporarily use 5W/20 gasoline oil in cold areas in winter, and 10W/30 gasoline oil in summer and spring. A lot of residuum type gear oil still use in China. The viscosity of this kind of oil is high, the fluidity is bad, the friction resistance and the anti-oxidation stability are all poor. The performance of the new gear oil has been improved greatly.

（5）30 号 CC-1 柴油机润滑油。30 号 CC-1 柴油机油具有较好的抗氧化抗腐及分散性，优良的高温清净性（形成积炭少）及抗磨性。

No. 30 CC-1 diesel engine oil. No. 30 CC-1 diesel engine oil has good oxidation resistance, corrosion resistance and dispersion, high temperature cleanness (less carbon deposition) and abrasion resistance.

6.1.3.4 润滑脂（Lubricating Grease）

润滑脂是在润滑油（基础油）中加入能起稠化作用的物质（稠化剂）把油液稠化成具有塑性的膏状的润滑剂。兼有液体和固体润滑剂的优点。基础油一般为各种黏度的石油润滑油或合成润滑油。稠化剂是各种金属的脂肪酸皂、地蜡、膨润土、硅胶和某些新型合成材料。不过用得最多的还是各种脂肪酸金属皂。

Grease is a lubricant that a thickener is added to a lubricating oil (base oil) and make lubricating oil become a plastic paste lubricant. Grease has both of advantages of liquid and solid lubricant. Base oil is generally petroleum or synthetic oils of various viscosities. Thickener is a something like fatty acid soap for various metals, ozocerite, bentonite, silica gel and some new synthetic materials. However, fatty acid metal soap is most commonly used.

与润滑油比较，使用润滑脂的主要优点如下：

The main advantages of grease comparing with lubricating oil are as follows:

（1）在摩擦表面的黏附性好，不易流失或飞溅，不会产生漏油现象。

Good adhesion on the friction surface, not easy to lose or splash, and no oil leakage.

（2）可起到密封作用，防止尘土进入摩擦面。

Can play a sealing role, prevent dust from entering the friction surface.

（3）比润滑油的减振性强，可减少噪声和振动。

Better vibration damping than lubricating oil, which can reduce noise and vibration.

（4）特别适用于滚动轴承的润滑，而且由于无需经常加油，从而减少了设备维护的工作量。

Especially suitable for lubrication of rolling bearings, and because the need for frequent refueling, thus reducing the workload of equipment maintenance.

润滑脂的缺点：散热能力差，输送性能不好，对大部分滑动轴承不适用，由于润滑的黏滞强，设备起动力矩增大。

The disadvantages of grease include poor heat dissipation capacity, poor transmission performance, not suitable for most sliding bearings. The starting torque of the equipment increase due to strong lubrication viscosity.

A 润滑脂的主要物理化学性能（Physical and Chemical Properties of Lubricating Grease）

润滑脂的主要物理化学性能指标有：针入度、滴点和稠化剂。

The main physical and chemical properties of grease are penetration, drop point and thickener.

（1）针入度。针入度标志润滑脂的软硬程度。它是利用150g的标准圆锥体在5s内沉入温度为25℃的润滑脂试样中的深度来表示，以0.1mm为单位来计量针入度。针入度是选择润滑脂的一项重要指标。

Penetration. Penetration is the degree of hardness and softness of the grease. It is represented by the depth of a 150g standard cone sinking into the 25℃ grease sample within 5s. Penetration is an important index for selecting grease.

根据针入度的大小，润滑脂分为9类，表6.1介绍了品种等级号所对应的针入度范围。常用的品种等级号为0~4号。等级号越小，针入度越大，润滑脂越稀。

According to the size of needle penetration, the grease is divided into 9 classes. Table 6.1 shows the penetration range corresponding to the variety grade number. The commonly used variety grade number is 0~4. The smaller the grade, the greater the penetration, the thinner the grease.

表6.1 润滑脂分类等级号所对应的针入度范围

Table 6.1 Penetration range corresponding to the variety grade number

品种等级号（Classification class number of grease）	工作针入度范围（25℃，0.1mm）（Scale of penetration（25℃，0.1mm））	品种等级号（Classification class number of grease）	工作针入度范围（25℃，0.1mm）（Scale of penetration（25℃，0.1mm））
000	445~475	3	220~250

续表 6.1

品种等级号（Classification class number of grease）	工作针入度范围（25℃，0.1mm）（Scale of penetration（25℃，0.1mm））	品种等级号（Classification class number of grease）	工作针入度范围（25℃，0.1mm）（Scale of penetration（25℃，0.1mm））
00	400~430	4	175~205
0	355~385	5	130~160
1	310~340	6	85~115
2	265~295		

注：经过 60 次机械剪切试验后的针入度称为工作针入度。

Note: Penetration after 60 times mechanical shear tests is called working penetration.

（2）滴点。润滑脂在试管中按规定方法加热，当开始熔化滴下第一滴油时的温度叫做滴点。滴点反映润滑脂的耐热能力，选用润滑脂时，应使滴点高于工作温度 20~30℃。

Dropping point. Dropping point is the temperature when grease is heated in the test tube by the regulation method, the first drop of oil starts to melt and drop. The dropping point reflects the heat resistance of the grease. When selecting the grease, the dropping point should be higher than the working temperature of 20~30℃.

（3）稠化剂。稠化剂对润滑脂的性能影响极大。稠化剂是润滑脂命名的依据。如钙基脂、钠基脂、锂基脂等。

Thickener. Thickener has a great influence on the performance of the grease. Thickeners are the basis for naming grease. For example, calcium based grease, sodium based grease, lithium based grease, etc.

（4）润滑脂的添加剂。为了改善润滑脂的性能以满足不同条件下机械对润滑性能的要求，常常加入各种添加剂，这些添加剂通常在石油化工厂就已经按配方加入，有时需要在现场添加。

Additives of grease. In order to improve the performance of grease to meet the mechanical lubrication requirements under different conditions, a variety of additives are often added. These additives are usually added in accordance with the formula in petrochemical plants. Sometimes oil additives need to be added on site.

添加剂有油性添加剂、极压抗磨添加剂、消泡剂、黏度指数改进剂、抗氧化添加剂、防锈添加剂、抗乳化剂、抗凝剂（起降低凝点的作用）和清净分散剂（起抑制油中漆膜生成和防止金属表面积垢）等。

There are a wide variety of additives, including oil additives, extreme pressure and anti-wear additives, defoaming additives, viscosity index improvers, antioxidant additives, anti-rust additives, anti-emulsifiers, anticoagulants (to reduce the freezing point), detergents (to inhibit the formation of paint film in oil and prevent metal surface scaling) and so on.

油性添加剂的品种有猪油、鲸鱼油、油酸、三甲酚磷酸酯及硬脂酸等。极压抗磨添加剂主要有硫化鲸鱼油、二甲基磷酸酯、氯化石蜡、二烷基二硫代磷酸锌等。

Oil additives include lard, whale oil, oleic acid, trimethylphenol phosphate ester and stearic acid. Extreme pressure anti-wear additives mainly include sulfurized whale oil, dimethyl phosphate, chlorinated paraffin, dialkyl dithiophosphate zinc and so on.

含硫、磷、氯等活性元素的化合物在较高温度条件下，可以在金属表面生成化学反应膜，起到润滑作用。但是，这种反应是不可逆的，所以对金属有腐蚀作用，随着使用时间的增长，油中极压添加剂含量减少，润滑性能下降，有时需要在现场定期补加。

Compounds containing sulfur, phosphorus, chlorine and other active elements can form a chemical reaction film on the metal surface at higher temperatures and play a role in lubrication. However, this reaction is irreversible, so it has a corrosive effect on the metal. With the increase of service time, the extreme pressure additive content in the oil decreases, and the lubrication performance decreases. Sometimes, it is necessary to add it regularly on the site.

在循环润滑系统中，由于泡沫会使油路发生断油故障，通常加入二甲基硅油或苯甲基硅油消泡。一般加入量为0.0001%~0.001%，加入时，先用煤油稀释后再加入搅匀。

In the circulating lubrication system, dimethyl silicone oil or polymethylphenyl siloxane fluid is usually added to defoam because the foam will break the oil circuit. Generally, the amount of dimethyl silicone oil or phenyl silicone oil should be added is 0.0001% ~ 0.001%. Dimethyl silicone oil or phenyl silicone oil should be diluted with kerosene before being added and stirred.

B　润滑脂的分类及用途（Classification and Usage of Grease）

目前国内多数企业使用的润滑脂还是以钙基、钠基等低性能润滑脂为主，这两类润滑脂虽然相对来说价廉，但性能差，润滑效果不好，换油周期短。其结果是耗量大、设备寿命降低，从而降低了设备的作业率，增加了零配件的消耗，从经济角度来衡量是不合算的。

At present, the lubricating grease used by most Chinese enterprises is mainly based on calcium base, sodium base and other low-performance lubricating grease. Although these two types of grease are relatively cheap, they have poor performance, poor lubrication effect and short oil change cycle. As a result, the consumption is large and the service life of the equipment is reduced. Thus, the operation rate of the equipment is reduced and the consumption of spare parts is increased. It is not economical.

锂基脂的性能则比较好，特别是采用12-羟基硬脂酸锂皂稠化的锂基脂，滴点较高、抗水性好，对各种添加剂的感受性也好，尤其是机械安定性优异，而且对金属表面的黏附力较强，由于针入度适中，因而泵送性好，是目前较为理想的一种多用途长寿命润滑脂。由于锂基脂具有上述优点，所以在西方工业化国家润滑脂的消耗中，锂基脂的比例逐年上升，而钙基、钠基的比例逐年下降。

The performance of the lithium base grease is better. Especially, the lithium base grease thickened by 12-Lithium hydroxystearate soap has high dropping point, good water resistance, high susceptibility to various additives, especially excellent mechanical stability, and strong adhesion on metal surface. The pumping ability is good due to the penetration is moderate. It is the ideal grease with a multipurpose and long-life. The proportion of lithium base grease increases year by year due to the above advantages of lithium base grease and the consumption proportion of calcium base and sodium base decreases year by year in western industrialized countries.

除锂基脂外，合成复合铝基脂与膨润土脂在使用中也显示出较好的润滑性能。

Besides lithium base grease, synthetic compound aluminum base grease and bentonite base grease also show good lubrication performance in use.

高性能的合成润滑脂是以润滑油作基础油加入合成皂基或普通皂基稠化剂制得的高性能润滑脂。

High performance synthetic grease is made by adding synthetic soap base or ordinary soap base thickener in lubricating oil as base oil.

合成润滑油是近似油的中性润滑材料，它并不直接由矿物处理得到，而是用有机合成的方法制得。由于它具有特定的分子结构，所以具有比矿物油更好的润滑性能。而且用作合成润滑油的化合物或聚合物有一个突出特点，就是可以根据需要而调整其有关性能。

Synthetic lubricating oil is a neutral lubrication material which is similar to oil. It is not obtained directly from mineral treatment, but obtained by organic method. Because of its specific molecular structure, it has better lubrication properties than mineral oil. Moreover, compounds or polymers used as synthetic lubricants have a prominent feature that their properties can be adjusted according to needs.

合成润滑油种类繁多，能适应多种多样的特殊用途，如极高温、极低温或宽温度范围的润滑油，抗燃润滑油，极压润滑油和能抗辐射耐高真空环境的润滑油等。但生产成本比较昂贵，所以价格远高于矿物润滑油。

Synthetic lubricants come in many varieties and can be used for a wide range of special applications, such as high, low or wide temperature range lubricants, fire resistant lubricants, extreme pressure lubricants, radiation resistant and high vacuum resistant lubricants, etc. But the production cost is expensive, so the price is much higher than the mineral oil.

部分国产合成润滑脂有通用润滑脂，高低温润滑脂，7018 高转速润滑脂，7019、7019-1 高温润滑脂，7020 窑车轴承润滑脂，高温极压轧钢润滑脂，7407 齿轮润滑脂和 BLN 半流体脂。

Some Chinese synthetic grease includes general grease, high temperature grease, 7018 high speed grease, 7019 or 7019-1 high temperature grease, 7020 kiln car bearing grease, high temperature extreme pressure rolling steel grease, 7407 gear grease and BLN semi-fluid grease.

在工矿企业的设备事故中，润滑事故占很大比重，而润滑材料选用不当又是引起这些事故的一个重要因素。如某厂压力机的大型轴承，由于润滑脂选用不当，一年就损坏八套，换用适合的润滑脂后，一套轴承两年还未见到显著的磨损。可见正确选用润滑材料的重要性。

In the equipment accidents of industrial enterprises, lubrication accidents account for a large proportion. The improper selection of lubrication materials is an important factor causing these accidents. For example, the large bearing of a press machine in a certain factory will be damaged eight sets in one year due to improper selection of grease. After replacing the suitable grease, a set of bearings has not seen significant wear after using two years. Therefore, the importance of proper selection of lubrication materials can be seen.

在各种润滑材料中，润滑油的摩擦较小，形成油膜比较均匀，优先选择润滑油。

In a variety of lubrication materials, lubricating oil friction is small, the formation of oil film is more uniform, the priority choice of lubricating oil.

润滑油脂的选择原则有以下几点：

The selection principles of lubricating grease are as follows：

（1）润滑材料的性能。

Properties of lubrication materials.

（2）负荷大小及负荷特性。

Value and characteristics of load.

（3）运动速度。

Speed of movement.

（4）工作温度。

Operating temperature.

（5）周围环境。

Surrounding environment.

（6）摩擦副的结构特点及润滑方式。

Structural characteristics of friction pairs.

（7）润滑模式。

Lubrication mode.

6.1.4 稀油润滑系统（Thin Oil Lubrication System）

习惯上把润滑脂润滑称为干油润滑，润滑油润滑称为稀油润滑。

Grease lubrication is usually called dry oil lubrication and oil lubrication is called thin oil lubrication.

6.1.4.1 稀油润滑的分类（Classification of Thin Lubrication）

根据润滑剂供往摩擦副的方式，润滑可划分为分散润滑与集中润滑，间隙润滑和连续润滑，无压润滑和压力润滑。根据对润滑剂的利用方式，润滑可分为流出式润滑和循环式润滑。

According to the way the lubricant is supplied to the friction pair, the lubrication can be divided into dispersed lubrication and centralization lubrication, clearance lubrication and continuous lubrication, non-pressure lubrication and pressure lubrication. According to the use of lubricants, lubrication can be divided into flow-out lubrication and circulating lubrication.

除了干油、稀油两种传统的润滑方式外，还有油雾、油气润滑（属于稀油润滑）和干油喷溅润滑等方式。

Besides dry oil and thin oil lubrication methods, there are oil mist, oil and gas lubrication（belongs to thin oil lubrication）and dry oil spray lubrication.

6.1.4.2 常用稀油润滑装置（Thin Oil Lubrication Devices）

常用稀油润滑装置包括：油孔和油杯润滑，油环、油链及油轮润滑，溅油及油池润滑。

Common thin oil lubrication devices in common use include：lubrication of oil hole and oil cup, lubrication of oil ring, oil chain, oil wheel, and spilled oil, lubrication of oil bath lubrication.

稀油集中循环润滑系统是指用油泵加压将润滑油泵送到各润滑点进行连续的强制润滑的系统，系统中设置了油箱、油泵、冷却器、过滤器、安全阀和各种控制仪表。

Thin oil centralization circulating lubrication system is a kind of lubrication system in which the oil is pumped to high pressure and sent to all lubrication points for continuous pressure lubrication. System contains tanks, pumps, coolers, filters, valves and various control instruments.

6.1.5 干油润滑系统（Dry Oil Lubrication System）

干油润滑虽然比稀油润滑的阻力大，但由于密封简单，不易泄漏和流失，所以在稀油容易泄漏和不适宜稀油润滑的地方，特别具有优越性。如轴承、开式齿轮传动、链条、某些导轨和机械上各种不适于稀油润滑的部位，特别是滚动轴承上用得最多。

Although the resistance of dry oil lubrication is more than that of thin oil lubrication, dry oil lubrication is not easy to leak and loss because of simple sealing. So dry oil lubrication especially has advantages where thin oil easy is not suitable. For example, thin oil lubrication is not suitable for bearings, open gear transmission, chains, some guide rail and mechanical parts. Dry oil lubrication is especially used in rolling bearings.

6.1.5.1 干油润滑的分类（Classification of Dry Oil Lubrication）

干油润滑的分类方法有很多种。按润滑方式可分为干油分散润滑和干油集中润滑。按压油的动力来源可分为手动干油润滑站和自动干油润滑站。按给油器的结构可分为单线和双线干油润滑系统。按主输油管的布置方式，可分为环式和流出式干油集中润滑系统。

Dry oil lubrication can be divided in many ways. Dry oil lubrication can be divided into dry oil dispersed lubrication and dry oil centralization lubrication according to the lubrication method. Dry oil lubrication can be divided into dry oil lubrication station manual dry oil lubrication station and automatic dry oil lubrication station. Dry oil lubrication can be divided into single-line dry oil lubrication system and double-line dry oil lubrication system according to the structure of the oil feeder. Dry oil lubrication can be divided into flow-out dry oil centralized lubrication system and circulating dry oil centralized lubrication system according to the main pipeline arrangement.

6.1.5.2 干油润滑的用途（Usage of Dry Oil Lubrication）

近年来，由于新型润滑脂的研制和润滑方法的改进，在闭式齿轮、蜗轮传动中，使用带抗磨、极压添加剂（如 MoS_2）的润滑脂也日益增多。

In recent years, greases with anti-wear and extreme pressure additives (such as MoS_2) are increasingly used in closed gear and worm gear transmission due to the development of new greases and the improvement of lubrication methods.

根据摩擦副的不同，有的采用单独分散的润滑方式（即由人工定期用加脂枪向润滑点或油脂杯添加润滑脂），有的则因摩擦副的数量多、工作条件的限制、用人工加脂有一定的困难（如高温、润滑点多、人工加脂忙不过来、人工加脂不易接近润滑点），则必须采用干油集中润滑系统定期加润滑脂。

Separate dispersed lubrication is a kind of lubrication of adding grease periodically to the lubrication points manually, using grease cup or grease gun according to the different friction pairs. If there are some limits such as adding grease manually is difficult because of many reasons, such as high temperature, many lubrication points, adding grease manually is not easy to approach lubrication points, workers are too busy to finish adding grease manually, dry oil concentrated lubrication system must be used.

手动干油分散润滑主要靠人工加脂，使用的装置主要有手动加脂的旋盖式干油杯和用脂枪加脂的压注油嘴。手动干油集中润滑系统由手动干油站、滤油器、给油器、主油管和支油管组成。单线流出式干油集中润滑系统主要由单线电动干油泵、滤油器、主油管、支油管和片式给油器组成。

Manual dry oil dispersion lubrication mainly relies on adding grease manually. Dry oil cup with screw cap and grease nozzle with grease gun can be used in manual dry oil dispersion lubrication system. Manual dry oil centralized lubrication system consists of manual dry oil station, oil filter, oil feeder, main oil pipe and branch oil pipe. Single-line flow-out dry oil centralized lubrication system is mainly composed of single-line electric dry oil pump, oil filter, main oil pipe, branch oil pipe and chip oil feeder.

6.1.6 典型零部件的润滑（Lubrication of Typical Parts）

滑动轴承、滚子轴承和齿轮属于典型的零部件，广泛应用于各种机械中，需要适当的润滑才能执行所需的功能。

Sliding bearing, roller bearing and gears are widely used in many kinds of machines. They need proper lubrication to perform the required functions.

6.1.6.1 滑动轴承的润滑（Lubrication of Sliding Bearing）

滑动轴承的润滑，主要是正确确定轴承的润滑方式、润滑材料、耗油量及润滑周期等。

Sliding bearing lubrication mainly considers correctly determining the bearing lubrication method, lubrication materials, the consumption flow and lubrication cycles, etc.

A 滑动轴承润滑方式的选择（Lubrication Method Selection of Sliding Bearing）

滑动轴承的润滑方式与轴承的载荷、速度、温度、轴承间隙及结构有关，通常用经验公式（6.1）和式（6.2）估算。

The empirical Formula (6.1) and Formula (6.2) are usually used to estimate sliding bearing lubrication method.

$$k = \sqrt{p_m v^3} \tag{6.1}$$

$$p_m = P/(dl) \tag{6.2}$$

式中，p_m 为轴颈投影面上的平均单位压力，Pa；P 为轴承的载荷，N；d 为轴颈直径，m；l 为轴颈长度，m；v 为轴颈圆周速度，m/s。

Where, p_m is average pressure shaft journal, MPa; P is the load on bearing, N; d is the diameter of the neck of sliding bearing, m; l is the length of the neck of sliding bearing, m; v is linear velocity of shaft journal, m/s.

$k \leqslant 6$ 时，用润滑脂润滑。$k = 6 \sim 50$ 时，用润滑油，采用针阀油杯或油绳润滑。$k = 50 \sim$

100 时，用润滑油，油环或飞溅润滑，但应对润滑油采取冷却措施。$k>100$ 时，用润滑油，集中压力循环润滑。

When $k \leqslant 6$, grease is suitable for lubrication. When $6<k<50$, lubricating oil is suitable for lubricating sliding bearing. Needle valve oil cup or rope lubrication is usually used. When $50<k<100$, lubricating oil is suitable for lubricating sliding bearing. Oil ring or splash is usually used. But cooling measures for lubricating oil should be taken. When $k>100$, lubricating oil is suitable for lubricating sliding bearing. A concentrated pressure circulation system should be used.

一般滑动轴承，转数不高、载荷较大及加工精度低，通常处于边界润滑状态，除选择合适的黏度外，还应注意油性和抗压性。

For general sliding bearing with low ratio of revolution, large load and low machining accuracy which is always in the state of boundary lubrication, oil performance and anti-pressure should be paid more attention in addition to the choice of suitable viscosity.

B 滑动轴承用润滑油的选择（The Choice of Lubrication Oil for Sliding Bearings）

滑动轴承选用润滑油的步骤是确定黏度、选择品牌、确定耗油量。

The steps of choosing sliding bearing oil is determination of viscosity, selection of lubricating oil brand, determining the oil consumption.

（1）确定黏度。有两种方法可以获得黏度。一种方法是在图表中查找，根据轴承的工作速度、载荷大小和直接工作温度，通过查表来选择所需的润滑油品种。一些高质量的特种润滑油也可以直接参考手册中的经验表或表格来选择。另一种方法是用式（6.3）经验公式计算和查图法。

Determination of viscosity. There are two ways to get viscosity. One way is looking up in the chart. According to the working speed of the bearing, and the size of the load, and direct operating temperature, the varieties of lubricating oil required is chosen by looking up in the chart. Some special lubricating oil with high quality can also be selected referring to the experience tables or forms of the manual directly. Another way is calculation and looking up chart method by empirical Formula (6.3).

$$\eta = \frac{10^{-9}P}{3.4d^2nl} \tag{6.3}$$

式中，η 为黏度；n 为轴承轴颈的转速，r/min。

Where, η is viscosity; n is rotate speed of shaft journal, r/min.

（2）选择润滑油的品牌。选择合适的黏度后可以选择润滑油的品牌。高油性通用机械润滑油用于低速中负荷滑动轴承。中极压工业齿轮油适用于中速、重负荷的滑动轴承。高黏度的极压工业齿轮油，可用于低速、重载的滑动轴承。

Selection of lubricating oil brand. The lubricating oil brand can be chosen after choosing suitable viscosity. Good oily general-purpose machine lubricating oil is used in low-speed medium load sliding bearing. The medium extreme pressure industrial gear oil can be used for the sliding bearing of medium speed and heavy loading. The extreme pressure industrial gear oil of high viscosity can be used for the sliding bearing of low speed and heavy loading.

（3）耗油量。人工加油、滴油和线芯润滑的滑动轴承的耗油量，主要根据轴颈直径、转速、轴承长度与轴颈直径的比值 l/d 而定。

Oil consumption of sliding bearing. Oil consumption of sliding bearing with manual adding oil, drops of oil, and line core lubrication can be determined by the diameter of shaft journal, rotation speed, the length of bearing over the diameter of shaft journal l/d.

如果从轴承流出的油量很小，说明供油不足会导致轴承温度上升，加重轴颈和轴承衬层的磨损，因此，应适当增加润滑油供应。如果所有流出的油都是新油，那就意味着已经给了太多的油，这会造成润滑油浪费，因此，应减少给油量。

If the amount of oil flowing from the bearing is very small, it indicates that the lack of oil supply will cause the temperature of the bearing to rise, aggravating the wear of the journal and the bearing lining. So, the oil supply should be appropriately increased. If all the oil flowing out is new oil, it means that too much oil has been given, which will cause oil waste. Therefore, the amount of oil given should be reduced.

C 滑动轴承润滑脂的选择（Lubrication Grease Selection of Sliding Bearing）

当 $k \leqslant 6$ 时，不宜使用或不方便使用润滑油的地方应使用润滑脂。滑动轴承的供油方式和润滑周期与润滑脂用量有关。

When $k \leqslant 6$, grease should be used where lubricating oil is not suitable or inconvenient. The grease supply method and lubrication period of sliding bearings are related to the amount of grease used.

滑动轴承的干油润滑制度可以根据滑动轴承轴颈的转速、工作温度和工作连续状况来确定加脂的润滑间隔周期，也可以按供油方式或装置以及工作条件制定，可参考表 6.2。

The dry oil lubrication system of sliding bearing can be determined according to the speed of the journal of sliding bearing, the working temperature and the working continuous state to determine the grease lubrication interval period, or according to the oil supply mode or device and working conditions. Refer to Table 6.2.

表 6.2 滑动轴承干油润滑制度

Table 6.2 Dry oil lubrication system of sliding bearings

供油方式或装置（Method or device for supplying oil）	工作条件（Running conditions）	润滑制度（Lubrication system）
旋盖干油杯（Dry oil cup with screw cap）	重载荷，间歇工作（Heavy load, intermittent operation）	8h 给脂一次（Give the fat once for 8h）
压力球阀油杯（Oil cup with pressure ball valve）	正常温度，经常运转（Normal temperature, regular operation）	8h 给脂 1~2 次（Give the fat 1~2 times for 8h）
	重载荷，高温下经常运转（Heavy loads, often running at high temperatures）	8h 给脂 2~3 次（Give the fat 2~3times for 8h）
集中润滑系统（Centralized lubricating system）	小载荷，间歇工作（Small load, intermittent operation）	1~2 天给脂一次（Give fat once a day or two）
	偶尔运转，不经常工作（Occasionally running not working very often）	4~6 天给脂一次（Give fat once every 4~6 days）

D 轴承的引油方法和润滑油沟(Oil Drainage Method and Oil Groove of Sliding Bearings)

轴承的润滑还涉及正确布置向摩擦表面引油的油孔，应具有适当的润滑油沟。

Lubrication of bearings also involves the proper arrangement of oil drainage holes to the friction surface and the proper lubricating oil groove.

轴承的润滑油应该在油膜上负荷最小的地方被引入摩擦表面。

Lubricating oil shall be introduced to the friction surface where the load on the oil film is minimal.

将油导入滑动轴承的最佳方法是在滑动轴承衬套顶部铲成镰刀弯。润滑油流进轴承座的镰刀弯形腔内。当轴转动时，一部分润滑油从下轴瓦的孔流入楔形间隙中，产生支承负荷的流体压力形成油膜。另一部分油流入轴径上轴瓦间环形间隔中，用以强烈冷却轴径，然后从轴承两端流出。

The best approach to guide oil into sliding bearing is shoveled a camber on the top of the sliding bearing bushing. Lubricating oil flow into camber cavity of the bearing block. When the shaft rotates, part of the lubricating oil flow into the wedge clearance from the bearing block hole and bearing load of fluid pressure and oil film is formed. Another part of the oil flow into the annular clearance between bearing diameter of shaft, for cooling diameter of shaft strongly. Then the oil flows out from both ends of the bearing.

对于滚动轴承，引入轴承上部空腔中，当轴承转动时，形成油膜。

For rolling bearings, lubricating oil should be introduced into the upper cavity of the bearing. Oil film is formed as the bearing rotates.

油沟不应开在油膜承载区，不应该有尖锐的边缘，也不应该太接近边界。对于开式滑动轴承，为了保证润滑油输入负荷部分和具有一定的冷却效果，在下轴瓦部采用其他形状的油沟。采用润滑脂的油沟尺寸应比采用润滑油的尺寸大一些，因为润滑脂流动性较差。

Oil grooves should not be opened in the film loading area, should not have sharp edges, and should not be too close to the boundary of bearing block. For the open-type sliding bearing, in order to ensure the input load of lubricating oil and a certain cooling effect, other shape of oil groove should be adopted on the lower of bearing bushing. The size of grease groove should be larger than that of the lubricating oil because grease has poor fluidity.

6.1.6.2 滚动轴承的润滑（Lubrication of Rolling Bearing）

滚动轴承的润滑主要是正确确定轴承的润滑方式和润滑剂。滚动轴承的润滑方法与轴承的类型、尺寸和运转条件有关。一般滚动轴承可使用润滑油和润滑脂进行润滑。在特殊情况下可使用固体润滑剂。目前，80%的滚动轴承采用润滑脂润滑。

The lubrication of rolling bearing is mainly to determine the correct lubrication mode and lubricant of the bearing. Rolling bearing lubrication method has relationship to the type, size, and operation condition of bearings. General rolling bearing can be lubricated with oil and grease. Solid lubricant can be used under special circumstances. Today, grease is adopted in 80% rolling bearings.

A 滚动轴承用润滑油的选择（Lubricating Oil Selection for Rolling Bearing）

滚动轴承用润滑油不仅要有合适的黏度，还要有良好的抗氧化安定性、热氧化安定性，不含机械杂质和水分。另外，滚动轴承和滚道存在较多滑动摩擦的滚动轴承（如球面滚子轴承），在载荷较重条件下，可采用加入极压添加剂的润滑油。

Lubricating oil for rolling bearing should not only have the suitable viscosity, but also has good anti-oxidation stability, thermal oxidation stability, not mechanical impurities and moisture. In addition, extreme pressure additives of lubricating oil can be used for rolling bearings combined with many sliding friction (such as spherical roller bearing), under heavy load conditions.

润滑油的黏度可以按内径、转数，对应工作温度从经验图上选择。

Viscosity can be chosen according to inner diameter, number of revolution, and temperature on the empirical figures.

一般载荷选用普通机械油、机床工业用润滑油。重载或冲击载荷选用黏度相同，但有油性和极压添加剂的润滑油。

Common mechanical oil and oil for machine tools are usually used for general load. Oil with same viscosity, oily and extreme pressure additives is fit for the machine with heavy or impact load.

耗油量可以用经验公式（6.4）计算。

Oil consumption of rolling bearing can be calculated by Formula (6.4).

$$Q = 0.00075db \tag{6.4}$$

式中，Q 为润滑油的消耗量，g/h；d 为轴承内径，mm；b 为轴承宽度，mm。

Where, Q is oil consumption, g/h; d is inner diameter of bearing, mm; b is the width of bearing, mm.

滚动轴承在正常工作条件下（有良好的密封、没有灰尘及水分侵入、工作温度不超过45℃），每三个月换一次油。如果轴承在温度较高（50℃以上）或有灰尘及水分侵入的环境中工作，则每半个月至一个月换一次油。

In normal working condition without dust and water, good sealing, and lower temperature than 45℃, change oil per three months. If the bearing is operating at a high temperature (above 50℃) or in an environment invaded by dust and moisture, the oil shall be changed every half month to once a month.

B　滚动轴承用润滑脂的选择（Lubricating Grease Selection for Rolling Bearing）

滚动轴承润滑脂的选择包括针入度的选择、品牌的选择、润滑脂消耗量的选择。

The choice of lubricating grease for rolling bearing includes the choice of penetration, the choice of brand, the choice of grease consumption.

（1）针入度的选择（The choice of penetration）

1）一般转速和载荷选 2 号润滑脂（针入度为 265~295）。

No. 2 grease (penetration is 265~295) is usually used for the machine with general rotate speed and load.

2）速度高时，选 1 号、0 号润滑脂（针入度分别为 310~340，340~380）。

No. 1 and No. 0 grease (penetration are respectively 310~340, 340~380) are usually used for the machine with high rotate speed.

3）低速重载时，选 3 号、4 号润滑脂（针入度分别为 220~250，175~205）。

No. 3 and No. 4 grease (penetration are respectively 220~250, 175~205) are usually used for the machine with low speed and heavy load.

4）干油集中润滑时，选择用1号或2号润滑脂（针入度分别为310~340，265~295）。

No. 1 and No. 2 grease (penetration are respectively 310~340, 265~295) are usually used for the machine with dry oil concentrate lubrication.

（2）润滑脂品种的选择（The choice of brand）

1）低速低负荷时，选用钙基脂和钠基脂。

Na (sodium) base grease and Ca (calcium) base grease are usually used for the machine with low speed and low load.

2）低速重载冲击时，选用二硫化钼锂基脂。

Li (lithium) base grease-molybdenum disulfide is usually used for the machine with low speed and heavy or impact load.

3）高温时，选用耐高温润滑脂。

Anti-high-temperature grease is usually used for the machine with high temperature.

4）高速（5000r/min左右）时，选用7018高转速润滑脂。

7018high speed grease is usually used for the machine with high speed (5000r/min).

（3）润滑脂消耗量选择（The choice of grease consumption）

n=1500r/min以上，装入量为空间的30%~50%。

When the rotation speed n of the machine is higher than 1500r/min, grease consumption over all space is 30%~50%.

n=1500r/min以下，装入量为空间的60%~70%。

When the rotation speed n of the machine is lower than 1500r/min, grease consumption over all space is 60%~70%.

低速时，装入量为空间的100%。

When the rotation speed n of the machine is very low, grease consumption over all space is 100%.

6.1.6.3 齿轮的润滑（Lubrication of Gear）

开式齿轮、闭式齿轮和蜗杆的润滑方式不同。

Lubrications for open gears, closed gears and worm are different.

A 开式齿轮的润滑（Lubrication of Open Gears）

人工涂抹润滑脂，或二硫化钼半干膜润滑再用干油喷溅方式保护膜是较好的润滑方式。

Manually adding grease, or technical service molybdenum disulfide semi-dry film lubrication and dry oil spray method is a better lubrication method for the lubrication of open gear transmission.

B 闭式齿轮的润滑（Lubrication of Closed Gears）

闭式齿轮的润滑包括润滑方式和润滑油的选择。

（1）润滑方式的选择。油浴润滑适用于低速中小型齿轮箱。高速、大功率齿轮箱采用循环润滑。高精度、小功率齿轮箱采用喷雾润滑。中低速减速机采用BLN半流体脂和

7407 脂的效果不错。二硫化钼半干膜润滑或干油脂飞溅也用于重载和低速大齿轮。

The choice of lubrication methods. The lubrication of closed gear requires the selection of lubrication mode and lubricating oil. Oil bath lubrication is used for medium and small gear boxes with low speed. Recycle lubrication is used for those gear boxes with high speed and higher power. Mist lubrication is used for those gear boxes with high precision and small power. BLN semi-fluid grease and 7407 grease have good effect for medium and low speed reducer and worm gear reducer. Molybdenum disulfide semi-dry film lubrication or dry grease splash is also used for heavy duty and low speed large gears.

（2）润滑油的选择（The choice of lubrication oil）

1）确定黏度。黏度可以按经验公式（6.5）选择。

Determination of viscosity. Viscosity is usually calculated by empirical Formula（6.5）.

$$°E_{50} = \frac{kp}{20} \tag{6.5}$$

$$p = \frac{N}{vL} \tag{6.6}$$

式中，k 为速度系数；见表 6.3；p 为单位齿宽负荷，kN/m；N 为传动功率，kN；v 为速度，m/s；L 为齿宽，m。

Where, k is the factor of speed, shown in Table 6.3; p is the load per tooth width, kN/m; N is transmission power, kN; v is speed, m/s; L is the width of tooth, m.

表 6.3　速度系数

Table 6.3　The factor of speed

v/m · s^{-1}	8	8~16	16~25
k	1.6	1.2	0.85

上述计算只适用于不带飞轮传动，如果带飞轮，还应考虑飞轮释放出来的能量，在齿轮上增加的压力。

The above calculation only applies to those drive devices without the flywheel. If the flywheel is used, the energy released by the flywheel and the added pressure on the gear should also be taken into account.

对于圆弧齿轮减速器，选用润滑油的黏度不低于渐开线齿轮减速器所用油的黏度。

For circular-arc gear reducer, the viscosity of lubricating oil is not lower than that of the lubricating oil used for the involute gear reducer.

2）润滑油牌号的选用。润滑油牌号选用的一般原则是：凡是生产有专用油，最好采用专用油。

Selecting lubricating oil brand. General principles of selecting lubricating oil brand is selecting specialized oil if there is specialized oil.

通用型机床工业用润滑油适用于中速轻载，中极压工业齿轮油（推荐产品）适用于低速重载。

General industrial lubricating oil for machine tools is generally suitable for the machine with medium speed and light load. Medium extreme pressure industrial gear oil (recommended) is suitable for the machine with low speed heavy load.

C 蜗轮蜗杆传动的润滑（Lubrication of Worm）

根据螺杆蜗轮的啮合特性，普通蜗轮的啮合滑动面不能形成动压油膜，因此不能根据黏度选择润滑油。根据输送功率和滑动速度选择合适的抗磨油是必要的。而润滑油中的润滑油添加剂是非常主要的，建议选用高黏度的油（金属切削机床上的蜗杆和蜗杆传动不属于这个范围）。

According to the meshing characteristics of screw worm wheel, the meshing sliding surface of ordinary worm wheel cannot form a dynamic pressure oil film, so the lubricating oil cannot be selected according to the viscosity. It is necessary to select the suitable anti-wear oil according to the transmitted power and sliding speed. And the oil additive in lubricating oil is very main, recommend to choose the oil of high viscosity (the worm and worm drive in metal cutting machine tool does not belong to this range).

无添加剂，低速低功率蜗轮可选用油缸或齿轮油。黏度应大于 $20mm^2/s$。可选用 680号、1000 号气缸油、460 号、680 号 CKE 蜗轮油、半流体润滑脂等。

No additives cylinder oil or gear oil can be chosen for low-speed low power worm gear. Viscosity should be greater than $20mm^2/s$. For example, No. 680 and No. 1000 cylinder oil, No. 460 or No. 680 CKE worm gear oil, and semi-fluid grease can also be chosen.

CKE 型 460 或 680 型蜗轮油适用于中速、中功率蜗轮。

No. 460 or No. 680 CKE worm gear oil can be chosen for medium speed and medium power of worm gear.

高速、大功率蜗轮可选用 460 或 680 CKE/P 型蜗轮油。

No. 460 or No. 680 CKE/P worm gear oil can be chosen for high speed and high-power worm gear.

6.2 密封的装配（Assembly of Sealing）

为了防止润滑油脂从机器配置设备接合面的间隙中泄漏出来，并不让外界的脏物、尘土、水和有害气体侵入，机器设备必须进行密封。

In order to prevent lubricating grease from leaking out from the gap of the connection surface of the machine configuration equipment, the machine and equipment must be sealed to prevent the invasion of external dirt, dust, water and harmful gas.

密封性能的优劣是评价设备的一个重要指标。由于油、水、气等的泄漏，轻则造成浪费、污染环境，对人身、设备安全及机械本身造成伤害，机器设备失去正常的维护条件，影响其寿命。重则可能造成严重事故。因此必须重视和认真搞好设备的密封工作。

The sealing performance is an important index to evaluate the equipment. The leakage of oil, water, gas, etc., will cause waste and pollution of the environment, harm to the safety of human beings, equipment, and mechanical equipment itself, and cause the loss of normal maintenance conditions of machines, affecting its service life. Serious accidents may result. Therefore, we must pay attention to and do a good job of sealing equipment.

机器设备的密封主要包括固定连接的密封（如箱体的结合面、连接盘等的密封）和活动连接的密封（如填料密封、轴头油封等）。采用的密封装置和方法种类很多，应根据密封的介质种类、工作压力、工作温度、工作速度、外界环境等工作条件以及设备的结构和精度等进行选用。

The sealing of the machine mainly includes the sealing of fixed connection (such as the sealing of the connection surface of the box, the sealing of the connecting disc, etc.) and the sealing of movable connection (such as packing seal, shaft head oil seal, etc.). There are many kinds of sealing devices and methods, which should be selected according to the type of sealing medium, working pressure, working temperature, working speed, external environment and other working conditions as well as the structure and precision of the equipment.

密封材料的性能是保证有效密封的重要因素，选择密封材料，主要是根据密封元件的工作环境，如使用温度、工作压力、所使用的工作介质及运动方式等。

The performance of sealing materials is an important factor to ensure effective sealing, the sealing materials should be selected mainly according to the working environment of sealing components, such as the use of temperature, working pressure, the use of the working medium and movement mode.

6.2.1 固定连接的密封（Sealing of Fixed Connection）

固定连接的密封包括密封胶密封、密合密封和衬垫密封。

The sealing of fixed connection includes sealant sealing, tight sealing and gasket sealing.

（1）密封胶密封。为保证机件正确配合，在结合面处不允许有间隙时，一般不允许只加衬垫，这时一般用密封胶进行密封。密封胶具有防漏、耐温、耐压、耐介质等性能，而且具有效率高、成本低、操作简便等优点，可以广泛应用于许多不同的工作条件。

Sealant sealing. In order to ensure the correct matching of the parts, there should be no gap in the connection surface. It is not allowed only adding gasket. Sealant is generally used to seal at this time. Sealant has many performances of leakproof, temperature resistance, pressure resistance, medium resistance and other performance, and has high efficiency, low cost, simple operation and other advantages, can be widely used in many different working conditions.

（2）密合密封。由于配合的要求，在结合面之间不允许加垫料或密封胶时，常常依靠提高结合面的加工精度和降低表面粗糙度进行密封。这时，除了需要在磨床上精密加工外，还要进行研磨或刮研使其达到密合，其技术要求是有良好的接触精度和做不泄漏试验。机件加工前，还需经过消除内应力退火。在装配时注意不要损伤其配合表面。

Tight sealing. Due to the requirements of the fit, when the bonding surface is not allowed to add padding or sealant, the sealing which can improve the machining accuracy of the bonding surface and reduce the surface roughness can be used. At this time, the connection surfaces need to be not only machined with high precision in the grinder, but also ground or scraped to reach close connection. The technical requirements the connection surfaces should have good contact precision and leakage test should be done. Before the machine parts or components are machined, they need to be annealed to eliminate the internal stress. Be careful not to damage the fitting surface during assembly.

（3）衬垫密封。承受较大工作负荷的螺纹连接零件，为了保证连接的紧密性，一般要在结合面之间加刚性较小的垫片，如纸垫、橡胶垫、石棉橡胶垫、紫铜垫等。垫片的材料根据密封介质和工作条件选择。

Gasket sealing. For the thread joins part bearing bigger work load, rigid lesser gasket, wait like pad of paper, rubber pad, asbestos rubber pad, red copper pad should be used between connection surfaces in order to assure the tightness of connection surfaces. Gasket material is selected according to sealing medium and working conditions.

衬垫装配时，要注意密封面的平整和清洁，装配位置要正确，应进行正确的预紧。维修时，拆开后如发现垫片失去了弹性或已破裂，应及时更换。

When assembling gaskets, attention should be paid to the smooth and clean sealing surface. The assembly position should be correct and the correct pre-tightening should be carried out. During maintenance, if the gasket is found to have lost elasticity or broken after disassembling, it should be replaced in time.

6.2.2 活动连接的密封（Sealing of Active Connection）

活动连接的密封包括填料密封、油封密封、密封圈密封、机械密封、干气密封和液膜密封等。

Active connection seal generally has packing sealing, oil sealing, ring sealing and mechanical sealing.

6.2.2.1 填料密封（Packing Sealing）

填料密封（如图6.10所示）的装配工艺要点有以下几点：

The key points of packing sealing (as shown in Figure 6.10) assembly are as follows:

（1）软填料可以是一圈圈分开的，各圈在轴上不要强行张开，以免产生局部扭曲或断裂。相邻两圈的切口应错开180°。软填料也可以做成整条的，在轴上缠绕成螺旋形。

Soft packing can be separated in circles. Each circle on the shaft should not be forced open to avoid producing local distortion or fracture. The incisions of the two adjacent circles should be staggered by 180°. Soft packing can also be made into whole strips and wrapped around the shaft in a spiral shape.

（2）当壳体为整体圆筒时，可用专用工具把软填料推入孔内。

When the box is a whole cylinder, special tools can be used to push soft packing into the hole.

（3）软填料由压盖压紧。为了使压力沿轴向分布尽可能均匀，以保证密封性能和均匀磨损，装配时，应由左到右逐步压紧。

The soft packing is compressed by the gland. In order to ensure the even distribution of pressure along the axial direction as far as possible, to ensure the sealing performance and uniform wear, the assembly should be gradually compressed from left to right.

（4）压盖螺钉至少有两只。必须轮流逐步拧紧，以保证圆周力均匀。同时用手转动主轴，检查其接触的松紧程度，要避免压紧后再行松出。软填料密封在负荷运转时，允许有少量泄漏。

There are at least two cover screws. Cover screws must be tightened step by step to make sure that the circular force is uniform. At the same time, turn the spindle by hand to check the tightness of the contact, and avoid loosening after tightened. A small amount of leakage is permitted for soft packing sealing during operation with load.

运转后继续观察，如泄漏增加，应再缓慢均匀拧紧压盖螺钉（一般每次再拧进 1/6~1/2 圈）。但不应为争取完全不漏而压得太紧，以免摩擦功率消耗太大或发热烧坏。

Continue to observe the leakage after running. If leakage increases, tighten the gland screw slowly and evenly (usually screw 1/6~1/2 turn each time). But the gland should not be screwed too tight and pressure is too high to strive for complete no leakage so as to avoid friction power consumption too large or burning by high temperature.

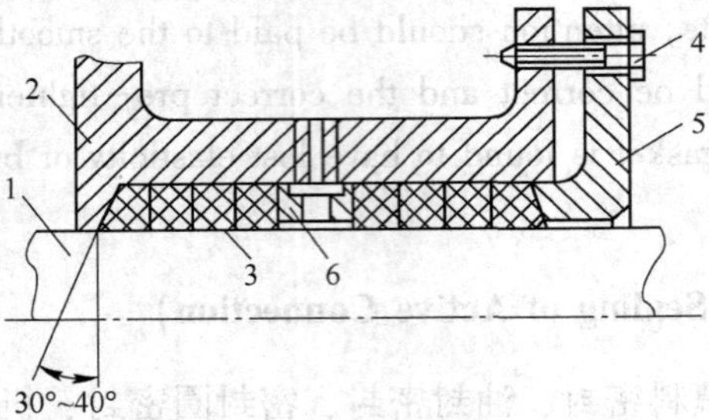

图 6.10 填料密封

Figure 6.10 Packing sealing

1—主轴（shaft）；2—壳体（shell）；3—软填料（soft filling material /soft packing）；
4—螺钉（screw）；5—压盖（gland）；6—孔环（air hole）

6.2.2.2 油封密封（Oil Sealing）

油封是广泛用于旋转轴上的一种密封装置，如图 6.11 所示，其结构比较简单，按结构可以分为骨架式和无骨架式两类。

Oil sealing is a kind of sealing device widely used on rotating shaft, as shown in Figure 6.11. Its structure is relatively simple, which can be divided into skeleton type and non-skeleton type according to the structure.

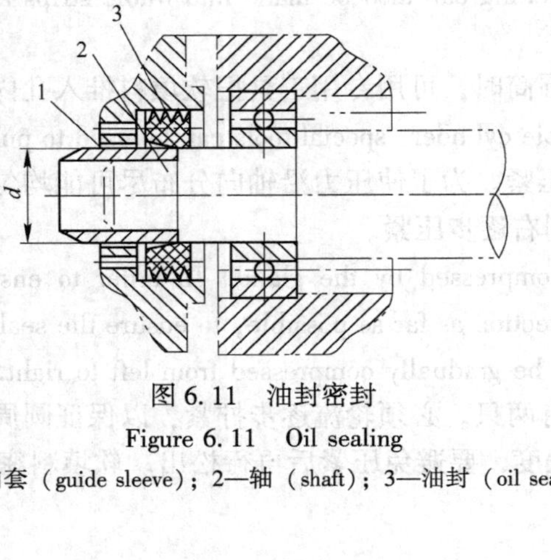

图 6.11 油封密封

Figure 6.11 Oil sealing

1—导向套（guide sleeve）；2—轴（shaft）；3—油封（oil sealing）

装配时应使油封的安装偏心量和油封与轴心线的相交度最小，要防止油封刃口、唇部受伤，同时要使压紧弹簧有合适的拉紧力。

When assembling, the installation eccentricity of the oil sealing and the intersection degree of the oil sealing and the axis line should be the minimum to prevent the oil sealing cutting edge, lip injury from destruction. The compression spring should have appropriate tension.

6.2.2.3 密封圈密封（Ring Sealing）

密封元件中最常用的就是密封圈，密封圈的断面形状有圆形（O形）和唇形，其中用得最早、最多、最普遍的是O形密封圈（如图6.12所示）。

The most commonly used sealing element is the ring sealing, the section shape of ring sealing are usually the circle (O shape) and the lip shape (as shown in Figure 6.12). The O-rings are the earliest and widely used.

O形密封圈是压紧型密封，故在装入密封沟槽时，必须保证O形密封圈有一定的压缩量，一般截面直径压缩量为8%~25%。O形密封圈对被密封表面的粗糙度要求很高，一般规定静密封零件表面粗糙度 Ra 值为6.3~3.2μm，动密封零件表面粗糙度 Ra 值为0.4~0.2μm。

O-rings are kind of compression type sealing. It must be ensured that the O-rings have a certain amount of compression in the sealing groove, general section diameter compression of 8%~25%. O-rings sealing have high requirements for the roughness of the sealed surface. Generally, it is stipulated that the surface roughness Ra value of static sealing parts or components is 6.3~3.2μm, and that of dynamic sealing parts is 0.4~0.2μm.

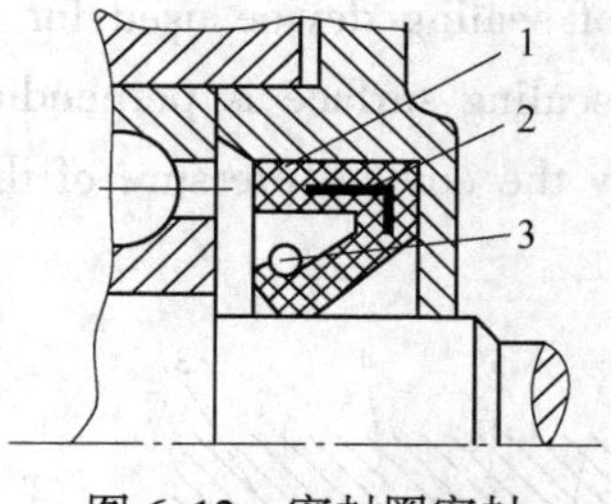

图6.12 密封圈密封

Figure 6.12 Ring sealing

1—密封圈（ring sealing）；2—金属骨架（metal bone）；3—压紧弹簧（press spring）

装配O形密封圈时应注意以下几点。

The following points should be noted when assembling O-rings seals.

（1）装配前须将O形圈涂润滑油，装配时轴端和孔端应有15°~20°的引入角。当O形圈需要通过螺纹、键槽、锐边、尖角等时，应采用装配导向套。

The O-rings must be lubricated before assembly. Assembly shaft end and hole end should be 15°~20° angle. Assembly guide covers shall be used when O-rings need to pass through threads, keyways, sharp edges, sharp corners, etc.

（2）当工作压力超过一定值（一般10MPa）时，应安放挡圈，需特别注意挡圈的安装方向，单边受压，装于反侧。

When the working pressure exceeds a certain value (generally 10MPa), retainer ring should be placed. Special attention needs to be paid to the direction of retainer ring installation. The unilateral ring is compressed and installed on the opposite side.

（3）在装配时，应预先把需装的 O 形圈如数领好，放入油中，装配完毕，如有剩余的 O 形圈，必须检查重装。

In the assembly, O-rings need to be assembled, should be installed in advance, and put in oil. After assembled, if there are any O-rings left, all O-rings must be reassembled.

（4）为防止报废 O 形圈的误用，装配时换下来的或装配过程中弄废的 O 形圈，一定立即剪断收回。

In order to prevent the misuse of the scrapped O-rings, the replaced and waste scrapped O-rings after assembly or in the assembly process must be immediately cut and recycled.

（5）装配时不得过分拉伸 O 形圈，也不得使密封圈产生扭曲。

Do not overstretch the O-rings during assembly, and do not distort the sealing ring.

（6）密封装置固定螺孔深度要足够，否则两密封平面不能紧固封严，产生泄漏，或在高压下把 O 形圈挤坏。

The depth of sealing device fixed screw hole is deep enough. Otherwise the two sealing plane cannot be tightly sealed. Leakage will occur or the O-rings will be squeezed under high pressure.

6.2.2.4 机械密封（Mechanical Sealing）

机械密封是旋转轴用的一种密封装置，如图 6.13 所示。它的主要特点是密封面垂直于旋转轴线，依靠动环和静环端面接触压力来阻止和减少泄漏。

Mechanical sealing is a kind of sealing device used for rotating shaft, as shown in Figure 6.13. Its main feature is that the sealing surface is perpendicular to the rotating axis, and the leakage is prevented and reduced by the contact pressure of the end face of the moving ring and the static ring.

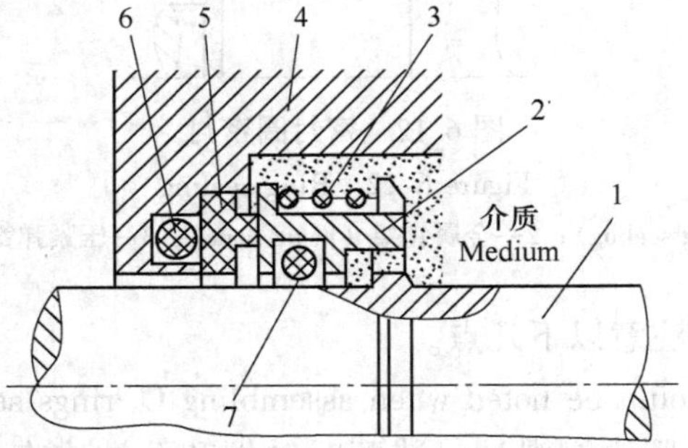

图 6.13　机械密封装置

Figure 6.13　Mechanical sealing

1—轴（shaft）；2—动环（moving ring）；3—弹簧（spring）；4—壳体（shell）；5—静环（static ring）；6—静环密封圈（static ring seal ring）；7—动环密封圈（moving ring seal ring）

对机械密封装置在装配时，必须注意如下事项。

When assembling the mechanical sealing device, the following points should be noted.

(1) 按照图样技术要求检查主要零件，如轴的表面粗糙度、动环及静环密封表面粗糙度和平面度等是否符合规定。

The main parts of mechanical sealing device should be checked according to the drawing technical requirements, such as shaft surface roughness, moving ring and static ring seal surface roughness and flatness, etc.

(2) 找正静环端面，使其与轴线的垂直度误差小于0.05mm。

Find the end face of the static ring, so that its vertical error with the axis is less than 0.05mm.

(3) 必须使动、静环具有一定的浮动性，以便在运动过程中能适应影响动、静环端面接触的各种偏差，这是保证密封性能的重要条件。浮动性取决于密封圈的准确装配、与密封圈接触的主轴或轴套的粗糙度、动环与轴的径向间隙以及动、静环接触面上的摩擦力的大小等，而且还要求有足够的弹簧力。

Must make the dynamic and static ring have a certain floating, so that in the process of motion the mechanical sealing device can be adapted to various deviations affect the dynamic and static ring end contact, which is the important conditions to ensure the performance of sealing. Floating depends on the accurate assembly of the sealing ring, the roughness of the spindle or shaft bushing in contact with the sealing ring, the radial clearance between the moving ring and the shaft, and the size of the friction force on the contact surface of the moving and static ring, and enough spring force is also required.

(4) 要使主轴的轴向窜动、径向跳动和压盖与轴的垂直度误差在规定范围内，否则将导致泄漏。

To make the error of the spindle axial movement, the error of radial run-out and gland and the error of shaft verticality in the specified range. Otherwise it will cause leakage.

(5) 在装配过程中应保持清洁，特别是在主轴装置密封的部位不得有锈蚀，动、静环端面应无任何异物或灰尘。

In the assembly process should be kept clean, especially the main shaft device sealing parts should not be corroded, moving, static ring end face should be free of any foreign matter or dust.

(6) 在装配过程中，不允许用工具直接敲击密封元件。

In the assembly process, hitting the sealing element directly with the tool is forbidden.

6.2.2.5 气膜密封(Gas Film Sealing)

气膜密封一般指依靠几微米的气体薄膜润滑的机械密封，也称为干气密封，即“干运转气体密封”，是将开槽密封技术用于气体密封的一种新型轴端密封，属于非接触密封。

Gas film sealing generally refers to mechanical sealing lubricated by a few microns of gas film, also known as dry gas sealing, also known as “dry running gas seals”, are a new type of shaft end sealing that use slotted sealing technology for gas sealing and are non-contact sealing.

气膜密封是一种新型的、先进的旋转轴密封，由它来密封旋转机械中的气体或液体介质。与其他密封相比，气膜密封具有泄漏量少、磨损小、寿命长、能耗低，操作简单可靠，被密封的流体不受油污染等特点。在压缩机领域、特殊泵领域正获得越来越多的应用。气膜密封使用的可靠性和经济性已被许多工程应用所证实。

Gas film sealing is a new and advanced rotary shaft sealing that seals gas or liquid media in rotating machinery. Compared with other sealing, gas film sealing have the characteristics of less leakage, less wear, long service life, low energy consumption, simple and reliable operation, and the sealed fluid is not contaminated by oil. More and more applications are being obtained in the fields of compressors and special pumps. The reliability and economy of using gas film sealing have been confirmed by many engineering applications.

气膜密封是20世纪60年代末期在气体动压轴承的基础上通过对机械密封进行根本性改进发展起来的一种新非接触式密封，通过在机械密封动环上增开了动压槽以及设置辅助系统而实现密封端面的非接触运行。英国的约翰克兰公司于20世纪70年代末期率先将干气密封应用到海洋平台的气体输送设备上并获得成功。干气密封最初是为解决高速离心式压缩机轴端密封问题而出现的，由于密封非接触式运行，因此密封摩擦副材料基本不受 pv 值的限制，特别适合作为高速高压设备的轴端密封。

Gas film sealing is a kind of non-contact sealing developed in the late 1960s on the basis of gas dynamic pressure bearing, through the study of the fundamental improvement of mechanical sealing. In fact, non-contact operation of the sealing face is realized through increasing dynamic pressure groove on the mechanical sealing rotating sealing ring and then set the corresponding auxiliary system. John Crane company in the UK was the first to successfully apply dry gas sealing to gas conveying equipment on offshore platforms in the late 1970s. Dry gas sealing was originally developed to solve the problem of shaft end sealing of high-speed centrifugal compressor. Due to the non-contact operation of the sealing, the sealing friction pair material is basically not limited by pv value, which is especially suitable for shaft end sealing of high-speed and high-pressure equipment.

当端面外侧开设有流体动压槽的动环旋转时，流体动压槽把外径侧（称之为上游侧）的高压隔离气体泵入密封端面之间，由外径至槽径处气膜压力逐渐增加，而自槽径至内径处气膜压力逐渐下降，因端面膜压增加使所形成的开启力大于作用在密封环上的闭合力，在摩擦副之间形成很薄的一层气膜从而使密封工作在非接触状态下完成。所形成的气膜完全阻塞了相对低压的密封介质泄漏通道，实现了密封介质的零泄漏或零逸出。

When the end of the lateral opening of fluid dynamic pressure groove ring rotates, high pressure isolation gas of diameter side (called upstream side) is pumped between the seal faces through the fluid dynamic pressure groove. The pressure on gas film increases from outside diameter to groove diameter, and declines gradually from groove diameter to inside diameter. Because of the mask opening which is formed by the pressure increase is greater than the closing force on the sealing ring, a thin gas film is formed between the friction pairs to enable the sealing to work in a non-contact state. The formed gas film completely blocks the leakage channel of the relatively low pressure sealing medium sequentially realizing zero leakage or zero escape of the sealing medium.

气膜密封与一般机械密封的平衡型集装式结构一样，但端面设计有所不同，表面上有几微米至十几微米深的沟槽，端面宽度较宽。与一般润滑机械密封不同，干气密封在两个密封面上产生了一个稳定的气膜。这个气膜具有较强的刚度使两个密封端面完全分离，并保持一定的密封间隙，这个间隙不能太大，一般为几微米。密封间隙太大，会导致泄漏量增加，密封效果较差。而密封间隙较小，容易使两密封面发生接触，因为干气密封的摩擦热不能及时散失，端面接触无润滑，将很快引起密封变形、端面过度发热从而导致密封失效。这个气膜的存在，既有效地使端面分开又使相对运转的两端面得到了冷却，两个端面非接触，故摩擦、磨损大大减小，使密封具有长寿命的特点，从而延长主机的寿命。

The structure of gas film sealing is the same with balance packaged structure of general mechanical seal, but the end face design is different. There are several microns to more than ten microns deep groove on the surface, moreover, the end face width is wider. Unlike general lubricated mechanical sealing, dry gas sealing produce a stable gas film on both sealing surfaces. The gas film has a strong stiffness to completely separate the two sealing faces and maintain a certain sealing gap, which can not be too large, generally a few microns. Too large sealing clearance will lead to increased leakage, poor sealing effect. Small sealing gap is easy to make the two sealing surfaces contact. Because the friction heat of the dry gas sealing can not be lost in time, the end contact without lubrication will soon cause sealing deformation, end excessive heating, resulting in sealing failure. The existence of the gas film, both effectively separate the end face and make the relative operation of the two end face cooling, two end face non-contact, friction, wear is greatly reduced, so that the seal has the characteristics of long life, so as to extend the life of the host.

开槽的密封面，分为两个功能区，外区域和内区域。气体进入开槽的外区域时，这些槽将压缩进入的气体，在槽根部形成局部的高压区，使端面分开，并形成一定厚度的气膜，为了获得必要的泵送效应，动压槽必须开在高压侧。开槽的密封间隙内的压力增加对干气密封的工作是至关重要的，它将保证即使在轴向载荷较大的情况下，密封也能形成一个不被破坏的稳定气膜。密封的内区域（即坝区）是平面的，靠它的节流作用而限制了泄量。密封工作时端面气膜形成的开启力与由弹簧和介质作用力形成的闭合力达到平衡，从而实现了非接触运转。干气密封的弹簧力是很小的，主要目的是在密封不受压或不工作时能确保密封的闭合，防止意外发生。

The sealing surface of the slotted surface is divided into two functional areas, the outer area and the inner area. While entering the outer area of the grooved area, the gas will be compressed into the groove root to form a local high pressure area, so that the end face is separated and a certain thickness of the gas film is formed. In order to obtain the necessary pumping effect, the dynamic pressure groove must be opened on the high pressure side. Increased pressure in the grooved seal gap is critical for the operation of dry gas sealing, which will ensure that the sealing will form a stable film that will not be destroyed even under high axial loads. The enclosed inner area (the dam area) is flat and limits the discharge by its throttling action. When the sealing works, the opening force formed by the end air film and the closing force formed by the spring and the medium force reach the balance, so as to realize the non-contact operation. The spring force of the dry gas sealing is very small. The main purpose is to ensure the closure of the sealing when the sealing is not pressed or not working, to prevent accidents.

气膜机械密封在泵、搅拌器等低速转动设备上已开始获得应用，展现了非常广阔的应用前景。值得注意的是，为了保障干气密封的稳定可靠运行，需要一套阻封气体的供应和控制系统。它一般具有气体过滤、流量调节与控制、压力调节与控制等功能。

Gas film mechanical sealing has been applied to pump, agitator and other low speed rotating equipment, showing a very broad application prospect. It is important to note that in order to ensure the stable and reliable operation of the dry gas sealing, a set of choke gas supply and control system is required. It generally has gas filtration, flow regulation and control, pressure regulation and control and other functions.

6.2.2.6 液膜密封（Liquid Film Sealing）

与气膜密封相对应，液膜密封一般指全液膜润滑非接触机械密封。减少或排除机械密封的泄漏，同时改善密封端面的润滑状况和操作稳定性，是密封使用者和研究者追求的目标。气膜密封在气相环境中获得了成功应用，但具有气体泄漏率较大的特点，将它直接用于液相环境，将可能出现不能接受的大泄漏量。近年出现的“上游泵送”密封概念有效地解决了这一难题，实现了非接触机械密封的低泄漏率，甚至零泄漏。

In contrast to gas film sealing, liquid film sealing generally refers to non-contact mechanical sealing with full liquid film lubrication. To reduce or eliminate the leakage of mechanical sealing and improve the lubrication and operation stability of sealing face is the goal pursued by sealing users and researchers. The gas film sealing has been successfully used in the gas phase environment, but it has the characteristics of large gas leakage rate. If it is directly used in the liquid phase environment, there may be an unacceptable large leakage amount. In recent years, the concept of “upstream pumping” sealing effectively solves this problem and achieves low leakage rate or even zero leakage of non-contact mechanical sealing.

上游泵送液膜润滑机械密封，就是普通机械密封的端面由一具有低流量、高压力“端面泵”所代替，该“泵”把少量的隔离流体沿着密封端面输送到密封腔。该密封端面的“泵”效应通过在端面开各种流体动压槽来实现，最常见的是螺旋槽。由于密封腔的液体压力比隔离流体的压力高，而隔离流体的流向是从低压的隔离流体腔流向高压的密封腔，故常被称为向“上游”泵送。与气膜密封不同的是，液膜密封的开槽区处于低压侧，槽与低压流体相通。未开槽的密封坝区靠近被密封的高压过程流体。

Upstream pumping liquid film lubrication mechanical sealing is that the end face of the ordinary mechanical sealing is replaced by a low flow, high pressure “end pump” and pump a small amount of isolation fluid to the sealing chamber along the sealing end face. The “pump” effect of the sealing face is realized by opening various fluid dynamic pressure grooves on the face. The most common is the spiral groove. Because the pressure of the liquid in the sealing chamber is higher than that of the isolating fluid, isolating fluid flows from the low-pressure chamber to the high-pressure sealing chamber, so it is often referred to as “upstream” pumping. Different from the gas film sealing, the slotting area of the liquid film seal is in the low pressure side, and the groove is connected with the low pressure fluid. The ungrooved sealing dam area is close to the sealed high pressure process fluid.

上游泵送机械密封通过端面螺旋槽的作用，在密封端面间建立了主动而明确的膜压分布，此膜压的最大值稍微超过被密封介质的压力，两者之差构成了液体上游泵送的推动力。端面间流体动压作用产生的动压力使两端面稍微分离，以便将下游流体送入上游，端面间为 100%的液膜，极大地改善了端面间润滑状况。

Upstream pumping mechanical sealing establishes an active and clear film pressure distribution between the sealing end faces through the end face spiral groove function. The maximum value of the film pressure is slightly more than the pressure of the sealed medium. The difference between the two constitutes the driving force of the liquid upstream pumping. The dynamic pressure of the fluid between the faces separates the two faces slightly in order to feed the downstream fluid upstream. The liquid film between the ends is 100%, which greatly improves the lubrication condition between the ends.

由于上游泵送液膜润滑非接触机械密封能通过低压隔离流体对高压的工艺介质流体实现密封，可以代替密封危险或有毒介质的普通双端面机械密封，从而双端面机械密封的高压隔离流体系统变成极普通的低压或常压系统，降低了成本，提高了设备运行的安全可靠性。

Due to the upstream pumping liquid film lubrication non-contact mechanical sealing realize high pressure technology medium fluid sealing by low pressure isolation fluid, it can replace the sealing dangerous or toxic medium of ordinary double mechanical sealing, and make double mechanical sealing system become very common low voltage or high voltage isolation fluid atmospheric system so as to reduce the cost, improve the safety and reliability of the equipment operation.

上游泵送机械密封已在各种场合获得应用，如防止有害液体向外界环境的泄漏、防止被密封液体介质中的固体颗粒进入密封端面、用液体来密封气相过程流体，或者普通接触型机械密封难以胜任的高速高压密封工况等。目前，上游泵送机械密封已应用于螺杆压缩机、污水泵、溶剂泵和液化石油气泵等特殊工况。

Upstream pumping mechanical sealing has been applied in various occasions. For example, it can used to prevent the leakage of harmful liquid to the external environment, to prevent the solid particles in the sealed liquid medium from entering the sealing end face, to seal the gas phase process fluid with liquid, or the ordinary contact type mechanical sealing is not competent for high-speed and high-pressure sealing conditions. At present, upstream pumping mechanical sealing have been used in screw compressors, sewage pumps, solvent pumps and liquefied petroleum gas pumps and other special conditions.

这类密封的不足是增加了在端面加工微米级槽的成本，少量的隔离气体会进入被密封的过程流体中。有较宽的密封环端面以布置螺旋槽，密封只能朝一个方向旋转等。

The disadvantage of this type of sealing is that it increases the cost of machining micron grooves on the end face, and a small amount of isolation gas can enter the sealed process fluid. There are wide seal ring face to arrange spiral groove. Seal can only rotate in one direction.

思考题
(Questions)

6.1 名词解释（Nouns Explanation）：

流体润滑（Fluid lubrication），流体动压润滑（hydrodynamic pressure lubrication），流体静压润

滑（hydro-static lubrication），闪点（flash point），凝点（condensation point），抗乳化度（anti-emulsification），抗氧化安定性（anti-oxidation stability），热氧化安定性（thermal oxidation stability），抗磨性（abrasion resistance），针入度（penetration），滴点（drop point），稀油润滑（thin oil lubrication），干油润滑（dry oil lubrication），动力黏度（dynamic viscosity），运动黏度（kinematic viscosity），恩氏黏度（Engler viscosity）.

6.2 润滑如何分类？

How to classify the lubrication?

6.3 润滑材料有哪些种？

How many classifications of lubrication materials are there?

6.4 润滑油有哪些物理化学性能及主要质量指标？

How many physical and chemical properties and main quality indexes of lubricating oil are there?

6.5 润滑脂有哪些物理化学性能及主要质量指标？

How many physical and chemical properties and main quality indexes of grease are there?

6.6 常用稀油润滑装置有哪些？

What are the commonly used thin oil lubrication devices?

6.7 稀油集中循环润滑系统由哪些零件和部件组成？

How many parts and components are there in thin oil centralized circulation lubrication system?

6.8 干油润滑分为哪些类？

How many classifications of dry oil lubrication are there?

6.9 滑动轴承的润滑，主要是正确确定哪些参数？

How many parameters should be determined correctly for sliding bearing lubrication?

6.10 滚动轴承的润滑方法与哪些参数有关？

How many parameters should be determined correctly for rolling bearing lubrication?

参考文献（References）

[1] Liping Wang, Guoqiang Zhang, Jingwei Piao. Engineering of mechanical equipment maintenance [M]. Beijing: Metallurgical Industry Press, 2019.

[2] 王立萍. 机械设备维修工程学 [M]. 北京：冶金工业出版社，2014.

[3] Hein Z P. Bloch, Fred K. Geitner. Mechanical equipment component maintenance and repair [M]. Elsevier, 2005.

[4] John Piotrowski. Shaft Alignment Handbook [M]. Second Edition. Revised and Expanded, President Turvac, Inc. Cincinnati, OH, 1995.

[5] Hein Z P. Bloch, Fred K. Geitner. Mechanical equipment Failure Analysis and Troubleshooting [M]. Elsevier, 1999.

[6] Lindley R, Higgins, P E, Keith Mobley R, et al. Maintenance engineering handbook [M]. Sixth edition. McGraw-Hill, 2001.

[7] Andrew K S, Jardine, Albert H C Tsang. Maintenance, Replacement, and Reliability theory and applications [M]. Talory & Francis, 2005.

[8] Jonathan David Morales Méndez, Ramon Silva Rodriguez. Total productive maintenance (TPM) as a tool for improving productivity: a case study of application in the bottleneck of an auto-parts machining line [J]. The International Journal of Advanced Manufacturing Technology, 2017, 92 (1/2/3/4): 1013-1026.

[9] Chlebus E, Helman J, Olejarczyk M, et al. A new approach on implementing TPM in a mine - A case study [J]. Archives of Civil and Mechanical Engineering, 2015, 15 (4): 873-884.

[10] Teonas Bartz, Julio Cezar Mairesse Siluk, Ana Paula Barth Bartz. Improvement of industrial performance with TPM implementation [J]. Journal of Quality in Maintenance Engineering, 2014, 20 (1): 2-19.

[11] 孙家骥. 矿冶机械维修工程学 [M]. 北京：冶金工业出版社，1994.

[12] 谷士强. 冶金机械安装与维护 [M]. 北京：冶金工业出版社，1995.

[13] 陈冠国. 机械设备维修 [M]. 2 版. 北京：机械工业出版社，2005.

[14] 费敬银. 机械设备维修工艺学 [M]. 陕西：西北工业大学出版社，1999.

[15] 李士军. 机械维护修理与安装 [M]. 北京：化学工业出版社，2004.

[16] 闫嘉琪，李力. 机械设备维修基础 [M]. 北京：冶金工业出版社，2009.

[17] 杨永军. 温度测量技术现状和发展概述 [J]. 计测技术，2009，29 (4)：62-65.

[18] Rajesh Attri, Sandeep Grover, Nikhil Dev. A graph theoretic approach to evaluate the intensity of barriers in the implementation of total productive maintenance (TPM) [J]. International Journal of Production Research, 2014, 52 (10): 3032-3051.